SPATIAL IMAGININGS IN THE AGE OF COLONIAL CARTOGRAPHIC REASON

This volume explores how India as a geographical space was constructed by the British colonial regime in visual and material terms. It demonstrates the instrumentalisation of cultural artefacts such as landscape paintings, travel literature and cartography, as spatial practices overtly carrying scientific truth claims, to materially produce artificial spaces that reinforced power relations. It sheds light on the primary dominance of cartographic reason in the age of European Enlightenment which framed aesthetic and scientific modes of representation and imagination. The author cross-examines this imperial gaze as a visual perspective which bore the material inscriptions of a will to assert, possess and control. The distinguishing theme in this study is the production of India as a new geography sourced from Britain's own interaction with its rural outskirts and domination in its fringes.

This book:

- Addresses the concept of "production of space" to study the formulation of a colonial geography which resulted in the birth of a new place, later a nation;
- Investigates a generative period in the formation of British India c. 1750–1850 as a colonial territory vis-à-vis its representation and reiteration in British maps, landscape paintings and travel writings;
- Brings Great Britain and British India together on one plane not only in terms of the physical geospaces but also in the excavation of critical domains by alluding to critics from both spaces;
- Seeks to understand the pictorial grammar that legitimised the expansive British imperial cartographic gaze as the dominant narrative which marginalised all other existing local ideas of space and inhabitation.

Rethinking colonial constructions of modern India, this volume will be of immense interest to scholars and researchers of modern history, cultural geography, colonial studies, English literature, cultural studies, art, visual studies and area studies.

Nilanjana Mukherjee is Assistant Professor at the Department of English, Shaheed Bhagat Singh College, University of Delhi. Her earlier publications include *Mapping India: Transitions and Transformations 18th–19th Centuries* (co-edited with Sutapa Dutta, 2019). She has also received the Meenakshi Mukherjee Memorial Prize for the year 2014 from the Indian Association of Commonwealth Languages and Literature for her article titled "Drawing Roads/Building Empire: Space and Circulation in Charles D'Oyly's Indian Landscapes" published in *South Asia: Journal of South Asian Studies,* Vol. 37. She is also a former Charles Wallace India Trust Visiting Fellow at the Institute of Advanced Studies in the Humanities and Centre for South Asian Studies, University of Edinburgh, UK.

SPATIAL IMAGININGS IN THE AGE OF COLONIAL CARTOGRAPHIC REASON

Maps, Landscapes, Travelogues in Britain and India

Nilanjana Mukherjee

LONDON AND NEW YORK

First published 2021
by Routledge
4 Park Square, Milton Park, Abingdon, Oxon OX14 4RN
605 Third Avenue, New York, NY 10017

Routledge is an imprint of the Taylor & Francis Group, an informa business

British Library Cataloguing-in-Publication Data
A catalogue record for this book is available from the British Library

Library of Congress Cataloging-in-Publication Data
A catalog record has been requested for this book

ISBN: 978-0-367-43018-4 (hbk)
ISBN: 978-0-367-50574-5 (pbk)
ISBN: 978-1-003-00070-9 (ebk)

Typeset in Sabon
by Taylor & Francis Books

DEDICATED TO MAA AND BABA

CONTENTS

CONTENTS

FIGURES

ACKNOWLEDGEMENTS

A list of all those who supported and contributed to this book can never be exhaustive. There have been innumerable people who have directly or indirectly rendered direction and motivation to this book in significant ways.

The research work for this book was undertaken in two phases. The first was during my doctoral research at the Jawaharlal Nehru University on which the book is based. Since then, of course, it has undergone substantial revisions. I am extremely grateful to my supervisors, Prof. G.J.V. Prasad and Prof. Neeladri Bhattacharya for providing constant support and encouragement which did not end with the submission of the dissertation. I am fortunate to have been taught by Prof. Shirshendu Chakravarty, Prof. Udaya Kumar, Prof. Gautam Chakravarty and Prof. Majid Siddiqi, whose courses proved to be eye openers for a young graduate student, and whose generosity in providing keen and valuable insights from time to time prodded me whenever and wherever I fell short. I shall remain forever indebted to King's College London for providing me with a most timely doctoral fellowship and funding in 2009. I express my gratitude to Dr Ruvani Ranasinha and Dr Jon Wilson for arranging this and for taking pains to make my stay in London most productive and fruitful. I also thank Dr Jane Cunningham of The Courtauld Institute of Art and Dr Jim Fowler of the Victoria and Albert Museum for letting me access their valuable archives and taking extraordinary interest in my work. I also thank Dr Emma Lauze and Dr Martin Postle of the Paul Mellon Centre for Studies in British Art for helping me locate archival materials and polishing my ways of interpreting them.

I sincerely value various libraries both in India and the UK and appreciate the labour behind maintaining well organised knowledge resources and towards making them conducive for a young researcher at that time. I am much obliged to the staff of the British Library, the Department of Prints and Drawings, the British Museum, Victoria and Albert Museum holdings at Blythe House, Olympia, The Courtauld Institute Library, the Paul Mellon Centre Library and the Maugham Library in London, the Lalit Kala Akademi Library and the Teen Murti Museum and Library, New Delhi, the National Library, Kolkata and the National Archives, New Delhi.

When I planned the conversion of the dissertation into a book, things had changed in my life. I found myself caught between parental care and responsibilities: on the one hand towards my new born daughter and on the other, towards my ailing father fighting cancer. Obviously, the progress of the work did not go as desired. Soon after, my father passed away and I was almost on the verge of shelving the revision. The anxiety of keeping my work relevant in a fast paced age of information circulation was getting the better of me. However, I managed to win the Charles Wallace India Trust Fellowship for a research stay for a different project at the Institute of Advanced Studies, University of Edinburgh. Freedom from the rigmarole of teaching and evaluation, and spoilt with a fully equipped centrally located Victorian office space, issued forth a new journey in the life of both this book and my academic work in general. Along with an allowance to borrow 60 books at a time from the University central library alone, apart from e-resources, e-books and books lying everywhere in the Institute lounge, to the archives at the National Library of Scotland situated not very far away, I also suddenly had all the time to read them as well! This, along with daily interactions with distinguished researchers from across the world would be enough to energise even a nonchalant scholar. Here, I met those people who I hope shall remain lifelong friends and mentors. Prof. Yearley, then newly appointed as the Director of the Institute, along with the staff were ever so kind, always ready to extend a helping hand, taking a keen interest and establishing connections wherever necessary. Here I met Prof. Charles Withers, whose works I had earlier followed closely, and who now proffered fresh perspectives to my work. I must also thank Radha Adhikari, Jeevan and Mona Sharma, Paul and Lotte, Arkotong Longkumer, Emile Chabal, Rakesh Ankit and Donald Ferguson for an extremely satisfying and academically rich time spent there.

My friends and colleagues who gave me company through my solitary journey as an academic in India are many. Among them, Amit, who I have known since my M.A. days in JNU, has been a constant source of intellectual support, never flinching from my requests to read and respond on writings, always readily supplying resources not available easily in India. Honestly, I have now grown quite superstitious and prefer not to send off any writing for review before making him read parts or the whole of it. The India International Society of Eighteenth-Century Studies has pulled together a small but energetic intellectual community of like-minded academics from diverse institutes and disciplines, enabling many pleasant hours of stimulating discussions and exchanges. I am thankful to Sutapa, Deeksha, Amba, Shivangini and Ridhima for this. I am also grateful to Shaheed Bhagat Singh College and the University of Delhi for relieving me from my duties for them for periods which enabled me to write my dissertation and avail fellowships with considerable ease. I also thank the fraternity of Shaheed Bhagat Singh College which has become my extended family in an otherwise alienating city. The students whom I have taught here over the years have contributed to my learning more than they could imagine.

This book has benefited from many thoughtful comments on earlier presentations of papers in national and international conferences which have later been accommodated in chapters here. My thankfulness should also extend to the many anonymous reviewers who have read my work and shared comments over the years. I admire the efficiency of the team of Routledge which has adeptly managed to expedite publication of an earlier volume and this monograph with precise planning.

This is a wonderful opportunity also to express my heartfelt gratitude and regards for my family and the person who has always been taken for granted: Arjun has been my wellspring for intellectual sustenance ever since I have known him. His unstinted faith in my capabilities, even at moments when I lacked it myself, has been my source of confidence and conviction. There are so many life skills I have learnt from Kouroki, with whom I have learnt to enjoy simple little matters in life. Maa and Baba have been a source of unwavering support and inspiration and had it not been for them, I would not have ventured as far as I have. I miss Baba but to Maa, (who keeps complaining I spend less time with her than with my books and laptop and everybody else) I dedicate this book.

INTRODUCTION

Maps, landscapes, travelogues: spatial articulation and the imperial eyes

> Just as none of us is outside or beyond geography, none of us is completely free from the struggle over geography. That struggle is complex and interesting because it is not only about soldiers and cannons but also about ideas, about forms, about images and imaginings.
>
> Edward Said: *Culture and Imperialism*

> But remember that words are signals, counters. They are not immortal. And it can happen – to use an image you'll understand – it can happen that a civilisation can be imprisoned in a linguistic contour which no longer matches the landscape of ... fact.
>
> Brian Friel: *Translation*[1]

It is some years since I was forced to rethink the beginnings of geographical identity of places and especially of nations. Questions about spatiality and spatial origins first occurred to me when reading Brian Friel's plays *Translations* (1980) and *Making History* (1988) in a Masters course.[2] That modern India as an insular geographical space born out of the imperial project of permanent possession is, in effect, a corollary to the spatial transformation in Great Britain, is an idea which crossed my mind on reading *Translations* years ago. Set in Donegal, Ireland in 1833, the play centres around the nodal points of spatial remapping and linguistic translation. While the Ordnance Survey "standardised" Irish-Gaelic place names, the newly introduced National Schools trained local students in English. In using the motif of mapping, it is Friel's agenda to show that, in effect, both the material and linguistic landscapes underwent a transformation and a translation of sorts from native Gaelic to English.[3] History is embedded in all kinds of local signs, symbols, artefacts, performances and practices. These also go to make the underlying codification "langue" which gets reflected in the "parole" or the spoken language.[4] A distortion affects the other, as a result of which autochthonous local place myths are subsumed under the meta structure of imperial history. As place names are

1

changed to English on the map, the consciousness of an entire culture is fractured by the transcription of one linguistic landscape, i.e. Gaelic by another, i.e. English. This completely and irrevocably destabilises and dispossesses the Irish of their language, heritage and history.[5] On the other hand, and closer home, another fictional character, Rudyard Kipling's Kimball O'Hara, a poor orphan of Irish descent, had to receive English education in the British School and imbibe survey techniques, paying particular attention to mapping and measuring, in order to be inducted in the prime colonial intelligence wing, the Ethnological Survey of India. These, almost ritual exercises, could ensure Kim the prestige and position of a true Sahib, a colonial master who could partake in the imperial project: the "Great Game" of the empire. Yet, his propensity to transgress cultural and racial differences in forging a friendly sympathetic relationship with the native, the Lama, (another representative character who traverses the margins of the British imperial heartland in India), disambiguates the inherent connection between the two: the subject status of both these individuals, pitching alongside each other, two different ways of making colonial analogies, one nationalist and the other, imperialist.[6]

In India, the late eighteenth and the early nineteenth centuries are crucial in relation to the colonial regime and its agenda for spatial production. The British, who looked upon the Indian landscape, its features, and its elements through a rigid disciplinary gaze, subjected its society, time and space to rigorous and hierarchical re-structuring.[7] In the following chapters, by clubbing the metropole and the colony in a "unitary field of analysis"[8] we shall see how European methods were applied to both England's first colony nearer home and overseas. Whereas the oppositional binary of the home state and the empire stands unaltered, it is essential to view the emergence of both the geographies as constructions of the same gaze. The geographical fashioning in both regions happened to be acts born out of an identical cartographic impulse that shaped the nation state of Great Britain and the British Empire.[9] The present work is a study of world views, of scopic patterns, viewpoints and visual paradigms which constructed Europe and the world, more specifically Great Britain and India, in the age of Enlightenment.

The aforementioned kind of representational structure through large scale mapping was not seen in India prior to the eighteenth century, although precolonial literary and ritual culture was ingrained with geographical awareness, as well as enumerative traditions.[10] One way to talk about pre-colonial imagined landscape is to see it as bearing imprints of meaning: the self-manifest eruptions of gods – the *swayambhu*, the footprints and trails of gods and epic heroes, the divine origins of rivers, or the body of the goddess etc.[11] This mental map is overlaid with layers and layers of stories and narratives. And one place is connected with another through stories forming an alternative network to that conceptualised by the British. On looking at native Indian/local/indigenous ideas of space and place making, (which is not the primary objective in this book) one is able to discern key themes of cultural representations of space and landscape through stylised aesthetic and narrative expressions of religious, historical and moral topophilia (i.e. a human being's affective ties with the material environment/habitat).[12] Through this study, one hopes to trace the transit of space from mythos to logos: from constructions of space and place through ritual and mythological narratives which memorialise space in a particular way, to disenchanted hodological space, i.e. space which is homogeneous, measured and regular, as opposed to sacred place.[13]

This book promises to similarly examine the process of discursive construction of "India" as a geographical category and as a specific place on the globe. This it will do by tracing the linkage in practices in Britain's internal and external colonies. Important here, are the simultaneous aspirations of fortification and exclusion (within the nation) and connection (overseas) in England's spatial figuration. This was crucial for the onset of global capitalism which led to lopsided spatial growth leading to spatial hierarchies.[14] The representation of the land as text, as a picture or an image transforms space into a place with mensurable exactitude identifiable with a proper noun. Where these are the first steps towards construction of a nation in Great Britain, they also form the foundation stones to imagining the British colony in India. On the flip side, we can go a step further to assert that colonial relations could have been employed as models for disciplining and controlling the body politic of England itself.[15] This book is about the construction of space, leading from imaginative to concrete contours, and ultimately attaining another meaning within the context of the British imperial enterprise.

Imaginative geographies of India

While studying the British invasion and refashioning of space in its colony, one should base one's understanding on the premise that this was, in effect, an encroachment of a pre-existing habitat with its native and lived ideas of space. Indigenous cultures of space making include the proto religious mandala which translated into the sacred Vaastu on the micro level and which provided models of state layouts on the level of polity.[16] Studies of monastic governmentality in the Himalayan border regions reveal an obliterated past of continuities in regional formations, successfully overwritten by the post Enlightenment colonial culture of border making.[17] Native scribes often collaborated with colonial Britons in writing about regions and setting them as distinct from erstwhile hegemons which led to hardening of boundaries and identities.[18] Sanjay Subrahmanyam has called for the necessity to challenge the colonial archive which has scripted spaces as distinct and disparate whereas he feels we today need to see them in terms of "connected histories".[19] That would effectively mean unpacking notions of space and territory that exist in contemporary times.

Arguments about India's distinct geographical past, have attained a validation from scholars of the Purana who cite the cosmography outlined in the source texts such as the Vishnu-Purana, and passages of Bharatavarsha-varna-nam "to draw connections between genealogy and space"[20]:

> Uttaram yat samudrasya Himadrascaiva daksinam
> varsam tad Bharatam nama[21]

On a mythological and narrative level, the epic journeys of Ram in the Ramayana and the Pandavas in Mahabharata are sited as the basis for integrity in the region. Folkloric narratives of the 51 *Sati Peethas* testify to and substantiate this spatial consciousness. Similarly, medievalists have cited the travelogues of Al Birauni and Ibn Battuta as demonstrating the cultural distinctness of the sub-continent from the regions lying to its west.[22] Yet, questions about the pre-colonial existence of geographical expressions in India demarcating a consolidated unit, have had diverse and sharply different scholarly trajectories in the hands of historians over the years. Where some have talked of the existence of Bharatvar-sha where a fundamental cultural linkage in practices formed the thrust,[23] some have outright refuted it as the "the imaginary institution of India":

> India ... the reality taken for granted in all attempts in favour and against, is not an object of discovery but of invention. It was historically instituted by the nationalist imagination of the nineteenth century.[24]

Manu Goswami has shown how India gets imagined as a territorial unit under the British. More importantly, she demonstrates how this new spatial con-ceptualisation welds with the language and nativist imagination of a chron-otopical Bharat emerging as a nation state worthy of being fought for on claims of independence and sovereignty.[25] Both Goswami and Chris Bayly have extensively talked about Western purveyors of science debating and actively collaborating with both Sanskritic and Mughal Islamic traditions of mathema-tical and astral practices to give rise to hybrid spatio-temporal configurations and new geopolitical realities.[26]

This new geographical consciousness percolated through the freshly estab-lished pedagogical processes of schools, textbooks and periodicals, naturalising the new state space.[27] The Vedic-Shastric-Puranic cosmographies intermixed with local mythologies found a way to meld with European Enlightenment geographies in creating an altogether new spatio-temporal paradigm. The con-cept of *Bhugol* (the term derived from the concept of the globe, which today is the equivalent of geography in most Indian languages) evolved out of two phases of transformations: first, from the cross fertilisation between Hindu medieval traditions of Surya Siddhanta, Jyotishastra and the Islamic astral calendar under Akhbar and Abul Fazl, and subsequently from interactions with Western sciences based on observations and exactitude. As Bayly points out, the intellectual debates and exchanges between the two and more cultural traditions were harbingers of the modern Indian nation.[28]

4

Production of space

Thus the nation space of today may be said to have embraced a multitude of intersections in the processes of its production. And if the national space is a product, our knowledge should extend to its processes of production. This act of excavating, identifying, analysing and explaining that, what continues to exist in the present, is called "architectonics" by Henri Lefebvre. This is similar in manner to Foucault's method of "archaeology" premised on the idea that epistemes and discursive formations are governed by rules, beyond those of grammar and logic, that operate beneath the consciousness of the individual subject defining set practices and conceptions in a given period.

Between the sixteenth and the nineteenth centuries, the status of space was reduced to one of reading. The underlying code of this legibility being:

> at once architectural, urbanistic, and political, constituting a language common to country people and townspeople, to the authorities and to artists – a code which allowed space not only to be "read'" but also to be constructed.[29]

A code of this kind must be correlated with a system of knowledge which brings an alphabet, a lexicon and a grammar together within an overall frame- work, situating itself within a syntagmatic sphere comprising of language, ordinary discourse, writing, reading, literature and so on. Western occupation in this region heralds in a logocentrism and bibliocentrism which did not exist in that specific form. The power of Logos hereby becomes the foundation of knowledge, technology, authority and finally through these, of colonisation and imperialism. Space is formed with, by and in language – for space is formed conceptually through a chain of verbal and nonverbal signifiers and hence a particular space can be treated as a message and representation as code. Lan- guage in a logocentric world has the ability to establish a realm of certainty which gradually extends its sovereignty. Space reduced to signs, becomes part of the commonly shared knowledge system which defines and determines the world. Lefebvre's "spatial practice" refers to this dialectical process of both creating and commemorating space, which produces as well as masters and appropriates space. It is the way in which space is seen, understood and intui- tively recognised. "Representations" of space are the ways in which any space is thought about, grasped, imagined, conceived and formulated.

Where Lefebvre makes use of the Marxist idea of the capitalist mode of production of space, Harvey too talks of spatial arrangements under succes- sive capitalist epochs facilitating the growth of production, the reproduction of labour-power and the maximisation of profit.[30] Most importantly, geo- graphical knowledge which was usually conveyed through representations, became a valued commodity in an increasingly profit conscious economy. This was an age of transitions involved with the transformation of the pre-existing feudal order into a capitalist one and was accompanied with all the disruptive effects of the changing economy, the most important of these being the monetisation of land. The accumulation of wealth and power got linked to knowledge and individual mastery over space as private property. Conceptualised space identifies lived space with what is perceived, and conceives it through a system of verbal and intellectually worked out signs.[31] As numericals, moduli and measurements play a dominant role in this mode of production of space, science was the other tool with which the goal was achieved. The conquest and control of space required it to be first imagined as usable and convertible through human action. Spatial representations such as maps and landscapes were some such technologies of control of physical space. Since space can be thought of as socially constructed, so is "the spatiality of social life, a practical theoretical consciousness that sees the lifeworld of being creatively located not only in the making of history but also in the construction of human geographies, the social production of space, and the relentless formation and reformation of geographical landscapes".[32]

The questions of space and place have today become central in studies in the humanities. The confidence in a "spatial turn" in epistemology from the late twentieth century onwards is evident in recognition of the importance of questions on "where" along with "how" and "why".[33] Following Lefebvre, just as space helps to constitute social relationships and meanings, space is also lent meaning by these. Although meanings are locally cast, these can change across time and space, and just as "knowledge in transit", both the local specificity of meanings and the nature of material forms such as maps can change or mutate as it travels from one place to another.[34] The dual processes of production and reception of spatial artefacts tell us about their power and purpose.

Places, therefore, are not just seen but are grappled with in diverse senses, which in turn are modified over time. Memories of a place are shared and are constructed over years and mostly through institutional commemoration which silence alternative memories of the past and of the place.[35] John Urry traces the articulation and discourse of spatial memory usually "organised around artefacts such as buildings, rooms, machines, walls, furniture and so on" calling them "place myths".[36] Nature itself is perceived, associated and appreciated as it is culturally ordered.[37] Images of places are fashioned according to the selective memories surrounding the place. Studying space should not be limited to a study of the material land alone, but involves a whole array of cultural products such as writing, architectural designs, paintings, guide books, literary texts, films, post cards, advertisements, music, travel patterns, photographs and so on. The consumption of such products is very much part and parcel of the processes of production and consumption of space. At another level, both the surveyor and the tourist map the land with the sense of possessing it, where the surveyor's quadrant shares similarities with the Claude glass both of which are designed with optic space in mind rather than haptic space, a space generated through physical contact with land.[38]

Imaginative geographies: discursive construction of spaces

Forays into the study of imperial practices of spatial representation cannot ignore the works of Edward Said, whose *Orientalism* (1980), as is well known, explores the ways in which racial identities are born through a Eurocentric geographical essentialism:

> the notion that there are geographical spaces with indigenous, radically "different" inhabitants who can be defined on the basis of some religion, culture or racial essence proper to that geographical space.[39]

Geography is a recurrent motif in Said's writings and what he engages us in is an act of "rethinking geography".[40] His work introduces us to an understanding of the creation of geographies through recognition of imagined and symbolic territories discursively consolidated into real spaces. This highlights the intersectionality of power, knowledge and geography inscribed within the imperial practices. What he calls the "imaginative geography", also was the key to the consolidation of the two polarities, the "Orient", or the East and the "Occident" or the West, as concepts and also spaces. This it does "by dramatising the distance and the difference between what is close to it [Europe] and what is far away".[41] In *Orientalism*, Said traces the triangulation of power, knowledge and geography through Foucault's conceptual analytic structures of the carceral archipelago and the order of things in the "historicogeographical specificity of the congruences between bodies and spaces put in place by particular constellations of Orientalism".[42] This depends on the ordered, systematic and differentiated arrangement of place. The similarities between the methods of Foucault and Said can be further developed through the identification of three points: division, detail and visibility. These three methods in European fields of scholarship formulated the exclusionary scopic regimes of the West and the East. The construction of the bipolar world depended heavily on a discursive construction of exclusionary and divisive geographies of the Orient and the Occident. The construction of each of these spaces is a constitutive function of the construction of the other, for, discourses of Orientalism not only essentialised the Orient but also essentialised the Occident. Categorisation and detailing, which were primary characteristic methods in scientific discourses, framed the two conceptual spaces into definitive grids of identification. Inventories were drawn up in order to specify features of the East as against the West in format of tabulation, which formulated these spaces. The concept of the scopic regime with its detailing is dependent on visibility, or the way spaces are designed to be seen in particular ways. It is Said's project to show how in the nineteenth century the European representations of the Orient as a theatre of magic gradually gave way to the Orient as "tableau, a museum and a disciplinary matrix".[43]

Though postcolonial scholarship has moved beyond Said's framework of essentialisation of the East by the West to accommodate narratives of agency and collaboration of colonial subjects in the making of the colonial world, Said's view of imaginative geographies is particularly useful here. The mapping impulse attains its apotheosis in the nineteenth century when colonies under European occupation were subject to routine scrutiny through the institutions of census, maps and museums. The colonial imaginary profoundly shaped the way in which the land and its people were understood by the colonial state at the onset of print capitalism.[44] More importantly, as Anderson points out, these became representational of how the colonised visualised themselves and managed to induce an amnesia, successfully superseding earlier cosmologies, cartographic traditions and older local methods of meaning generation. The new boundaries consolidated a fresh territorial cognisance of the geographic space of the nation "parallel and comparable to those in Europe."[45] Winichakul's work on early and modern cartographic reproduction of Siam has proven to be influential in this respect, and provided a cue to understanding the construction of post-colonial geographies of new nations in Asia and other continents across the world.[46]

In this model, the prime aim of the new rulers in their expansionist efforts was to bring under sway all territory not already theirs within the natural frontiers. Small nationalities which had failed to develop as states were swallowed up. Similar structures of spatiality in both internal and overseas colonial system could be seen as the outcome of the same social forces in both. The peripheral is made dependent and subordinate to the core mainly through trade. Hence, cities in this spatial scheme serve as way stations in the trade between colonial hinterlands and metropolitan ports. Hence, cities and colonial capitals tend to be located on coasts with direct access to the metropolis. The Celtic fringes of Britain's internal colony is similarly forced into complementary development to the core and thus became dependent on external markets. Therefore, great migration and mobility is rampant in these fringes and it is not merely a coincidence that these skilled and unskilled personnel from the fringes were employed in large numbers to lay base and prepare the grounds for colonial expansion at the frontiers overseas.[47]

Contributions of colonial personnel who engaged with foreign cultures in the moment when political and cultural boundaries were hardening, have a causal connection with imaginary spatial configurations taking tangible forms. Studies of lives and legacies of travellers, painters and cartographers bring coherence to the seemingly disparate practices of forging the contact zone. Piecing together their individual stories of self-fashioning and fortune seeking, unfold the narrative of an entire region acquiring its modern form.[48] This book will approach these practices by studying the lives of certain people as agentive in making this new meaning. The distinguishing theme in this study is the production of India as a new geography sourced from Britain's own interaction with its rural outskirts and domination in its fringes.[49] Therefore,

I read the persons and their actions as subjects shaped first by the influence of an internal colonisation, and then by the shared imperial consciousness.[50] The spatiality and re-centring that emerges in South Asia is a direct derivative of the cartographic rationale in Great Britain itself. What is differently done in this book is, perhaps, the tracking of a uniform formulaic gaze among a crude (even haphazard) assemblage of varied careering individuals, working with ostensibly disjunct media, performing apparently different tasks, who in actuality work in unison to unfurl the cartographic web across diverse regions bringing them under a conceptual scopic compass. This calls for a reassessment of what counts as knowledge and creativity in every field when these subscribe to a worldview which, in a concerted and sustained manner, supplants all other existing knowledge systems and are historically responsible for disempowering alternate views and imaginations. Therefore, this also calls for a questioning of clear definitions of the idea of India, as inherited in this world view.

The ocular space of geography and its quest for accuracy

In India, the eighteenth and the nineteenth centuries are of special significance in relation to the colonial regime and its agenda for spatial production. According to Bernard Cohn:

> metropole and colony have to be seen in a unitary field of analysis. In India the British entered a new world that they tried to comprehend using their own forms of knowing and thinking. There was widespread agreement that this society, like others they were governing, could be known and represented as a series of facts.[51]

The British looked at the Indian landscape, its features, and its elements through a rigid disciplinary gaze, which subjected its society, time and space to rigorous partitioning and hierarchical structuring. "Gaze" also bifurcates space between foreground and the distant panorama – the space of the observer, in this case the coloniser, and the space of the observed, i.e. of the colonised. Matthew Edney points out the two distinctive "gazes" employed by the British in looking at India: the "scientific gaze" and the "picturesque gaze". However, both were engaged in achieving a certain amount of precision and were part of what Bernard Cohn calls the "investigative modality".[52] There were of course, according to Cohn, a criss-crossing of several modalities, the "historiographic", "observational/travel", "survey", "enumerative", "museological", "surveillance" and "investigative", all of which were employed by the British in constructing an epistemological territory. The scientific gaze involved the examination, recording and the graphic representation of the alien landscape. As Edney explains, it was driven by two assumptions:

9

That an exact image of the real world is impressed onto the mind; and, that the viewer cannot help but be "intently engaged by the aggressive identity of a particular object".[53]

The scientific gaze with its quest for accuracy and precision, was most obvious in the colonial enterprise of surveys.[54] When conjoined with the picturesque gaze or a principle of aesthetics, the object of representation, the landscape, results in topographical drawings. The proliferation of representations of the landscapes of India during this time took different forms. The empirical sciences including geography and anthropology were employed to this end. Lord Curzon, the erstwhile Viceroy of India, addressing the Royal Geographical Society in 1912, stated that geography had been promoted from being a "dull and pedantic" science into "the most cosmopolitan of all sciences ... the hand-maid of history".[55] Michel Foucault, in his essay, "On Geography", points out the inevitable inter-linkage between the discipline of geography and historical processes of establishment, institutions and power:

> to trace the forms of implantation, delimitation and demarcation of objects, the modes of tabulation, the organisation of domains meant the throwing into relief of processes – historical ones, needless to say – of power.[56]

Clearly, thus, geography served as the primary means of the colonial purpose of accumulation of information, wealth and power. To return to Said, then, geography needs to be resituated within a political and strategic sphere of the colonial mission. Much has been made recently of the nature of geographical inscription. The Greek roots of geography as "writing the earth" has been taken up in an interrogation of the nature of geographical texts and other writing/etching/ representation of space, place and landscape.[57] The texts talked of here are, as is discernible, not only the traditional written bit, but following Roland Barthes[58] and other literary theorists and cultural anthropologists, the expanded concept of the text, including all other cultural productions such as paintings, maps, landscapes, journals, as well as social, economic and political institutions.[59] Lefebvre's "spatial architec-tonics" is thus relevant here in order to reassemble knowledge disseminated by partial disciplines of ethnology, ethnography, human geography, anthropology, history and sociology so as to excavate processes of spatial production at this juncture in history. In other words, these can be seen as constitutive of reality rather than as re-presenting it. It is imperative to put down here, that this book is an attempt to critique primarily the figurative, tropic and narrative representations in the wake of colonial assumption of power in the Indian subcontinent.

Maps, in this respect, are a form of discourse in that they represent viewpoints, aspirations and statements to their readers: for maps are scarcely only representation of facts but carry ideological and rhetorical devices. Maps are "part of a visual language by which specific interests, doctrines, and even world views were communicated" and "spatial emblems of power in society".[60] Even before cartography was imposed on colonies to articulate imperial space, maps and related iconography first concretised the idea of Europe, which was rather loosely developed in the Middle Ages. The emancipatory force of Renaissance humanism first permitted the triumphal celebration of Europe in the written texts and graphic images. New discoveries, the paradigm shift from a Ptolemaic to a Copernican conception of universe, new sea and land routes, new continents, all these provided the impetus towards an enhanced accuracy for maps of both Europe and the world. Theological and ideological changes also allowed an emancipation from the metaphysical strait-jacket of medieval cartography. Somewhat paradoxically, these changes initially appear rather tiny in comparison to the rest of the world. But new mathematical techniques were employed to alter projections through which Europe was placed in the central and most dominant position on the map. Europe also appeared larger as its border with Asia shifted eastward. To explain the iconography attached with the map of Europe, I quote from Michael Wintle's brilliant research in this area:

> Habsburg ambitions lent a new triumphalism to the portrayal of Europe, which was increasingly depicted as a unitary figure, usually a young woman, more noble and regal than the female figures which were also regularly used to personify other continents. The iconography of Europe was enthusiastically adopted and refined by Renaissance cartographers who incorporated the map of Europe into the physical shape of a queen, and decorated geographical texts with images of the "noble" continent lording it over her "sisters". Europe's tentacles stretched around the earth, and the other parts of the world brought her willing tribute on bended knee.[61]

This was the beginning of Eurocentrism. And it is interesting to see how the very royal figure of the image of Europe was, before long, transferred onto the map of Britannia, playing a key role in garnering the idea of British nationalism in the eighteenth and nineteenth centuries. In what John Pickles calls propaganda maps and Judith Tyner calls "persuasive cartography",[62] the map is seen as a subjective form aiming to persuade large groups of people to believe something or act in a way that they otherwise would not. Maps, such as the one talked about above, are a sophisticated technique in seeking to manipulate relationships in order to persuade people about particular truth claims or in other words, to perpetuate hegemony.[63]

The early maps of uncharted territories like South America, Africa and Australia, were often works of art and imagination rather than of objective scientific fieldwork of later times. These maps were populated with drawings of imaginary monsters, pygmies and animals as substitutes for gaps in topographical knowledge. From the Renaissance onwards, maps came to be valued for their objectivity, because claims to accuracy had now become politically

11

and economically important as new trade routes opened up. But this did not make the maps "neutral" in any way. For example, as mentioned earlier, the Mercator projection, which was used from the sixteenth century, distorted the size of the land masses so that Europe and North America appear bigger than they really are. The huge proliferation of maps in circulation in the seventeenth and the eighteenth centuries was directly connected to the colonisation of the rest of the world by European powers. The process of "discovering" land and promptly appropriating it as part of colonial territory, were accomplished with the help of cartography. Cartography became a way of textually appropriating spaces and renaming them as it naturalised culturally created boundaries and power arrangements such as occupation of adjacent spaces by rival European nations. A range of work in human and historical geography has discussed naming and mapping as reflections of social power. Some critical attention has been paid to authority and naming in an historical context, in relation to the extension of British colonial power and the mapped expression of that power. In the case of North America, for example, several scholars have considered the contact between European settlement and colonial expansion in terms of exclusion of native names, even native peoples, from the toponymic and mapped representation of the "new" colonial space. K.G. Brealey has discussed the ways in which maps helped "actualise the territorial dispossession" of the Nuxalk and Ts'ilhqot'in peoples of what is now British Columbia and demonstrated how European maps both "contained" and "represented" natives according to the world view of the coloniser.[64] The same has been the trajectory of many other native groups in North America such as the Sioux of the northern United States and the Inuits of the Arctic. In his spatial and cultural history, Paul Carter seeks the origins of Australian civilisation in journals, letters home, unfinished maps and other narratives by explorers and soldiers and emigrants, including hearsay and tales of escapes of convicts who helped settle the new place. The transformative power of naming is demonstrated here in Captain Cook's renaming of Botany Bay in 1770; the bay that was initially called Stingray Bay. Cook's revision altered the bay's meaning placing it within the context of Cook's entire journey and his experience.[65] Researches undertaken in Australia, New Zealand, Hawaii and the south Pacific Islands have discussed the cartographic inscription of what was, purportedly, Terra Nullius, the empty expanse of wilderness first encountered by colonial explorers and mapmakers who named and transformed them into inhabitable space. Within Britain, linguistic and cultural "others" like Scotland, Wales and Ireland were mapped through surveys, with the objective to incorporate them into the larger geo-body of "Great Britain". Charles Withers shows how the Ordnance Survey in the nineteenth century Scottish Highlands, undertaken by the English, transformed the Gaelic landscape of the natives. An examination of the Ordnance Survey's Original Object Name Books is used to explore how the landscape was "authorised" or rewritten.[66]

The empire of the gaze: surveys and maps

With reference to India, Matthew Edney points out the complex process of compilation of data, where locations were to be fitted to the rigid mathematical framework of longitudes and latitudes. This involved a process of translation of native astrological vocabulary into a European system of meridians and parallels. The translation itself was dovetailed with the incorporation of native knowledge into the British archive.[67] This signified not only the conquest of the physical space of the Indians, but also, the appropriation of an epistemological space, and what Cohn calls "epistemological conquest".[68] The famous cartouche inscribed on the first edition of James Rennell's *Map of Hindoostan*, (1782), depicts this surrender of native knowledge in the form of the Brahminical sourcebook, *the Vedas*, as tribute to the superior imperial nation of Britannia, personified as a queen. The map symbolised the combined act of military, geographical, cultural and epistemological conquest, for the British made themselves the intellectual masters of the Indian landscape. Side by side, most importantly, cartography in India, as in other places, in defining the colonial territory, laid down the boundaries of a future nation state.

However, cartography did not merely pertain to the production or representation of spaces but, with the help of what can be called "anthropometric" or racial cartography, also charted the bodies which inhabited those spaces. Cartographic representations of race were situated within the context of late nineteenth- and early twentieth-century Britain. Such constructions of race tended to dwell on photographic and museum representations.[69] In trying to look beyond the apparent, natural sciences like botany and zoology, they tried to seek the sources of species, language and development. Mary Louise Pratt traces the prevalence of the scientific scrutiny in travel narratives as being a part of the emergent "planetary consciousness" during the time period 1750–1800. This attempted to systematise, catalogue and classify all existing species occupying the surface of the earth, "into sequence of a descriptive language".[70] In being engaged in discovering the so-long hidden system of nature, the natural historians, as it were, led to "a new field of visibility being constituted in all its density".[71] Early nineteenth century botany, zoology and anatomy cannot be understood without this attempt to seek out remote fields for the collection of as yet unknown specimens in order to bring them under the overarching classificatory system of natural history. The definition of a scientific "object" was usually central to the politics of colonial anthropology. Such anthropological knowledge production was not limited to discourses alone, but was fundamentally linked to vision and to exhibition. In this context the importance of photography as document privileged this artefact, which had a clearly articulated role as embodiment of cultures, catering to the visualist bias of Western ethnological fieldwork. In India, colonial officials posing as amateur photographers took up a number of such photographic enterprises, including Herbert Risley, J. Forbes Watson and John William Kaye, to name a few. The colonial photographic enterprise in India can be seen as another "form of knowledge" working in tandem with British anthropological writings of the nineteenth century. Photography was taken up by colonial officials, sergeants and anthropologists with "scientific" agendas in studying the Indian people in their plurality and diversity. There was a need to record the exact number of human "specimens" to be found in India claiming adherence to race, tribe or caste, which could be clearly tabulated in photographic encyclopedias.

Surveying and mapping were important aspects of the technologies of state formation and administration even in early modern Europe, which were replicated during the period of colonial expansion all over the world. The eighteenth and nineteenth centuries saw an upsurge of surveying and mapping activities all across the colonies. The British in India in the late eighteenth century commissioned surveys which were meant as exploration of the natural and social landscape. The survey in India encompassed a wide range of activities apart from mapping, like collecting botanical specimens, recording of architectural and archaeological sites of historical significance and land surveys for revenue and taxation. It should be remembered, however, that this was the age of Enlightenment, the age of reason, science, progress and discoveries. As Nicholas Dirks observes:

> Reason made discovery the imperative of Western thought, [...] colonialism provided a critical theatre for the Enlightenment project, the grand laboratory that linked discovery and reason. Science flourished in the 18th century not merely because of the intense curiosity of individuals working within Europe, but because colonial expansion both necessitated and facilitated the active exercise of scientific imagination. It was through discovery – the siting, surveying, mapping, naming and ultimately possessing – of new territories of conquest: cartography, geography, botany, and anthropology were all colonial enterprises.[72]

The picturesque gaze

The systematic knowledge so collected, not only restricted itself to written documents but permeated into exhibitions and picturesque representations. Where maps were constructs that combined assorted observations into a unifying image of a space, the landscape paintings of the period added perspective to views, the images then standing in lieu of the world. Strangely, enough, science and the aesthetics joined hands together as the "Europeans were actively constructing both the picture and the picture-like quality of an East that was still for them, [...] a moving image".[73] Just as the surveys aimed to unfold universal and objective truths, so also:

the picturesque conventions of aesthetic transcription worked to constitute India as a place where reality itself was defined as presentation, all the while masking the brutality of conquest, justifying the British military presence, and in the end defining India as a place where time had stopped and culture had given way to nature.[74]

Arguably, landscape implies the denial of process. Reading landscape in this manner is based on understanding the concept's historical emergence that emphasises its visual and painterly dimensions. Landscape in this interpretation is a restrictive way of seeing, which privileges the "outsiders" point of view, while sustaining a radical split between the outsider or the seer and the insider or the seen: between those who relate directly to the land and those who relate to it as a form of exchange value.[75] Lefebvere's concepts of "representation" and "representational" space are useful here in talking about the artistic form, the gardens and estates and the space that exists as a part of everyday social practice. The scholarship in this field has generated an opening up or "unpacking" of the concept of the landscape in denying the related concepts of space/place, inside/outside, original, image and representation of a precise closure. The general idea here is that there is no "absolute" landscape and the relationship between place and space, inside and outside and image and representation are cultural constructs and dependent on historical processes. The organisation of space is always already coded in the way it is experienced. Similarly, the "visual", considered the primary property of pictures, too needs to be unchained from restrictive practices to turn our attention to "ways of seeing" that mould the experience of space and through which space itself is moulded.[76] Further, landscape paintings of the colonial moment involve a materially located process of perception and identification that works to shape forms of social identity.

In talking of the "world as picture", Christopher Pinney invokes Heideggar to speak of it as a typically modern capability. Heideggar traces a resultant alienation out of scientific pursuit from Descartes onwards which allowed the world to be seen as a picture – an object to be viewed and seen rather than to be lived and experienced. The "Cartesian perspectivalism", the outcome of Cartesian dualism, of that which is experienced by the body and that which is perceived by intuition, led to claims of certainty of representation. The modern epoch placed the world as a picture in the realm of man's knowing:

> There begins that way of being human that mans the realm of human quality as a domain given over to measuring and executing, for the purpose of gaining mastery over that which is as a whole.[77]

On the other hand, the act of reading a pictorial representation involves, as Louis Marin explains, the dual achievement of pleasure and power:

> pleasure, for in a representation, the theoretical desire of identification is achieved, the reflection of representation on itself where the subject

15

is constituted in truth; power, since by mastering the appearances painted on canvas, the subject dominates and appropriates for itself its vision in theory, occupies fictively and legitimately the position – without position – of God.[78]

To the same tune, Gillian Rose explains the colonial "frame" when she perceptively identifies what she calls "the uneasy pleasures of power" with the view of the Orient-as-woman reclining before the scopic virilities of the masculinist spectator.[79] As discussed before, the paradigm of visibility is quite obvious in British representations of their empire. It is, as it were, literally, what Foucault called "the empire of the gaze", as colonisation designed spaces to make things seeable in certain ways. Timothy Mitchell speaks of Orientalist knowledge as markedly pictorial. In the same vein, Said suggests that colonial inscriptions of the orient were constituted panoramically. In demonstrating this, Derek Gregory discusses the production of *Description de l"Egypte* (1809–22) by the orders of Napoleon.[80] In analysing the image in the frontispiece to the original French compendium, Gregory points out how in "a view through the portal of a stylised temple on to the monumentalised landscape of ancient Egypt, all signs of life or contemporary inhabitants of Egypt have been erased. This "memorialised landscape" signified "a union of power, knowledge and geography" which inscribed its surveyors in "positions of power and prominence". The *Description* itself was remarkable in its sheer detailing. At each site the inventory begins with an eagle's eye view with topographic maps locating the antiquities, which are then displayed in panoramic view giving way to perspective views, which in turn dissolve into the close-up detail of reliefs and inscriptions. The detailed representation and ethnography were not merely a way of claiming empirical authority in the sense of "being there" but invested the colonial regime with legitimacy of presence.[81] However, pictorially, it insists upon the "imaginative geographies" of Orientalism. It superimposed the abstracted space or Lefebvre's "represented space" of the coloniser onto the concrete place of the native in presenting it as an essentialised and timeless setting.[82]

The Indian picturesque is precisely imperial landscape. The semiotic features of landscape, and the historical narratives they generate, are tailor made for the discourse of imperialism as the very genre metaphorises an expansion of landscape which is understood here as an inevitable outcome of colonisation:

Empires move outward in space as a way of moving forward in time; the "prospect'" that opens up is not just a spatial scene but a projected future of development exploitation.[83]

The "picturesque" was an ideal; and only through art and the active agency of the artist/viewer could a natural scene be converted into the picturesque. The British found the "picturesque" to be the perfect intellectual tool for imagining the landscapes of South Asia. Just as India served as a laboratory for various disciplines, for painting and fine art, it provided the sites of naturally occurring beauty that could be carried back to England and owned by means of replicas as the Daniells, the professional painters to India remark:

> Science has had her adventures, and philanthropy her achievements; the shores of Asia have been invaded by a race of students with no rapacity but for lettered relics; by naturalists, whose cruelty extends not to one human inhabitant: by philosophers, ambitious only for the extirpation of error, and the diffusion of truth. It remains for the artist to claim his part in these guiltless spoliations, and to transport to Europe the picturesque beauties of these favoured regions.[84]

However, the picturesque in India was manipulated and manufactured. It was also highly selective and the natural had to fall into the rigid aesthetic schema of the "picturesque" as understood in Great Britain. The "picturesque" as a way of seeing was highly in fashion by the late eighteenth century. By 1780, "picturesque" beauty referred in general to scenes, which recalled two different kinds of paintings – on the one hand, the ideal classical landscapes of Claude Lorraine, Gasper Poussin and Salvator Rosa, on the other the naturalistic views of Ruysdael, Hobbema, Gyp and Van Goyen. During the 1790s three books provided a working aesthetic: Richard Payne Knight's didactic poem, *The Landscape* (1794), Uvedale Price's "An Essay on the Picturesque" (1794), and Humphry Repton's *Sketches and hints on picturesque gardening* (1795). Apart from these, there were the rules of the "picturesque" simplified and laid down by Dr William Gilpin who produced a series of guides between 1782 and 1809. These trained and educated new artists to look at landscapes in a certain way. And if that was not all, there were special scientific instruments to produce a picturesque scene, such as the special lenses called "artists' viewers", the convex mirrors called "Claude Glass", the "camera obscura" and the "camera lucida", all of which solved various problems of size, form, perspective and view.[85] R.R. Reinagle's letterpress to Turner's *Views in Sussex,* stresses the "science" which lies behind the apparently natural effects that master artists achieve:

> Science alone, aided by genius, can do this. These are the high qualities that can enslave and enchant the eye. It is the science of the art so little known, though never failing in the works of those who have been crowned by the praise of the world, and successive ages, that is constantly overlooked and mistaken for art only. The art is imitative; science produces choice; and entangles and entwines itself within the former so carefully, as to be unperceived.[86]

The cult followed with ardour in England, with outskirts bordering the English mainland, under the linguistic and cultural predominance of Celtic and

Gaelic dialects, after which artists flocked to India which served as suitably rich material, to be "captured" in paintings. Therefore, Gombrich observes that the Western tradition of painting has been "pursued as a science" through a process of "ceaseless experimentation".[87] As many Marxist critics have pointed out, landscape paintings were produced for use primarily in urban spaces and it was by far a class-based aesthetic invoking a feudal attitude, or at least a nostalgia towards land against the eighteenth and nineteenth centuries' backdrop of emergent capitalism which can be extended to imperial attitudes to colonial lands.[88]

Spatial metaphorics

The concepts of spatial production or of "imaginative geography" that are being talked of here involve the assignment of systematic and differentiated attributes to places in order to designate in one's mind a familiar space which is "ours" and an unfamiliar space which is "theirs" but could be "ours". This requires what Gregory calls the "spatial metaphoric" or the practice of "poetics of space" whereby a space acquires emotional and even rational sense by a kind of poetic process with the help of language, and the vacant or anonymous reaches of distance are granted meaning with an attachment of figurative or imaginative value. It is imperative for us to understand how emotions like anxiety, fear, desire and fantasy enter into the production of imaginative geographies. It is important to analyse topographies as represented through writing and literature, which describe them in metaphors. Richard Rorty claims that metaphors have no meaning other than their literal one, which is nonsensical. Precisely because metaphors are literally nonsensical, they are the "frisson" that makes us see the world in a different way, a way that could not be imagined before the metaphor was used. Metaphors create new angles on the world once they are "savoured rather than spat out".[89] Metaphors are socially and culturally constructed within the literary sphere as they act as tropes for reducing the unfamiliar to familiar, by translating the unknown through already known frames of references. For our concern, metaphors can be subdivided into "big" and "small". Big metaphors are those that lie behind general research methods and school of thought, while small metaphors are used as ornamentation in pieces of writing, and surely the two are connected.[90] Both are used as persuasive and rhetorical devices in representing the unknown in known language. Hayden White equates historiography to a literary activity when speaking of the historical narrative as extended metaphor which evokes the familiar and thereby persuades by likening the events reported in them to some form with which we are already familiar through the literary culture.[91] Writings which represent space, employ similar metaphors which invoke a recognisable narrative unity.

Geography and the craft of penmanship also go hand in hand. Connections between polite writing and geographical knowledge were (and still are), part of eighteenth- and nineteenth- century imperial and domestic politics. Travel literature, journals and travelogues are narrative accounts of eyewitness records during voyages and stays in foreign countries constitutive of this very geographical knowledge talked of. In Dennis Porter's words:

> From the beginning, writers of travel have more or less unconsciously made it their purpose to take a fix on and thereby fix the world in which they found themselves; they are engaged in a form of cultural cartography that is impelled by an anxiety to map the globe, centre it on a certain point, produce explanatory narratives, and assign fixed identities to regions and the races that inhabit them.[92]

Travel literature, however, is an inherently interdisciplinary form because of its cross-generic status in its intermixing of fiction, autobiography, history, reportage and natural history. Factually, the Britain of the eighteenth and the nineteenth centuries saw the highest proliferation of travel literature which is indicative of its connection with the empire. As Mary Louise Pratt observes, travel literature during this period is fundamentally linked to the colonial gaze. It is engaged with the metropolis's obsession with presenting and representing its peripheries and its others to itself. In this it constructs not only the space of the "contact zone" but also the metropolis itself.[93] Travel writing thus also involves a mapping, a production of an imaginative geography which turns the experience of an unfamiliar place into narrative. In this it employs all the literary devices and tropes, which grants it a complex mix of fact and fiction, of objective reality and subjective experience. The journal format of the travel account has attracted critics to reading it in terms of autobiography, as a technology of individuation or writing the self. The act of writing the self, of course, implies also, the act of construction of identity, and in the case of colonial travelogue, it is the imperial self that gets articulated through the construction or representation of the other. On the other hand, as a spatial practice, it tends to become the representational space that Lefebvre discusses as it employs descriptions of one's own experience and lived reality. However, the flux and the locomotion which characterises the travel narrative, where the writer–protagonist sequentially moves from one to another foreign land and places, seen in fleeting glances, restrains it from becoming a full-fledged representational space and remains, therefore, a representation of space. It is after all an outsider who visualises and writes of a space.

For the European Renaissance traveller, writing was an integral part of the activity. In the face of rivalry among European nation states, political or commercial documents and maps would have to abide by codes of secrecy. It was only through travel writing that public interest in foreign lands was kindled. As Peter Hulme and Tim Youngs point out, this public interest aroused by stories of faraway places was an important way of attracting investment and, settlers, once the colonies were established.[94] For Francis Bacon, who prescribed travel as a form of education for creating an ideal Renaissance man, the travellers of the Renaissance had discovered a "new continent" of empirical truth which defeated earlier speculations of the ancients. John Locke founded his observations on the world and nature of man on numerous travel accounts he collected about various places. Although travel in the sixteenth and seventeenth centuries was exclusively opportunist and commercial, from the eighteenth century onwards there emerged new kinds of travellers, implicated in the rhetoric of science. These were natural scientists such as biologists, botanists and geographers who were initially sponsored by the European nation states, but then gathered momentum on their own, with the Royal Geographical Societies spearheading the movements. The expeditions of geographers and natural scientists provided some initial instances of travel writing such as Darwin's *The Voyage of the Beagle* (1840) which also draw on the conventions of the "anti-conquest", by which term Pratt refers to the "strategies of representation whereby European bourgeois subjects seek to secure their innocence in the same moment as they assert hegemony".[95] While often disguised as innocent activities, the White explorer–spectator enforces colonialist assumptions in their romanticisation of the "exotic" or the "primitive". For example, travellers, explorers and missionaries talk of the lands of Africa, Caribbean islands and Australia in similar fashion as infested by strange animals and primitive black men. Likewise, the India of these fanciful travel accounts was:

> a land of "castellated elephants", "proud rajas" and "melodious bulbuls", of silks, muslins, precious gems, of "white cities", "gilded minarets", and "glittering scimitars", of "snorting arabs" … and the "dark-eyed daughters of the east". It was a place of "tall palm trees and browsing camels, rose gardens and citron-groves".[96]

David Arnold mentions a specific literary style, akin to British Romanticism, which evolved through these travelogues, which could at once be identified as Orientalist or associated with tropical writing. The land and environment of the spaces described, acquired a visibility through this language, which Gregory calls the "metaphorics of space".[97]

Paginal spaces

As Pinet says, "No one can be a good chorographer who is not a good painter".[98] The combination of maps, pictures and written documents not only play a definite role in providing geographical knowledge, education and entertainment, these also, most importantly, construct a territorial space, influencing future decisions and policy making. In the age of Enlightenment, this combination was characteristic of the technique of copper-plate engraving which, unlike letterpress printing, could readily and elegantly combine words and images on one plate in many different ways. The same combination carries on in map production in the sixteenth century, where the same copper-plate engraver's burin could inscribe both the linework and the lettering, along with elaborate pictorial cartouches, to achieve a unified stylistic effect where word and image coexisted. In the eighteenth century, too, engravings integrated text as accompanying poems, speech bubbles or identifying notations.[99] The same was noticeable in the depictions of newspapers, books, broadsides and manuscripts of the nineteenth century, continuing to the present. The combination and synchronisation of graphic and textual images constitute the organisation of knowledge of a space in vivid terms: a map shows the place in its totality, the landscape paintings depict individual scenes from vantage points, and the travelogue fabricates a story of a journey through these spaces, representing daily life, flora and fauna in its minute details. Thus, the collaboration of the "god's eye view", "the bird's eye view" and the "ant's eye view" has the power to conjure up a space which is ostensibly "real".

Most importantly, in writings, cartographic images and landscape paintings in the period of the eighteenth and nineteenth centuries, new views of local and national space began to be depicted. This is surely an outcome of the "implied augmentation and pluralisation" of the world, born due to new discoveries and eventual colonialism.[100] Fresh local and national idioms matured, replacing the dilemma over uncertain borders which had been given under the authority of Greek and Latin forebears. The simultaneity of the new topographical map with the new literary self-consciousness is clearly discernible as a hallmark of this re-centring. The topographer or the cartographer reveals to sight extraordinary varieties of terrestrial space that can be depicted in novel ways that exceed what eyes could normally see and had not seen so far. The resultant vision of nature is transported to the new world held under conquest drawn through the romance of the fantasy portrait of the individual traveller who is also the painter and the map maker, and who also doubles up as the harbinger of science. On the other hand, this scientific discourse is itself the product of the interaction between the core and periphery and the need to explain matters that is exposed to sight. Things unknown became singularities which it became the task of the traveller cartographer to eke out, in the foreseeable plot of distinguishing one geographical or cultural zone from another. In this manner, the new idiom acquired a descriptive status defining people and places in certain singular representations.

For the coloniser, the work did not stop, of course, at the level of conjuring a space in the colonial consciousness, but went further to transform the space for inhabitation and settlement. A reading of the town planning, sanitation and architecture in order to "modernise" native villages into colonial quarters, would complete the understanding of the process of production of space.[101] Through case studies in the following chapters, I shall study the process of evolution of the idea of India as a self-referential entity during the period of British rule. It shall justify my eschewal of a more socially relevant idea of space as a human artefact, encoded with numerous cultural, political and economic forces. To summarise: the main arguments of this book stand focussed on the dominance of cartographic reason in the age of European Enlightenment, which framed aesthetic and scientific modes of representations and imaginations such as maps, landscape paintings and travel narratives. These cultural expressions are spatial practices playing key roles in the processes of spatialisation. They produce spaces according to dominant terms of knowledge, which has tenable links with power. This cartographic reason is ocular centric, is expansive, and tries to create an integrated "one world view"[102] (see Figure I.1) which is a forerunner to today's globalisation; likewise, it often functions with a high-altitude aerial panopticon-like[103] point of view integrating it with finer in situ minutiae intending to familiarise the landscape, often leading to stereotypical essentialisations. I argue that colonialism embraces this gaze. This line of argument is fruitful to pursue because it helps us understand and critique spatial matrices constituted by the forces of colonialism. What came to be known as Great Britain, and what got conceptualised as British India were the products of an identical imagination: flipsides of the same coin. Therefore, we can say that the production of India as a new geography is sourced from Britain's own interaction with its rural outskirts and domination in its fringes. The making of colonial geographies were contingent upon the making of imperial Enlightenment persona within which key contributors wanted to fit their individual identities as respectable and worthy gentlemen of science and contributors to the colonial encyclopedic archive. In India, as in Britain, this search for knowledge acquired a remarkably repetitive route, (reminiscent of Rennell's first map project), travelling inwards from the colonial presidencies, especially the port city of Kolkata, towards the countryside and finally the high mountains in the northern side which stalled imperial expansionist ambitions and eventually turned into the frontier of the British colony, and thereafter, border to the country. Practices of these individuals, therefore, materially and visually shaped India.

The by-and-large repetitive structural pattern is integral to the central arguments and claims I make in this book regarding the rise of the cartographic rationale in the West and the fashioning of new global geo-political realities which were products of this same cartographic reason. The following chapters are clubbed under sections which identify and discuss three kinds of spatial practices, namely, maps, landscape paintings and travel writings. The first chapters in each of these sequences discuss the British, specifically, English spatial interventions where knowledge is produced at its own fringes. These chapters explore the deep interconnections between the development of each of these cultural forms as an industry, and the making of the national territory, identifying people and practices which have architected its metamorphosis. The second chapters in each section go on to expand the margins of both discourse and space to favour and fit the process of colonisation, specifically in the region which came to be known as India.

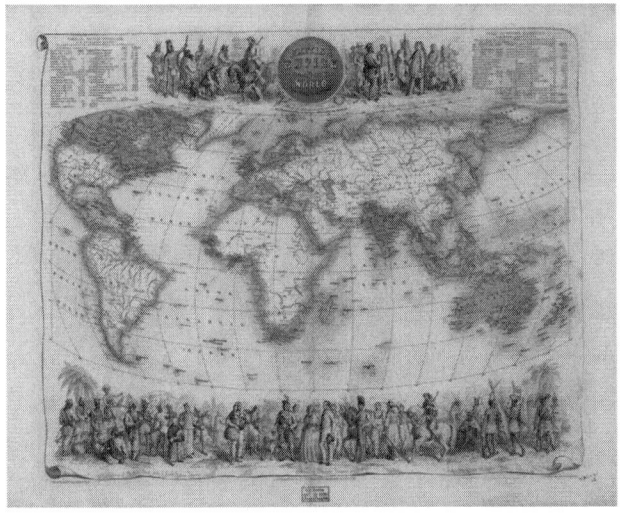

Figure 1.1 British Empire Throughout the World Exhibited in One View, John Bartholomew and Co. c. 1850
Image courtesy: Library of Congress, Geography and Map Division

Notes

1 Friel, Brian. Translations. In *Collected Plays*. Vols. 2 & 3. Ireland: The Gallery Press. 2016.
2 Ibid.
3 See Loomba, Ania. *Colonialism/Postcolonialism*. London: Routledge, 1998. pp. 87–88; Mukherjee, Nilanjana. "Translations/Representations/Colonisations", *Language Forum*, Vol. 30, No. 2. July–Dec. 2004.
4 de Saussure, Ferdinand. *Course in General Linguistics*. Trans. R. Harris. Chicago: Open Court Publishing, 1998. pp. 9–10.
5 See Cronin, Nessa. "Lived and Learned Landscapes. Literary Geographies and the Irish Topographical Tradition. In Marie Mianowski ed. *Irish Contemporary Landscapes in Literature and the Arts*. Palgrave Macmillan, 2012. pp. 106–18; Cronin, Nessa. "The Eye of History: Spatiality and Colonial Cartography" (Unpublished Ph.D. thesis), NUIZ, Galway, 2007. Also see Kiberd, Declan. *Inventing Ireland: The Literature of the Modern Nation*. London: Jonathan Cape, 1995.
6 Kipling, Rudyard. *Kim*. (1901). Oxford: Oxford University Press, 2008. Also see Nagai, Kaori. *Empire of Analogies: Kipling, India and Ireland*. Cork, Ireland: Cork University Press, 2007.
7 See Foucault, Michel. "Of Other Spaces: Utopias and Heterotopias". In *Architecture/Mouvement/Continuite*. Tr. Jay Miskowiec. 1967, for a discussion on hierarchies of space.
8 Cohn, Bernard. *Colonialism and its Forms of Knowledge*. Oxford: Oxford University Press, 1997. p. 4.
9 A recent work which tries to put the cultural assimilation of diverse regionalities into a unifying paradigm and the tensions within Britain and its overseas empire in

a comparative framework is Lamont, Claire and Michael Rossington eds. *Romanticism's Debatable Lands*. UK: Palgrave Macmillan, 2007.

10 Wagoner, Phillip B. "Precolonial Intellectuals and the Production of Colonial Knowledge". *Comparative Studies in Society and History*. Vol. 45, No. 4, 2003; Raj, Kapil. *Relocating Modern Science: Circulation and Construction of Knowledge in South Asia and Europe, 1650–1900*. Basingstoke: Palgrave Macmillan, 2007, especially the section "Circulation and the Emergence of Modern Mapping: Great Britain and Early Colonial India", pp. 60–93.

11 See Eck, Diana L. *India: A Sacred Geography*. New York: Harmony Books, 2012. Eck, Diana L. "The Goddess Ganges in Hindu Sacred Geography". In John Hawley and Donna Wulff eds. *Devi: Goddess of India*. Berkeley: University of California Press, 1996; Eck, Diana L. "The Imagined Landscape: Patterns in the Construction of Hindu Sacred Geography". *Contributions to Indian Sociology* 32:2, 1998, pp. 165–188; Sinha, Amita. *Landscapes in India: Forms and Meanings*. New Delhi: Asian Educational Services, 2011; Kramrisch, Stella. "Space in Indian Cosmogony and in Architecture". In Kapila Vatsyayan ed. Concepts of Space: Ancient and Modern. Indira Gandhi National Centre for the Arts, New Delhi: Abhinav Publications, 1991. pp. 101–4.

12 Tuan, Yi-Fu. *Topophilia: A Study of Environmental Perceptions, Attitudes, and Values*. New York: Columbia University Press, 1990.

13 See Bollnow, Otto Friedrich. *Human Space*. Tr. Christine Shuttleworth. London: Hyphen, 2011; Deleuze, Gilles. *Essays Critical and Clinical*. Tr. Daniel W. Smith and Michael A. Greco. London: Verso, 1998; Jameson, Frederick. *The Geopolitical Aesthetic – Cinema and Space in the World System*. London: BFI, 1992.

14 See for example, Helgerson, Richard. *Forms of Nationhood: The Elizabethan Writing of England*. Chicago: Chicago University Press, 1994; McLeod, Bruce. *The Geography of Empire in English Literature 1580–1745*. Cambridge: Cambridge University Press, 1999.

15 Netzloff, Mark. *England's Internal Colonies: Class, Capital and the Literature of Early Modern Colonialism*. New York: Palgrave Macmillan, 2003.

16 See for instance, Bergmann, Sigard. *Religion, Space and the Environment*. New Brunswick: Transaction Publishers, 2014. Also see Wolters, O.W. *History, Culture and Regions in South East Asian Perspectives*. SEAP Publications, 1999. For discussions on Vastu Shastras and its incorporation in maps and plans, see the section titled "Spatial Archetypes" in Sinha, Amita. *Landscapes in India: Forms and Meanings*. New Delhi: Asian Educational Services, 2011. pp. 125–66.

17 Chatterjee, Indrani. *Forgotten Friends: Monks, Marriages, and Memories of Northeast India*. Delhi: Oxford, 2013.

18 Chatterjee, Indrani. "Connected Histories and the Dream of Decolonial History". *South Asia: Journal of South Asian Studies*. Vol. 41, Issue 1, 2018. pp. 69–86.

19 Subrahmanyam, Sanjay. "Connected Histories: Notes Towards a Reconfiguration of Early Modern Eurasia". *Modern Asian Studies*. Vol. 31. No. 3 Jul 1997. pp. 735–62.

20 Chattopadhyaya, B.D. *The Concept of Bharatavarsha and Other Essays*. New Delhi: Permanent Black, 2017. p. 9.

21 Ibid. p. 10.

22 Habib, Irfan. "The Formation of India – Notes on the History of an Idea", *Social Scientist*, nos. 7–8 (1997), p. 8. See also Habib, Irfan. "The Envisioning of a Nation: A Defence of the Idea of India", *Social Scientist*, vol. 27 (nos. 9–19), 1999, pp. 18–28.

23 Chattopadhyaya, B.D. *The Concept pf Bharatavarsha and Other Essays*. New Delhi: Permanent Black, 2017. pp. 1–25. See Mookerji, R.K. *The Fundamental Unity of*

India. New Delhi: Chronicle Classics Series, 2003 (originally published in 1913); Mukherjee, B.N. *The Concept of India*. Calcutta: Sanskrit Pustak Bhandar, 1998.

24 Kaviraj, Sudipta. "The Imaginary Institution of India". In Partha Chatterjee and Gyanendra Pandey eds. *Subaltern Studies*, Vol. VII, New Delhi: Oxford University Press, 1999. pp. 1–39.

25 Goswami, Manu. *Producing India: From Colonial Economy to National Space*. Chicago: Chicago University Press, 2004.

26 Ibid. Specifically the chapters "Colonial Pedagogical Consolidation" pp. 132–53 and "Space, Time, and Sovereignty in Puranic Itihas" pp. 154–64. Bayly, Chris. *Empire and Information: Intelligence Gathering and Social Communication in India, 1780–1870*. Cambridge: Cambridge University Press, 1996. pp. 247–64.

27 Goswami, Manu. *Producing India: From Colonial Economy to National Space*. Chicago: Chicago University Press, 2004. pp. 132–53. Basu, Subho. "The Dialectics of Resistance: Colonial Geography, Bengali Literati and the Racial Mapping of Indian Identity". *Modern Asian Studies*. 44, 1. pp. 53–79. Seth, Sanjay. *Subject Lessons: Western Education of Colonial India*. London: Duke University Press, 2007.

28 Bayly, Chris. *Empire and Information: Intelligence Gathering and Social Communication in India, 1780–1870*. Cambridge: Cambridge University Press, 1996. p. 247. Bayly is a chief proponent in the field of research focussed on indigenous agency behind forging of new structures of information order. Here he differs from Cohn who, in his books like *An Anthropologist Among Historians and Other Essays*. New Delhi: Oxford University Press, 1987, and *Colonialism and Its Forms of Knowledge*, studies the nexus between Western knowledge and the power of the colonial state.

29 Lefebvre, Henri. *The Production of Space*. Oxford: Blackwell, 1991. p. 7.

30 Harvey, David. *Condition of Postmodernity: An Enquiry into the Origins of Cultural Change*. Oxford: Basil Blackwell, 1989.

31 Also see Soja, Edward. "The Trialectics of Spatiality". In *Third Space: Journeys to Los Angeles and Other Real and Imagined Spaces*. Oxford: Blackwell, 1996. pp. 53–82. The trialectics comprising of perceived, conceived and lived corresponds to what Soja refers to as First, Second and Third Space. The Third Space is the space of resistance to hegemony of the First and Second Spaces and is therefore the space of possibilities or hybridisation.

32 Soja, Edward. *Postmodern Geographies: The Reassertion of Space in Critical Theory*. London: Verso, 1989. pp. 10–11.

33 Brewer, Daniel. "Lights in Space". *Eighteenth Century Studies*. Vol. 37, No. 2, 2004. pp. 171–86.

34 Withers, Charles. "Art, Science, Cartography and the Eye of the Beholder". *Journal of Interdisciplinary History*. Vol. 42, No. 3, 2012. pp. 429–37.

35 Withers, Charles. "Memory and the history of geographical knowledge: the commemoration of Mungo Park, African explorer". *Journal of Historical Geography*. Vol. 30, 2004. pp. 316–39.

36 Ibid.

37 Low, Martina. "The Constitution of Space: The Structuration of Spaces Through the Simultaneity of Effects and Perception". *European Journal of Social Theory*. Vol. 11, Issue 1. 2008. pp. 25–49. The article suggests how space is constituted through a process of perception and ideation giving shape to a social structure based on specific orderings of living entities and social goods.

38 Broglio, Ron. *Technologies of the Picturesque: British Art, Poetry and Instruments 1750–1830*. Lewisburg: Bucknell University Press, 2008. p. 58.

39 Said, Edward. *Orientalism*. London: Routledge and Kegan Paul, 1978. p. 322.

40 Gregory, Derek. "Edward Said's Imaginitive Geographies." In Mike Crang and Nigel Thrift eds. *Thinking Space*. London: Routledge, 2000. p. 302.
41 Said, Edward. *Orientalism*. New Delhi: Penguin, 2001. p. 55.
42 Gregory, Derek. "Edward Said's Imaginitive Geographies." In Mike Crang and Nigel Thrift eds. *Thinking Space*. London: Routledge, 2000. p. 313.
43 Ibid. p. 317.
44 Anderson, Benedict. *Imagined Communities: Reflections on the Origin and Spread of Nationalism*. London: Verso, 1991. pp. 163–86.
45 Ibid. p. 192.
46 Winichakul, Thongchai. *Siam Mapped: A History of the Geo-Body of the Nation*. Honolulu: University of Hawaii Press, 1994.
47 See for instance, Bowen. H.V. *Elites, Enterprise and the Making of the British Overseas Empire, 1688–1775*. New York: St. Martins's Press, 1996; Wilson, Kathleen. *The Sense of the People: Politics, Culture and Imperialism in England, 1715–1785*. Cambridge: Cambridge University Press, 1998.
48 Recently historians of New Imperialism, a phase characterised by unprecedented territorial acquisition, have developed the analytic category of studying connections and networks between metropole and colony. See for example, Lambert, David and Alan Lester eds. *Colonial Lives Across the British Empire: Imperial Careering in the Long Nineteenth Century*. Cambridge: Cambridge University Press, 2009; Butlin, Robin A. *Geographies of Empire: European Empires and Colonies c. 1880–1960*. Cambridge: Cambridge University Press, 2009.
49 See Armitage, David. *Ideological Origins of the British Empire*. Cambridge: Cambridge University Press, 2000 for a rare study of the ideological basis underlying the history of cultural unity of England, Scotland and Ireland.
50 See Colley, Linda. *Britons: Forging the Nation, 1707–1830*. New Haven: Yale University Press, 1992; MacKenzie, John. *Propaganda and Empire: The Manipulation of Public Opinion, 1880–1960*. Manchester: Manchester University Press, 1984; MacKenzie, John ed. *Imperialism and Popular Culture*. Manchester: Manchester University Press, 1986. These books address the consolidation of British identity in the Isles based on ideological premises justifying overseas aggrandisement and trade, territorial acquisition and war with other European nations. MacKenzie's other works like his "Essay and Reflection: On Scotland and the Empire", *The International History Review*, 15: 4. 1993, pp. 714–39 and "Scottish Orientalists, Administrators and Missions: A Distinctive Scots approach to Asia?" in Devine, T. M. and Angela McCarthy eds. *The Scottish Experience in Asia c. 1700 to the Present: Settlers and Sojourners*. Palgrave Macmillan, 2017. pp. 51–73 discuss and trace specifically Scottish connections within British Empire.
51 Cohn, Bernard. *Colonialism and its Forms of Knowledge*. Oxford: Oxford University Press, 1997. p. 4.
52 Ibid. p. 11.
53 Edney, Matthew. *Mapping the Empire: The Geographical Construction of British India, 1765–1843*. Chicago: Chicago University Press, 1997. p. 5. Here, Edney quotes from Barbara Stafford. *Voyage into Substance*.
54 See Raj, Kapil. "Relocating modern science." *Circulation and the construction of knowledge in South Asia and Europe: 1650–1900*. London: Palgrave Macmillan, 2007. Also see Bravo, Michael T. "Precision and Curiosity in Scientific Travel: James Rennell and the Orientalist Geography of the New Imperial Age (1760–1830). In Jas Elsner and Joan-Pau Rubies eds. *Voyages and Visions: Towards a Cultural History of Travel*. London: Reaktion Books, 1999. pp. 162–83.
55 Quoted in Moran, Joe. *Interdisciplinarity*. London: Routledge, Indian reprint, 2007. p. 167.

56   Foucault, Michel. "Questions on Geography". In Colin Gordon ed. *Power/Knowledge: Selected Interviews and Other Writings 1972–1977*. Hempstead: Harvester Wheatsheaf, 1980. p. 70.

57   Ogborn, Miles. "*Geographia*'s pen: writing, geography and the arts of commerce, 1660–1760". *Journal of Historical Geography*, 30, London, 2004. pp. 294–315. p.

58   See Barthes, Roland. *The Responsibility of Forms: Critical Essays on Music, Art and Representation*. Tr. Richard Howard, Oxford: Basil Blackwell, 1986.

59   Barnes, Trevor J. and James S. Duncan eds. "Introduction" to *Writing Worlds: Discourse, Text and Metaphor in the Representation of Landscape*. London: Routledge, 1992. p. 5.

60   Harley, J.B. "Maps, knowledge and power". In D. Cosgrove and S. Daniels eds. *The Iconography of Landscape: Essays on the Symbolic Representation, Design and the Use of Past Environments*. Cambridge: Cambridge University Press, 1985. pp. 300–3.

61   Wintle, Michael. "Renaissance maps and the construction of the idea of Europe". *Journal of Historical Geography*, 25, 2, London, 1999. pp. 137–65. p. 159.

62   Tyner, Judith. "Persuasive Cartography". *Journal of Geography*, 81, 1982, pp. 140–4.

63   Pickles, John. "Texts, Hermeneutics and Propaganda Maps". In Trevor J. Barnes and James S. Duncan ed. *Writing Worlds: Discourse, Text and Metaphor in the Representation of Landscape*. London: Routledge, 1992. pp. 197–201.

64   Bearley, K.G. "Mapping them 'out': Euro-Canadian cartography and the appropriation of the Nuxalk and Ts'ilhqot'in first nations' territories, 1793–1916, *The Canadian Geographer*, 39, 1995. pp. 140–56.

65   Carter, Paul. *The Road to Botany* Bay: *An Essay in Spatial History*. New York: Knopf, 1988.

66   Withers, Charles W.J. "Authorizing landscape: 'authority', naming and the Ordnance Survey's mapping of the Scottish Highlands in the nineteenth century." *Journal of Historical Geography*, 26, 4, 2000. pp. 532–54.

67   Edney, Matthew H. *Mapping an Empire: The Geographical Construction of British India, 1765–1843*. Chicago: The University of Chicago Press, 1997. p. 96–7.

68   Cohn, Bernard. *Colonialism and its Forms of Knowledge: The British in India*. Oxford: Oxford University Press, 1997. p. ix-xi. This concept is engaged with by others such as C.A. Bayly, Nicholas Dirks and U. Kalpagam. See Dirks, Nicholas B. *Castes of Mind: Colonialism and the Making of Modern India*. Princeton: Princeton University Press, 2001; Kalpagam, U. *Rule By Numbers: Governmentality in Colonial India*. New York: Lexington Books, 2014. p. 22.

69   Winlow, Heather. "Anthropometric cartography: constructing Scottish racial identity in the early twentieth century. *Journal of Historical Geography*, 27, 4, 2001. pp. 507–28.

70   Foucault, Michel. *The Order of Things: An Archaeology of the Human Sciences*. (1966) London: Routledge, 2005. p. 149.

71   Ibid. p. 29.

72   Dirks, Nicholas B. "Guiltless Spoliations: Picturesque Beauty, Colonial knowledge, and Colin Mackenzie's Survey of India". In Catherine B. Asher and Thomas R. Metcalf eds. *Perceptions of South Asia's Visual Past*. New Delhi: Oxford and IBH Publishing Co. ltd., 1994. p. 211–12.

73   Ibid. pp. 227–8.

74   Ibid. pp. 229.

75   Hirsch, Eric. "Introduction Landscape: Between Place and Space". In Eric Hirsch and Michael O. Hanlon eds. *The Anthropology of Landscape: Perspectives on Place and Space*. Oxford: Oxford University Press, 1995. p. 22.

76 Green, Nicholas. "Looking at the Landscape: Class Formation and the Visual". In Eric Hirsch and Michael O. Hanlon eds. *The Anthropology of Landscape*. pp. 31–41.

77 Heideggar, Martin. "The Age of the World Picture", (1977). Quoted in Christopher Pinney. "Moral Topophilia: The Significations of Landscape in Indian Oleographs" in ibid. pp. 99–100.

78 Marin, Louis. *On Representation*. Stanford: Stanford University Press, 2001. pp. 313–14.

79 Rose, Gillian. "Looking at landscape: the uneasy pleasures of power". In *Feminism and Geography: The Limits of Geographical Knowledge*. Cambridge: Polity Press, 1993. pp. 86–112. However, critics have pointed out that the sexual politics implicated in Orientalism was, by far, more complicated than a simple equation between Orientalism and masculinism to be confined to a heterosexual imagery.

80 Description de l'Égypte (English: Description of Egypt) is a comprehensive scientific description of ancient and modern Egypt as well as its natural history. It is a publication that is the collaborated work of about 160 civilian scholars and scientists, known popularly as the savants, who accompanied Napoleon's expedition to Egypt from 1798 to 1801 as part of the French Revolutionary Wars, as well as about 2000 artists and technicians, including 400 engravers, who would later compile it into a full work.

81 Gregory, Derek. "Edward Said's Imaginative Geographies". In Mike Crang and Nigel Thrift eds. *Thinking Space*. London: Routledge, 2000. pp. 317–23.

82 Lefebvre, Henri. *The Production of Space*. Oxford: Blackwell, 1991.

83 Mitchell, W.J.T. ed. *Landscape and Power*. Chicago: The University of Chicago Press, 1994. p. 17.

84 Daniell, Thomas and William Daniell. *A Picturesque Voyage to India by the Way of China*. London, 1810. p. ii.

85 Archer, Mildred. *British Drawings in the India Office Library*. Vol. 1: Amateur Artists. London: Her Majesty's Stationary Office, 1969. pp. 1–56.

86 Cooke, W.B. *Views in Sussex, Drawn by J.M.W. Turner, R.A. and Engraved by W.B. Cooke*. [Text by RR. Reinagle], London, 1819. Quoted in Andrew Hemingway. *Landscape Imagery and Urban Culture in Early Nineteenth-Century*. Cambridge: Cambridge University Press, 1992. p. 77.

87 Gombrich, E.H. *Art and Illusion*. London: Phaidon, 1962.

88 See Mitchell, W.J.T. *Landscape and Power*. Chicago: The University of Chicago Press, 2002.

89 Rorty, Richard. *Contingency, Irony and Solidarity*. Cambridge: Cambridge University Press. 1989. p. 18.

90 Barnes, Trevor J. and James S. Duncan ed. "Introduction" to *Writing Worlds: Discourse, Text and Metaphor in the Representation of Landscape*. London: Routledge, 1992. p. 11.

91 White, Hayden. Metahistory: The Historical Imagination in Nineteenth Century Europe. Baltimore: Johns Hopkins University Press, 1973.

92 Porter, Dennis. *Haunted Journeys: Desire and Transgression in European Travel Writing*. Princeton: Princeton University Press, 1991. p. 20.

93 Pratt, Mary Louise. *Imperial Eyes: Travel Writing and Transculturation*. London: Routledge, 1992. p. 6.

94 Hulme, Peter and Tim Youngs eds. "Introduction" to *The Cambridge Companion to Travel Writing*. Cambridge: Cambridge University Press, 2002. p. 3.

95 Pratt, Mary Louise. *Imperial Eyes: Travel Writing and Transculturation*. London: Routledge, 1992. p. 7.

96 Quoted in Arnold, David. *The Tropics and the Traveling Gaze: India, Landscape, and Science 1800–1856*. Delhi: Permanent Black, 2005. pp. 120–1.

97   Gregory, Derek. *Geographical Imaginations*. Oxford: Blackwell, 1994, p. 355.

98   De Pinet, Antoine. *Plantz, Pourtraits et descriptions de plusieurs, villes et for-tresses, tant de l'Europe, Asie, Afrique que des Indes, et des Terres Neuves*. Cited in *The History of Renaissance Cartography*. p. 404.

99   Ogborn, Miles. *Geographia*'s pen: geography and the arts of commerce, 1660–1760. *Journal of Historical Geography*, 30, 2004, pp. 294–315.

100 *The History of Renaissance Cartography*. p. 404.

101 See Metcalf, Thomas R. *Imperial Vision: Indian Architecture and Britain's Raj*. (1989) New Delhi: Oxford University Press, 2002.

102 I have used this phrase as a concept from John Bartholomew's (1831–93) map titled "British Empire Throughout the World Exhibited in One View" published in the 1850s. Bartholomew and Co. was a map making firm established in Edinburgh which secured a commission to produce maps for the *Encyclopedia Britannica*. This map published for the first time around the 1850s stated: "The British possessions are engraved in a bolder character and coloured Red". The framing cartouches are supposed to depict friendly encounters with natives in different parts of the world.

103 Jeremy Bentham's eighteenth century coinage was used to describe the architecture of a special circular jail in which the inmates would be under surveillance but be unable to locate the single watchman. Michel Foucault in the 1970s used the term as a metaphor implying the essence of a modern disciplinary society where the location of power is invisible. See Foucault, Michel. *Discipline and Punish: The Birth of the Prison*. (1975). New York: Random House, 1989.

Part I

CARTOGRAPHIC IMAGINATION

With post-colonial scholarship as the backdrop, it is important to see the functional dimension of cartographic endeavours in scheming the metropolis, the colony and the empire in general. Whereas the oppositional binary of the home state and the empire stands unaltered, it is essential to view the emergence of both the geographies as constructions of the same gaze. The fashioning of both the geographies happened to be acts born out of an identical cartographic impulse that shaped both the nation state of Great Britain and the British Empire. In the chapters in this section, the first is an account of the process of cartographic formulation and formalisation of the British Isles, and the second deals with the extension of those ideas implemented on the Indian colony: the act involving a translation of sorts devised to appropriate complexities and peculiarities under a totalitarian all-encompassing gaze.

1

MAPS

The onset and dominance of cartographic reason

Only geometry can hand us the thread [which will lead us
through] the labyrinth of the continuum's composition.
Gottfried Leibniz (1676)[1]

Francis Bacon lays down some foundations in the political construction of territory from space: that territories be compacted and not dispersed; that the
heart and seat of the region or the centre of the state be sufficient to support
provinces and additions; that no part or province of the state be utterly unpro-
fitable and must confer some use or service to the state.[2] The map is the political instrument through which land can be converted to territory, by subjecting
it to rigorous geometrical rules. It outlines the process of consolidation, the
forced or wilful submission of dominions before a commonwealth for their
common good[3], or a merger of smaller king-heads into the larger geo-body of
the sovereign godhead.[4]

The map is a crucial technology of control over space. Map production over
centuries went through various transitory phases to arrive at what we equate
them with today – as images of spatial realities. Whether the places precede
maps or maps pre-empt geo-political realities is a debate which has been dealt
with and intrigued the minds of numerous scholars from across disciplines.
Benedict Anderson, for one, points out how indebted present-day nationalisms
are on enumerative and geographical artefacts, namely the census, map and the
museum. In this chapter, I shall discuss the instrumentality of maps in fashioning geo-political realities in concrete spatial terms. After exploring, briefly,
the historical evolution and transformation of maps in the post-Renaissance
West, I shall focus on cartographic practices and cultures of map production in
England. My objective in this chapter, remains, as in the latter ones, to see the
geographical artefact of the map as an ideologically constructed representation
of space, as are landscape paintings or travel writings. Maps, and such other
representative artefacts of modernity act as a demiurgic force that mediates
experiential reality of space and imbues it with a superficial structural value.
Further, it is the map, which is the visual apparatus which envisions space as
territory and inscribes it with power. The formation of Great Britain as a geo-

political reality, its existence as a nation, is as much chequered by and entrenched in such assumption of power as is its empire in the eighteenth and nineteenth centuries. I intend to argue in this chapter and the next, that the map and its acclaimed scientific principles were as much used in fashioning the spatial conglomerate of Great Britain as it was in forging the empire in tectonically visible terms.

Maps as ideological constructs

The map, which seems to be a simple and natural representation of landscape is, in fact, a highly artificial construction. Today, it is generally linked to power and control. The story of its progress and development from being indexical of emotional or moral state to its present connotation as a spatial representation is an interesting one. The English usage of the term "map" dates back to around 1527, but at this time the term usually recurred in poetry and drama as a conceit, shorn of the scientific and mimetic implications it has today.[5] The word used today derives from late Latin "mappa", by way of "mappa mundi", a cloth painted with the representation of the world. In some other European languages, it derives from late Latin "carta", which meant any sort of formal document. Hence the art/science of construction of maps is called cartography. The development of maps in Europe was more or less complete and standardised by the sixteenth century.[6] The ancient Greeks had developed a theory of geography allied to their knowledge of distant lands acquired by military expansion and commerce in the Roman Empire which culminated in the work of Ptolemy resulting in maps of the world as known to Europe. He was the first to devise a system of grids, later realised as faulty. According to Ptolemy, the goal of "chorography is to deal separately with a part of a whole", whereas the task of "geography is to survey the whole in its just proportion".[7] Simultaneously, he also asserted the theoretical distinction of the global and the local, the whole and its parts. Christendom produced very different maps later on, which were probably influenced by Babylonian sources, circular in shape and Jerusalem at the centre. Most of these early maps, however, were pictorial in form. The modern topographical map was quite a late development. The maturing of this technology heralded a new age of representations based on a scientific temperament of accuracy.

This chapter will deal with the constructed nature of maps. As numerous scholars have pointed out, maps are not neutral or disembodied representations. As they transformed the world into portable pictures, they have become embedded in the embattled domain of ownership and possession. They have been used for military and political purposes in being able to graphically represent a space. Its capability of providing knowledge about specific locations has helped in formulating strategies of defence or fortification through ages. Properties of scalar fixation and gratular accuracy and thus the map's equivalency with the world outside were additions which gradually took shape in post-Renaissance Europe. As David Harvey puts it, the Renaissance invention of perspectivism furnished an entire set of fundamental qualities such as objectivity, practicality and functionality to map making.[8] The cartographic image, then, is a visual reproduction based on a technology of signs invested with the unique power to imitate what is thought of as the real world:

> it manages, without any apparent effort, to replace a natural world beyond our physical control with the promise of mental order wrapped in the Euclidean rhetoric of "poynt, lyne, angle or measure'".[9]

In the later chapters, I will show how space is visualised and narrativised. This chapter will deal with the stage in the final conceptualisation of space in the process of its reproduction: that of measurement of space by mapping it. The grid provided the structure to absorb the inflow of new information which ordered the distribution of life, population and societies across the globe.[10] Landscape paintings and travelogues therefore become the correlatives which supplied information about the differential phenomena into the mathematically derived reconciliatory and totalitarian framework of the map.

Fine art to field science

Among the many questions opened up by scholarship within the history of cartography is that of the complexity of discourses between art and science which cartography mediates. As Denis Cosgrove points out, the obvious parallelism between map and pictorial art emerge from both practices being concerned with "technical questions of content selection and emphasis, medium, line, colour and symbolization, and both require similar decisions about form, composition, framing and perspective".[11] A number of historians of early modern art, such as Samuel Edgerton and Svetlana Alpers, have exposed the shared techniques and interests among Italian and Dutch painters and mapmakers.[12] On another note, Brian Harley sees the essential dichotomy between art and science as disabling and using the Foucauldian paradigm locates cartographic practice in the domain of cultural production and as a human artefact on a par with any other social practice. What Harley terms the "sacred dichotomy" between science and art is itself a late eighteenth-century construction and this is also the time when maps shed their last remnant of aesthetic colouring and emerged into the domain of science under the pervasive scientism of the age. However, the steady veering of map making towards accuracy and objectivity and thereby the deployment of specific instruments to optimise its verifiability definitely shows a transition.

According to Alpers, the link between maps and picture-making is an old one that dates back to Ptolemy's *Geography* (c. 150 A.D.). While Ptolemy invokes the analogy of picture making with both geography as well as chorography, he "connects the training and skills of the mathematician to geography and those of the artist to chorography".[13] Having laid this distinction, Ptolemy's own work and his maps were inclined towards the former and employed mathematical principles in working out planar projections of the earth's spheroid. Harvey reads Ptolemy's innovation of the system of grid, as connected to perspectivism in that "Ptolemy had imagined how the globe as a whole would look to a human eye looking at it from outside". This entails also, an epistemological possibility of perceiving the "globe as a knowable totality". Therefore, the seemingly infinite space once uninterpretable could be represented and scaled down to finitude following mathematical principles to appropriate the globular space onto a plane surface.[14] With the translation of Ptolemy's writings in Renaissance Europe, there came about a resurgence of his geometrical framework and ideas of linear perspective in projecting spaces and bodies onto a plane surface whether in art or cartography, elaborated by Florentine artists such as Brunelleschi and Alberti:

> It supplied to geography the same aesthetic principles of geometrical harmony which Florentines demanded of all their art.[15]

According to Harvey, however, though this elevated and distant view could provide an all-encompassing view, a territorial totality in effect, it lost much of the sensuous quality of medieval maps and pictures and generated images "completely out of plastic or sensory reach."[16] The dilution of boundary between what is generally deemed a view and what is deemed a map is furthered when taking into account that the specular position constructed within Western cartography is essentially the same for panoramas, picturesque views and other visual delights. All of them are constructed with the viewer at its centre, at a privileged vantage point with the framed image under surveillance. It posits the power of the mapmaker and the view made available comes mediated through his vision. One cultural construct is that the vision in maps is used to negotiate another cultural construct, the space.[17]

The use value of the map derives from its ability to convince its users of its accuracy, which too is constructed and culturally mediated by the mentioned relationship between the viewer and the viewed. Renaissance maps, while maintaining their aesthetic qualities, took on this quality of scientific precision and accuracy which were increasingly being valued for determining property rights, political boundaries and navigation. English estate maps belonging to the age were no less regarded than works of art. The decoration in these maps was an integral part of such maps. William Leybourne, in his instructive treatise to surveyors, *The Compleat Surveyor* (1653), lays down the format and aesthetic principles for cartographers to follow, for, "Your plot", he says:

> will be a neat ornament for the lord of the manor to hang in his study, or other private place, so that at pleasure he may see his land before him.[18]

To this effect, Leybourne instructs the mapmaker to draw "divers little Trees" in the most important places and to use lively colours for them and for the topographical features. Further, he points out, the coat of arms should be represented in the upper part of the map, "correctly coloured", together with a compass rose, scale bar and a picture of the manor house, along with a diagram to show where these should go.[19] English maps continued to include these features till the eighteenth century, when landscapes of estates and their maps eventually parted ways irrevocably, even though both of these artefacts, even in the eighteenth and the nineteenth centuries, occupied similar wall spaces in the lord's "study or other private places".

The steady transition of maps from the realm of art to its incorporation in the domain of science is also traced in the history of map colouring which was, during the Renaissance and later, a highly skilled, and specialised profession. Colour was an important selling point in the general appreciation of the map as an art item. Holland and what was called the Low countries were the hub of European cartographic culture. Alpers speaks about the overwhelming carto-graphic impulse among seventeenth-century Dutch painters and shows how artists at the time emerged from a previous profession as cartographers. Most of the maps in other European regions were engraved by Dutch and Flemish engravers. Needless to say, cartography picked up elaborate artistic traditions and innovations of the time and place. The profuse baroque decoration of the period allowed the production of superbly hand-coloured atlases. As baroque style succumbed to the spread of rococo, colouring became less flamboyant till such time that colour disappeared altogether from cartouches and vignettes and was used more sparingly on the map face. The bright elaboration of the six-teenth and early seventeenth century gave way to the refined, lighter, subtle colouring of the late seventeenth and early eighteenth century, before gradually evolving into the austere functionalism of the early nineteenth century. Map colouring underwent its technical revolution from the nineteenth century with the introduction of colour lithography especially after the Great Exhibition of 1851, which eventually rendered the map colourist redundant.[20]

The next phase in modernising mapping practices came in the eighteenth century with the revolution of surveying techniques. The use of numerous sci-entific tools and instruments in cadastral surveys and cartography marked its final transition from being a craft to a highly specialised field science. This was possibly a direct outcome of the process of stabilisation of military-fiscal nation states in Europe, and "their attempts to secure more complete and rational control over their own territories, was the mapping of those territories at ever larger scales and across ever more extensive areas."[21] Such large scale surveys could hardly be undertaken as individual initiatives any longer and required state support for financial and other logistic aid. One of the most extensive of such surveys during the period was France's famous Cassini surveys. It made use of a novel technology called triangulation which could function "simulta-neously at the smallest and largest of scales" capable of providing the most unambiguous of results, and it thus "represented a dramatic extension of state power, reaching deep down into different landscapes all across a state's territory".[22] The pervasive encyclopedism of the Enlightenment age incorporated and reconciled multiple viewpoints and generated a new idiom of systematic and disciplined observation, which triangulation characterised. Though triangulation as a practice was initiated in the late eighteenth century, most European states did not find it feasible, partly because of the heavy expenditure involved, till the mid-nineteenth century. This was also the time when cartography as an exclusive scientific discipline was accepted in its own right. As Edney points out, Enlightenment order successfully invested cartography with ideas of exactitude and accuracy as separate from other forms of geographical representations, be it textual or pictographic, although these too had their own claims to objectivity:

> Enlightenment is central to modern cartography's self-definition as a "science" because it was then that "art" was apparently purged from maps, thereby freeing cartography's pristine, scientific core to develop and to progress.[23]

Mapping, at this time, received its necessary practical adjunct of applied geometry or trigonometry. Significant advances were made in the area with rigorous systematisation of survey procedures. Surveys quantified space along geometrical criteria for purposes both of cartography as well as statistics, which determined not only form and extents of land but also relationships between soil and subject, between land and owner.[24] This also entailed the engagement of elaborate scientific instruments, the theodolite, the plane table, the globe which translated social space into sets of tables and diagrams.

Mapping property: private land and the state

Maps were the cultural apparatus of a space economy. Surveys, as Klein sees them, are a phenomenon attached to the larger story of the transition from feudalism to capitalism. Post-Reformation England saw land being subjected to new economic forces with the dissolution of Catholic churches and monasteries and with the collapse of the manorial system. This provided stimulation for private property by monetising land. The land itself rather than its produce, became a commodity. Therefore:

> The dynamics of a fluid land market affected the ways in which the whole practice of surveying was understood, and its main impact, in the latter part of the sixteenth century, was to gradually naturalise a perspective on agrarian space which foregrounded its status not as a social realm but as marketable commodity.[25]

Land was now required to be fervently mapped, as it frequently changed hands, to develop new and exclusive private rights of the owner over his property. So, as landlords and surveyors joined hands, space was produced as a coded order of legible signs. The estate map is the direct articulation of this collaboration. According to P.D.A. Harvey, the estate map was:

> a work the estate owner could consult for detailed information about the lands it showed; or he might point to it with pride, seeing it as a graphic epitome of his property, wealth and social position. Often it was clearly designed for display, beautifully coloured and elaborately ornamented. Often signs of wear, and many added corrections and annotations, testify to long service as a functional tool of estate management.[26]

The overriding reason behind the production of these estate maps was, of course, to indicate that all the land put on visual display belonged to one owner. The cartographic image inscribed with the estate owner's insignia and coat of arms together with a landscape view of his manor in the cartouche, therefore reflected his manorial authority. The map denoted his social stature accruing from his wealth and property. The map along with the topographical inventories were indexical of the lord's social standing, power and identity in general. Thus, the map acted as one badge of the owner's local authority. The family coat of arms added on the top of the map was much more than mere decoration, for "the right to these heraldic emblems also incorporated an individual's right, rooted in the past, to the possession of land".[27]

In England, while an estate map recorded the boundaries of land owned by a particular person or group of individuals or an institution, cadastral maps recorded property falling within a particular administrative unit. However, English cadastral maps often concerned not just a single individual but involved the local community and were thus called "enclosure maps", or they combined national interests in "tithe maps".[28] Most professional surveyors and cartographers of the sixteenth and seventeenth centuries whose role became crucial, in such a transitory economy, mapped estates first of all. The spur to the surveyors was not a metaphysical or intellectual rediscovery of the inhabited space correlative with Renaissance discovery of self: there was a material basis for the action. The economic imperatives of the age urged the landlord "to know his own" (a material counterpart of the metaphysical quest of "know thyself") – a phrase which did the rounds in the survey manuals. The surveyor not only drew up an inventory of the entire estate in order to "apportion objective rights of ownership over goods and land" but also for each new prospective buyer to know what he was purchasing.[29] As society became increasingly commercial, socially ambitious and litigious, the surveyor was required to expand his activities and improve his techniques in order to meet the new demands for maps of multifarious practical and symbolical motives. Therefore there was a need to make the techniques competent and scientifically advanced enough to cope with the varied needs of a quickly accelerating land market and to achieve accuracy in mensuration.

If maps were increasingly required for property assessment, taxation and revenue, on the other hand, they provided simultaneous impetus for their development from the state's end. Consequently, the practice of mapping Britain as a single unit emerged from regional mapping arising out of practices of mapping revenue regimes. The early administrative units in England called county or shire which derived their autonomous status from their ruling heads, usually called earls, bowed down to the sovereign jurisdiction of the Tudor court and its revenue regime. Even with the growing class of acquisitive gentry establishing itself on confiscated monastic land, these units remained unaltered. These counties became the units of basic division which constituted the nation and remained so for years to come.

In a study of English maps produced at the time of Queen Elizabeth I's reign, Helgerson shows how the English were exposed to an image through maps which "let them see in a way never before the country – both county and nation – to which they belonged and at the same time showed royal authority".[30] Maps had an ideological effect in the age in that they strengthened both local and national identity, dovetailed with an identity based on dynasties. The maps of this time show the gradual transition from universal Christendom, to dynastic state, to land-centred nation.[31] The county maps of England and Wales have an ancient history commencing in the sixteenth century. The oldest known series of regional maps of this kind is one compiled by Laurence Nowell (1559–76). They remained unprinted. A direct result of the state's increasing interest in administrative mechanisms, Nowell's *General Description of England and Wales* was a preliminary specimen intended to promote a larger project which purported to include a whole series of individual maps covering the area which Nowell describes in a letter as "our region". The scope of "our region" extended beyond English and Welsh counties to skirt Scottish lowlands and adjoining coastal areas of Flanders and France as well as Ireland. The composition therefore represented not individual but a collective ("our") view of the dominions under the Tudor monarchs. The map's title cartouche inscribed with the crowned Tudor arms confirms the legitimacy of royal control over the land portrayed. Through a semiotic reading of the maps, Bernard Klein unravels "the political statement of the map", which, according to him is "the description of a space that aspires to the collective vision of a fully anglicized terrain ... translated into an almost physical incorporation of those areas which, viewed from the English centre of power, constitute the outlying regions of an unevenly structured polity, the culturally diverse margins of the Tudor state."[32] The striking inclusion of Ireland, the itineraries noted on the verso etc., indicate a desire to visually master and therefore politically claim culturally diverse but contiguous spaces.

In 1579, appeared Christopher Saxton's collection of county maps as an atlas, which had been coming out individually during the preceding nine years. A "land-meater" by profession, Saxton's collection of 34 regional maps is instrumental in organising subdivisions of the nation. The collection opens with a general map titled *Anglia*, which exhibits the differing cultural composites i.e. England and Wales as a tightly enmeshed single unit, with parts of Scotland, France and Ireland hovering at the margins. Though in comparison to Nowell's *General Description*, Saxton presents a much more restricted version of national territory. He, however, devises a more consistent and homogeneous and thus a unified geography than Nowell's loose one. The 34 county maps in his atlas are therefore given equal status in the great partnership called the nation: "*Anglia* offers the central frame of reference to which the individual maps of the shires metonymically relate".[33] Queen Elizabeth's image on the frontispiece of the atlas points to its being undertaken under royal commission. In fact, his enterprise was achieved under the authority of Queen Elizabeth and patronised by Thomas Seckford, a government official and eminent lawyer, who bore the title of "Master of the Court of Requests and Surveyor of the Court of Wards and Liveries". Saxton's atlas not only established a tradition of county mapping which was to continue till the Ordnance Survey in the nineteenth century, but also promulgated the practice of funding by state patronage. Seckford stimulated official interest in the project. The Privy Council issued orders for the continuation of the survey in 1575 and 1576. The proof copies were used for tactical and defensive planning.

The late sixteenth and early seventeenth century, which were crucial years in framing the national imaginary, witnessed an expansion of map production. John Norden mapped a few counties, introducing roads and triangular distance tables detailing the distances between towns, generally known as *Guyde for English Travellers* (1625). He intended to call his work *The Speculum Britanniae*, "after the most painful and praiseworthy labors of Master Christopher Saxton, in the redescription of England".[34] Though his method would be adopted later in the history of English cartography, his project was doomed for lack of sponsor. His ambitious schema of recording a series of county histories was likewise stalled.[35] In 1607, an edition of Camden's *Britannia* came out complete with maps, which can be called the second atlas of England and Wales. This contained reductions of Saxton's and Norton's maps after fresh engravings.[36]

The culmination of Tudor cartographic advance was the publication of John Speed's *Theatre of the Empire of Great Britain*, though published in 1611 under the Stuart regime. Speed's atlas is dedicated to James I, the "Inlarger and Uniter of British Empire". Likewise, the *Empire of Great Britain* promotes the concept of spatial unity of the British Isles through visually translating the desire onto cartographic plane.[37] The spatial event is replaced by a historical stage, a spectacle – the theatre. The series consists of a separate map of each county in England Wales, besides general maps of each kingdom, one for Scotland and five for Ireland, depicting its four provinces. As scholars have pointed out, Speed's work was the first to cartographically integrate culturally diverse spaces within a single pictorial unit called the "empire". In Helgerson's analysis, the project of describing Britain's geography was an inherently political act. In his opinion, the land-based chorography which identified specific individuals or houses as the owners of spaces mapped, posited a challenge to the idea of the monarch as the sole repository of power and land ownership. There appears therefore, a conceptual split between the visual maps and the steadily gaining sentiment of that of the monarch as the body politic of the nation. Saxton, Camden, Norden, Speed, Drayton and other county chorographers played an inescapable part in creating the cultural and political entity they only pretended to represent, namely the nation.[38] However these maps are signifiers of emerging nationhood as they were the means of organising knowledge not only about the external frame but also about the internal configuration of the nation.[39]

Map literacy through popular culture

In the first half of the eighteenth century, along with atlases containing county, provincial and regional maps, maps were a frequent feature in periodicals and guide books in England. The geographical grammars, such as Herman Moll's *A System of Geography* of 1701 or Emanuel Bowen's *A Complete System of Geography* of 1747 were extremely popular. However, apart from these exclusively geographical sources, around the middle of the eighteenth century, cartographic works were often contained in popular literary periodicals. Maps became very much the object of curiosity for the elite and middle classes as were travelogues and other, fictional, content in the journals. As tools to cultivate polite culture in men and women, these journals were replete with diverse educative, scholarly and entertainment-based pieces. That maps of the British Isles should find a place alongside such diverse literary and other topics, points to the fact that maps as specialised field science was an idea generated later. Also such maps were effective in spreading map literacy and an awareness about one's own nation. Some of these were *The Gentleman's Magazine and Historical Chronicle* (fl. 1736–75); *The London Magazine or Gentleman's Monthly Intelligencer* (fl. 1747–60); *The Universal Magazine of Knowledge and Pleasure* (fl. 1747–97); *The General Magazine of Arts and Sciences* (1755–64); *The Universal Museum, or Gentleman's and Ladies Polite Magazine of History, Politicks and Literature* (fl. 1762–64); *The Universal-Museum, and Complete Magazine of Knowledge and Pleasure*; and *The Political Magazine and Parliamentary, Naval, Military and Literary Journal* (fl. 1782–90).[40] Many of the maps also accompanied and supplemented travelogues and antiquarian survey literature such as Kitchin's maps in Boswell's *Antiquities of England and Wales*, 1786. In fact, officially, maps were placed under the surveillance of the Committee of Polite Arts. Maps constantly impressed upon the people, making them alert to the recesses of their native counties which were, nevertheless, parts of a unified whole, the nation. This surfaced simultaneously with a rise in home tours in Britain, together with a new awakening towards the natural history and antiquities of Britain and generally towards its countryside.

It was recognised that good maps had many practical functions:

> A complete knowledge of the Situations, Bearings, Levels and other Topographical Circumstances of this Kingdom, being of great use in planning any scheme for the Improvement of Highways, making Rivers Navigable and providing other means for the Ease and Advancement of the National Commerce.[41]

While journals and pamphlets teemed with maps, inaccuracies in existing maps were steadily coming under criticism. County surveys were still based on road traverse, and angled with the circumferentor or surveying compass, processes which had continued since Elizabethan times. For example, a correspondent of the *Gentleman's Magazine* tartly referred to Herman Moll's popular small-scale pocket maps as "Moll's little erroneous trifles". Herman Moll, being a Dutchman, underwent a severe criticism for his highly ornamental style in the typically Dutch tradition. This was the time when the Dutch school of cartography had declined and British cartography was competing against steadily progressing French methods. William Baker reiterated the same indictment of British cartography:

> That branch of knowledge (though our pamphlet shops are full of boasted surveys) when it is examined accurately will be found ... excessively low, oppressed as it is with errors arising (not to mention the ill capacity of common Map-makers) from hasty observations without a variety of good instruments.[42]

William Borlase and Henry Baker were the two people who first brought to the notice of the Royal Society of Arts, the cause of national cartography. It was concluded that a map encompassing the whole of the nation was urgently required. As Harley mentions, the general climate of opinion in the country favoured cartographic improvement as a crucial signature of national triumph. France had already taken the lead in mapping an entire nation through an advanced scientific process called the trigonometrical survey. There was consciousness even in the provinces in England about Cassini de Thury's trigonometrical survey and map of France commissioned by Louis XV.

Therefore, the general demand was to formulate a map of the nation following more precise standards comparable to French advanced cartography. Therefore, after numerous debates and deliberations the Royal Society of the Arts, in November 1759, decided to give a premium of £100 and gratuity for a specifically trigonometrical survey. A draft of the advertisement which was to announce the remuneration also specified all the instruments which should be used:

> The Horizontal Distances of all places in the Map to be taken with the Theodolite or Plain Table and the roads to be measured with a Perambulator and noted down

Moreover, a clarification regarding the above advertisement specified that:

> The intention of the Words Theodolite or Plain Table ... was to guard against taking the Angles by the Circumferentor or such like uncertain Instruments; but that if the surveys of any Counties taken by the Candidates be done Trigonometrically by a new invented Instrument of known use and certainty, that such Surveys shall be entitled to the premium according to their merit.[43]

As a response to the Society's invitation, a large number of county maps were produced in the latter half of the eighteenth century.

Cartography excited the cultural imagination of the nation in most imaginative of ways. Not only did maps emerge as a favourite geographical metaphor in poetry, novel and drama, but the map consciousness extended to sport and pastimes.[44] A board game, The Royal Geographical Pastime Exhibiting a Complete Tour Thro' England and Wales, was published in 1770 by Thomas Jefferys, a cartographer, engraver and a geographer to George III. Another board game called Tour through England and Wales, made by John Wallis in 1794, was a hand-coloured engraved map on linen to be played with pyramid-shaped markers, an octagonal teetotum spinner and four markers (Figure 1.1). Players had to make their way through ricocheted lines closely resembling the nationwide triangulation, designating 117 playing spaces, beginning from Rochester and ending in London. The worst place to land was the Isle of Man, where a traveller is shipwrecked and had to leave the game. The places of interest highlighted their worth in trade.[45]

Mapping home and outside

Since the global map in the fifteenth and sixteenth centuries was practically a three-quarters empty canvas, it provided ample scope for inscribing on it images of distant and newly "discovered" lands. The first English incursions into the wider fields of foreign cartography began in the sixteenth century with the first known English world map being compiled in 1542 by John Rotz, geographer to Henry VIII. Anthony Jenkinson, a member of the Mercer's

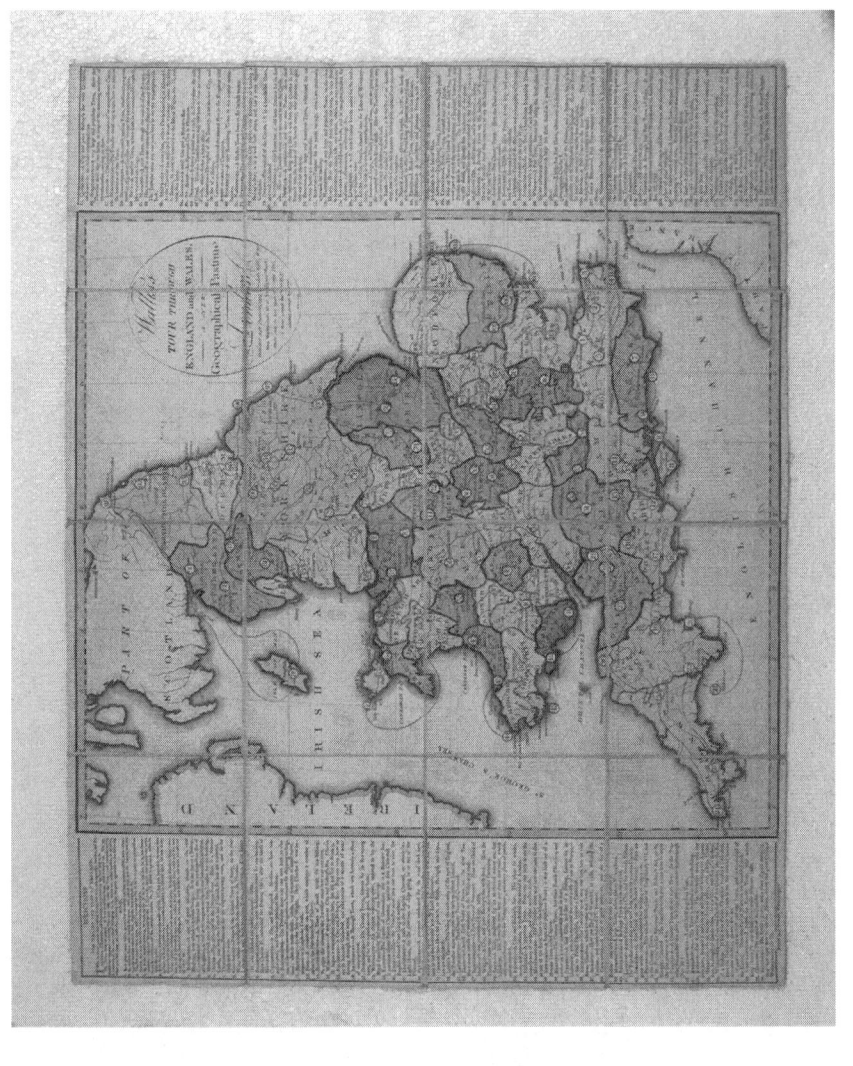

Figure 1.1 John Wallis "Tour Through England and Wales: A New Geographical Pastime" (1794)
Image courtesy: Victoria and Albert Museum of Childhood

Company, was appointed leader of an expedition sent to Russia in 1557 to open up the eastern trade routes for the Muscovy Company. He drew up a map of the country which was later also used by Ortelius in his *Theatrum*. Early English navigators, such as Chancellor, Drake, Gilbert, Frobisher, Davis and Raleigh, composed maps and charts from their explorations which caused lively discussions in the court and among the learned at the time. Another world map was compiled by Edward Wright in 1600 to accompany Hakluyt's *Voyages*. The seventeenth century saw further outpour of cartographic works which increasingly sought to bring the outside world into purview. With the English translation of Ortelius being made available in 1606, there appeared a range of maps based on Ortelius and the travels of numerous explorers into the external world. In 1612, Captain John Smith's map of Virginia appeared engraved by William Hole. Smith's *History of Virginia* had four new maps of "Oulde Virginia", Virginia, the Summer Islands and New England. In 1619, William Baffin compiled a map of the Mughal Empire based on the information given by Thomas Roe. In 1627, John Speed published his *Prospects of the Most Famous Parts of the World*, the first printed general atlas of the world in English. It was combined with the *Theatre of the Empire of Great Britain*, containing 22 maps in all including county maps. Reissued in 1631, 1662 and 1676, it later saw the addition of new maps of Virginia and Maryland, New York and New England, Jamaica and Barbados, Carolina and the East Indies, and Russia and Palestine.[46]

From 1698 to 1700, the famous English scientist, Edmund Halley sailed on the *Paramour*, on the first sea voyage undertaken for a purely scientific objective to the North and South Atlantic. The outcome of his travels were several scientific charts. Interestingly, the magnetic chart, in its first edition of 1701, dedicated to William III, shows the Atlantic only. The second edition of 1702 was dedicated Prince George of Denmark, and extended to include the whole world. Probably, this was a token gesture to acknowledge Dutch influences and contributions. However, in the eighteenth century, British cartography emerged with a strong fervour with increasing colonisation and in the face of rising competition from the French empire. By now, much of the once unknown lands had been conquered by European forces. This was also the age of colonising powers which emerged as naval and military forces. Knowledge about these newly acquired territories proved to be not only a fashion of the times, but an ineluctable necessity, in the face of a growing requirement of a certain perception of the geographical space and the extent of area thus possessed. Some of the most aggressive mapping in this century was done by surveyors in North America, the most famous of these being Thomas Jefferys and his successor, William Faden. James Rennell's work in India was not only appreciated, but impacted on British cartography on home soil as well. At around the same time, Cook's *Voyages* and atlas, published 1773–1784, supplied an impetus through an accurate mapping of considerable portions of the South and North Pacific and the coasts of Australia and New Zealand. With this, by the end of the eighteenth century, the outline of the world had been chalked and brought before the British public. The setting up of numerous academies and societies enabled the sponsorship of scientific expeditions and surveys as well as channelling scientific thinking regarding Britain's colonies. In its colony in South Asia, the establishment of the Asiatic Society provided a liminal creative space to scientists through which researches in the colonial periphery were made congruent upon those in Britain.[47]

While the unknown charmed and filled minds with curiosity, the spaces nearer home were also increasingly being brought under cartographic surveillance. The inhabited locality was as much part of geographic discovery as was the transoceanic world. Whereas the overseas explorers worked towards the larger cause of geodesy, chorography was the subject of domestic exploration producing landscapes of the inhabited space. The geographical and historical detailing of localities was part of the process of realisation of the larger space of the nation.

In fact, the trigonometric mapping in Britain was a response to the revolutionary spurts brewing next door. The cartographic articulation of revolutionary principles in 1790s France had a deep association with what was called the "*esprit geometrique*", exalting the connection of triangulation with rationality, realism and Enlightenment equality. This began with the move to adopt a standard system of measurement spanning the entire French nation, replacing all existing local and regional variants of weights and measures with the metric system. The metre was designed to eliminate all arbitrary physical authority like the foot and sought to homogenise land upholding stakes and claims of every citizen equally. On the one hand, while Britain imbibed this cartographic rationale as its underlying principle in mensuration as an Enlightenment impulse, on the other, the reason behind the nationwide military survey and the consequent hardening of borders was also a defence against the French Revolutionary Army.[48]

Triangulation and mapping of the national space

The national map of Great Britain was the outcome of years of labour of field surveyors. In the eighteenth century, cartography required the self-presence of the surveyor on the field as validation or proof for the map's authenticity and precision.[49] Chorography by the eighteenth century, had largely become an insignificant and obsolete field. Chorography, which employed merely vision, was not enough anymore. The inadequacy spurred a sharp distinction between viewing and observation/examination: as between an indiscriminate spectator and an insightful observer. At the wake of emergent national consolidation, vision became an immaculate perception configured increasingly by mechanisation and scientific sophistication. Thereby measurement became a surrogate to examination or observation. The survey map has been a technology of vision, which employed instrumentalised vision in conceptualising space. This

47

amounted to compiling an accurate measurement of that space which meant, in this case, the extent of the territory under the British monarch's command. The survey map's spatial structure was taken as a direct reproduction of the nation's political existence.

The surveys were political statements which attempted to fix imperial control over its territories. As Edney says, the perfection of the geographic panopticon and its archive was promised by the use of "triangulation". The system was based on measuring an accurate distance between two points on a longitude called the base line, and thus plotting a third point by calculating angles with the help of the optical instrument called the theodolite. The surveyor occupied high vantage points like the peak of a hill or the summit of towers. These were called the trig points. The adoption of triangulation as a method, marked a shift in the source of propagation of truth and certitude from the site of the office to the field. The cartographic practice was validated by its capability of creating rationally coherent and geometrically accurate geographic space. The "uninterrupted series of triangles" created an image of continuous and connected space. As geo-spaces got chequered by the ever-continuing triangles, they invariably got inscribed as imperial space, shaping new geo-political realities such as the nation or the empire.[50]

Scotland and Ireland were the first to be brought under the imperial gaze of the English monarchy. Though Scotland featured on maps of Britain from Ptolemy onwards, its mapping was catapulted into a new era with the Culloden suppression of 1746.[51] It came to notice in this campaign and the subsequent pacification of the region, that maps of the Scottish Highlands were inadequate and affected adversely the military operation. Lt. Col. David Watson, who was entrusted with the duty to plan and set up army posts and build roads in the newly pacified regions henceforth, found the unreliability and inadequacy of existing maps hugely crippling. It was Col. Watson who, under the orders of the government and the army, soon after the campaign, started mapping the Highlands in 1747. Col. Roy, who was once assistant quarter master on this project, later reflected in *Philosophical Transactions* for the Royal Society in 1785:

> The rise and progress of the rebellion which broke out in the Highlands of Scotland in 1745 ... convinced Government of what infinite importance it would be to the State that a country so inaccessible to nature, should be thoroughly explored and laid open, by establishing military posts in its inmost recesses, and carrying roads of communication to its remotest parts.[52]

While talking of the existing maps so far prepared, which were still in manuscript form, Roy talks of the requirement to recast these maps according to the advanced technological development, the time had seen:

Although this work ... answered the purpose for which intended; yet having been carried with instruments of the common, or even inferior kind, and the sum allowed for it being inadequate to the execution of so great a design in the manner, it is rather to be considered as a magnificent military sketch, than a very accurate map of a country.

The word "sketch" had a deeper significance, as much of the aesthetic quality was endowed by the Chief Draughtsman, Paul Sandby, who spent many years in Scotland drawing prospects for the military survey. Close, on discussing the maps of the Highlands produced after the campaign confirms the obsolescence of the maps as they were no better than a skilful military sketch asserting its undue nexus with academic art:

The work is clearly an elaborate compass sketch; the roads and some of the streams have been paced, and the mountains have been put in roughly by eye. Near the towns the work is carefully drawn. Cultivation is indicated by open diagonal hatching. The hill features are shown by rough, faint brush, sepia shading or hachuring; the higher mountains are shown by similar, only darker shading.[53]

This appears to be the first call for a trigonometrical survey of the region which initiated the process of the Ordnance Survey. For all its flaws and inaccuracies, the military survey was still the first important reference point to map the entire island in its entirety. It helped establish the Ordnance Survey of the later years. After the successful suppression of the Jacobite rebellion, cartography was routinely taken up as a civilising process into the nineteenth century. The processes and patterns of material transformation in the cultural region of the Highlands and its appropriation were enveloped in the language of improvement which ranged across economic, political and sociological models. Thus, the survey of Scotland helped to bring the country under a unified Parliament. With the Ordnance Survey and William Roy, began a new era of cartography in Great Britain based on accuracy, precision and mathematical calculations (see Figure 1.2).

Even at a very early cartographic stage, in a bid to reduce topographic complexities and regional and cultural differences, Wales was painted the same colour as England on the map, leaving differences and diversities within Wales itself unaddressed. Similarly, Ireland was subject to successive phases of mapping, often following violent suppression of rebellions on Irish soil. As far back as during the reign of Henry VIII, plantations were established on Irish soil. Land belonging to rebels were forfeited and redistributed among some English and Scottish families who then established plantations there. Thus also arose the need to clearly chalk out land ownership. Ireland directly came under an English survey-based imperial cartographic regime after a rebellion which took place in 1641.[54] In 1649, Oliver Cromwell trampled the revolt with the help of a

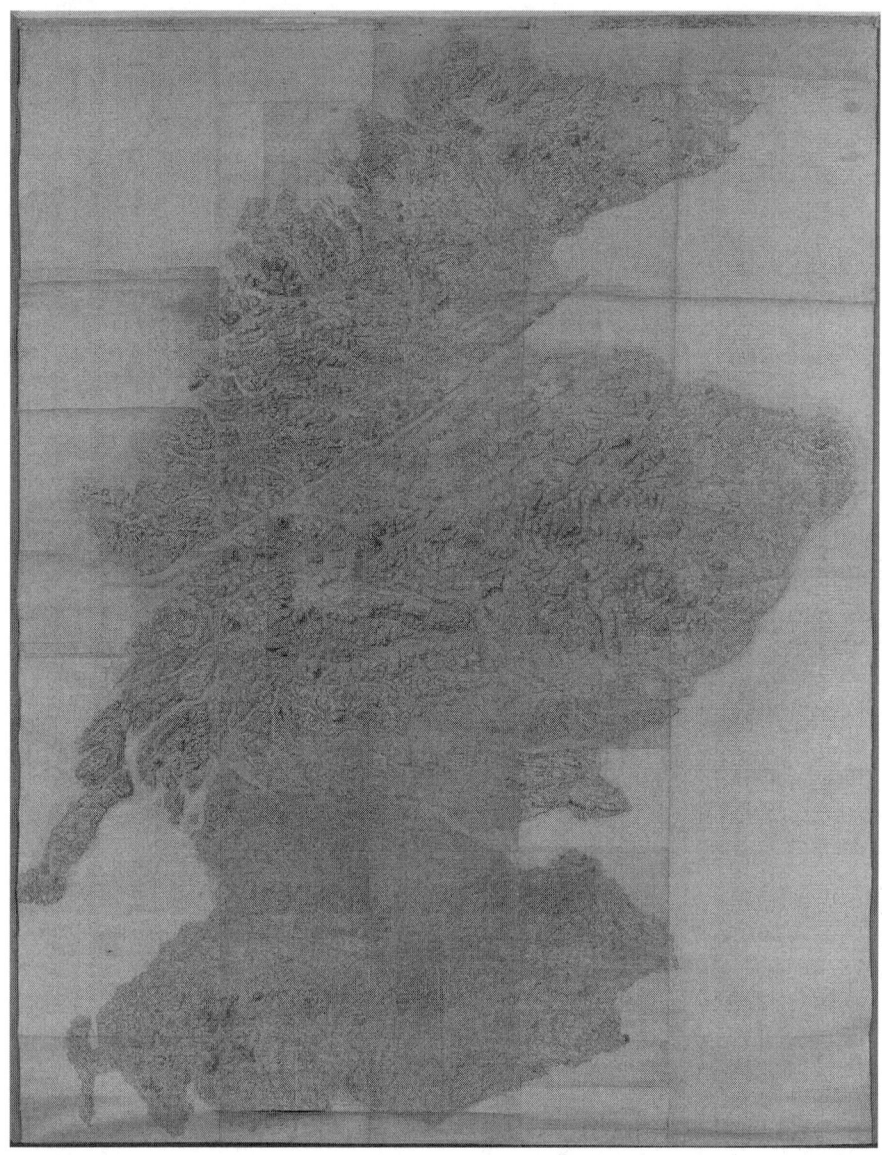

Figure 1.2 General Roy's Military Survey of Scotland, 1747–55
Reproduced with the permission of the British Library

disciplining New Model Army and later ordered what was called the "Down Survey". In 1653, an Act was passed under the Commonwealth, for the *Satisfaction of the Adventurers for Lands in Ireland, and of the Arrears due to the Soldiery there, and of other publique Debts.* In the Preamble, it stated:

> Whereas many well-affected persons, bodies politique and corporate, did subscribe and pay in, upon several Acts and Ordinances of the late Parliament, divers considerable sums of money by way of adventure towards suppression of the late horrid rebellion in Ireland, which said sums of money were, by the said Acts and Ordinances appointed to be satisfied by several proportions of the lands and the rebels there.[55]

Basically, this meant that the property of the Irish Catholic rebels who had opposed Cromwell, was confiscated in the name of meeting the expenses for suppressing and pacifying the rebellion. For this purpose, the Act appointed a committee to conduct "exact and perfect survey and admeasurement of all and every the honors, baronies, castles, manors, lands, tenements, and hereditaments forfeited".[56] These were officially called the "Down Survey" as they were set "down" on paper. Benjamin Worsley and an English natural philosopher Dr William Petty were jointly named surveyor generals for the operations to chalk out both the settlements of Cromwellian soldiery, and the settlement of "adventurers". Petty devised a form of governmentality to bring together economics, demography and geography as he felt that Ireland's location was crucial in terms of trade as it lay between England and its colonies and therefore its territorial appropriation was emergent.[57]

It was not till the nineteenth century that such extensive surveys would be re-initiated. The last official survey before 1824, when the survey of Ireland was resumed, was that made during the reign of William III of the "Forfeited Lands", but which were not very impressive. The Down Survey, in comparison, still stood as the most effective and accurate assessment. In 1815, the Select Committee met and reported in favour of rendering the assessments more equal. But it was only in 1824 that a fresh survey was recommended in view of more equal apportionment of local tax. It was decided on the insistence of Major Colby that a "central and effectual control is indispensible". The execution of the survey was decided to be given over to the Ordnance Survey which was already 33 years in progress. The Committee concluded its report with:

> all former Surveys of Ireland originated in forfeitures and violent transfer of property; the present has for its object the relief which can be afforded to proprietors and occupiers of land from unequal taxation ... In that portion of the empire to which it particularly applies, it cannot but be received as a proof of the disposition of the Legislature to adopt all measures calculated to advance the interests of Ireland.[58]

The chief object of the Ordnance Survey of Ireland was marking on paper the exact areas and boundaries of counties, baronies, parishes and townlands. Though Ireland had its own native tradition of map making, the British imperial cartographic integration of Ireland was complete by the late nineteenth century.[59] In view of the history of combative Anglo-Irish history, Lord Salisbury opined in 1883, "The most disagreeable part of the three kingdoms is Ireland, and therefore Ireland has a splendid map", underlining the long-standing British belief that rebellions provided the strongest impetus to cartographic fortification.[60]

Colby's scheme of covering Ireland through triangulation was approved by the Duke of Wellington. While the trigonometrical survey of Scotland was in progress, some of the hills in Ireland had already been marked and linked up by intersection to the Scottish hills. These hill stations extended from the Mourne Mountains in County Down to Malin Head in Donegal, 150 miles from the north-east coast of Ireland. For example, Colby mentions the importance of Divis, near Belfast:

> The triangles of which it is the apex cover a space of about one hundred and thirty miles in one direction and about eighty miles in the other – no less than two hundred Trigonometrical Points were observed from it.[61]

The "heliostat" and the "limelight", two novel innovations by Drummond were used for the triangulation. The artillery division was entrusted with the field work, who Colby found educated, intelligent and trained in practical geometry. Colby, however, refused to take responsibility for determining the boundaries on the ground. Therefore, the duty of ascertaining boundaries on the ground fell on the Boundary Department, under Richard Griffith, and special boundary surveyors were employed for the purpose.

The Ordnance Survey: the military take over

The word "ordnance" is a syncopated variant of "ordinance", which is derived from the old French word "ordenance", meaning a regulation and an arranging in order. In the seventeenth century, the word "ordnance" came to be exclusively applied to artillery and engineer personnel and material, and the services relating to these. The Board of Ordnance, a defunct organ of the English state, was constituted in 1683 to deal with questions of national defence, ordnance supplies, and the deployment of engineering corps and artillery regiments. Among other responsibilities, the board was also to act as the custodian of lands, depots, and forts and all overseas possessions. The phrase "Ordnance Survey" was first used in 1820, whereas the process of its coming into being had begun in the previous century.[62] Whereas it entailed ideas of fortification, defensible property and spaces, it also amounted to the military takeover of an episteme, that of cartography. Therefore, the term ordnance in the expression "Ordnance Map" retained meanings of both the original and the etymologically derived terms in that it ordered space, as well as transformed it into a military regime or a martially occupied and inscribed territory. There eventuated, thereby, a sharp distinction between civil and political geography, with maps being increasingly seen as part of the technology of administration and control. The transition was thus symptomatic of transformation of lived and experienced space into an ideologically architected and constructed space.

In the *Philosophical Transactions* of the Royal Society for the year 1785, Roy, in an account on the measurement of the Hounslow Heath base line, opens with a general observation that, if a country has not been surveyed, or is but little known, a state of warfare generally produces the first "improvements" in its geography. So in Britain, "on the conclusion of the peace of 1763, it came for the first time under the consideration of the Government to make a general survey of the whole island at a public cost". In this project, Roy's part was to make "the map of Scotland ... subservient, by extending the great triangles quite to the North extremity of the island, and filling them in from the original map".[63] This was the beginning of what Close calls the Accurate Surveys 1783–90. Another incident which expedited the survey was a proposal from France, based on observations by Cassini de Thury in his memoir, that the latitude of Greenwich was in doubt by some 15 seconds and that in order to determine the correct value it was desirable to connect geodetically with Paris. Though objection was naturally raised to such a statement, it gave an added impetus and was used as an effective lever by the missionaries of scientism and the enthusiasts of accuracy. The adoption of the proposal had deep-seated national, political, scientific and practical motivation. Both King George III and the Royal Society were willing to assist in this project to ascertain the difference between the longitude of the two observatories. However, Roy, to whom the responsibility of execution of this project was given, conceived the task in a different way. His principal scheme "has always been considered of a still more important nature [than the mere joining of the observatories], namely, the laying the foundation of a general survey of the British Islands".

It was decided to observe the angles with a large circular instrument called the "3 ft. Theodolite R.S." designed and constructed by Jesse Ramsden, the famous English instrument maker. The volunteers for assisting in the preliminary measurement declined the age-old method of measuring with wooden rods for it was accepted that variation in humidity affected their length. In its stead, steel chains were ordered at the outset but that too was deemed ineffective. Finally, the definitive measurement was carried out by the most unusual means, being used for the first time, namely glass tubes of fixed length and diameter. Ramsden was once again called upon to make them suitable for measurement on the field. Also, as a policy, chiefly with a view to the more effectual execution of the work, it was judged to be a right measure to obtain and employ soldiers, instead of country labourers in tracing the base, clearing the ground, and assisting in the subsequent operations.[64]

They would furnish the necessary sentinels for guarding the apparatus. This was a crucial change which was to remain for years to come and which irrevocably wrenched cartography away from the artist's hands. Accordingly, a party of the 12th Regiment of Foot, consisting of a sergeant, corporal and ten men were ordered from Windsor to Hounslow Heath.[65]

The French representatives were three distinguished members of the Academy of Sciences, Comte de Cassini (the fourth Cassini), Mechain and Le Gendre. They arrived at Dover and the two parties seemed to have amicably settled the details of the operations in 1787, the derived difference of longitude, Greenwich–Paris was 2°19′ 51◻ . According to Close:

> It was the first accurate triangulation carried out in this country and set a remarkably high standard; it amply fulfilled its original scientific purpose; it provided for the first time, a thoroughly reliable framework for map making; and it led directly to the formal founding of the Ordnance Survey.[66]

Another figure that influenced and aided the national survey to a considerable extent was Charles Lennox, the third Duke of Richmond who picked up the project and boosted it after a lapse of a few years due to Roy's death in 1790. In fact, it was from the Duke, who was greatly interested in matters of fortification and survey, that the national survey received its earliest and decided support. Bestowing a new confidence on the military, he corroborated "the claims of Ordnance officers to be looked upon as scientific men", and asserted their role in national defence. The existence of the Ordnance Survey itself can be traced to him. After Roy's death, he took the decisive step of putting the survey under the Board of Ordnance in which he held the post of Master-General from 1782–95, crucial years in British military history. Therefore, the name "Ordnance Survey", as mentioned earlier, was used much later. In its stead, expressions such as "British Survey", "British Trigonometrical Survey" or "General Survey of Great Britain" were used. The term which arose from its being regulated by the Board of Survey, finally came to mean and became synonymous with "survey of the highest accuracy".

An interesting corollary to the Ordnance Survey was the Royal Military Academy established in 1741. Notably, it was set up not only to train cadet officers or artillery engineers, but a contingent of surveyors and draftsmen. Engineers and gunners of the British Army had to amalgamate drafting architectural drawings with their martial and defence expertise. The curriculum consisted of multifarious subjects such as fortification and artillery along with presumably civil components such as geography, drawing, classics, writing, arithmetic, and the art of surveying and levelling.[67] Till Roy's death, the operations were carried under the general directions of the Royal Society, with assistance from the Board of Ordnance, while the cost of the instruments was personally borne by the King himself. While the reasons for the continuation of the survey after Roy's death remain doubtful, the military interest in the project was aroused by the possibility of war with France. The military authorities, persuaded by the Duke himself, were convinced of the necessity and value of a trustworthy map of southern England if such a war broke out.

Therefore, within the course of the year, the various negotiations with the financial authorities, with the Royal Society, and with Ramsden and the East India Company regarding the procurement of the second "3 ft theodolite", which was meant for the survey of India, were successfully concluded. The Duke of Richmond's plea to the Royal Military Academy, Woolwich to provide him the best officers of the artillery was also answered. Maj. Williams, Lt Mudge and Isaac Dalby were appointed for the job. The official constitution could therefore be taken as the 10th July 1791. Their first task was to remeasure the base at Hounslow Heath to verify Roy's assessment, which they began in the presence of Sir Joseph Banks, Dr Maskelyne, the Astronomer Royal and several other members of the Royal Society. This time measurements were done using steel chains fashioned by Ramsden. The mean of the two measurements was taken as the length of the base. What is remarkable is the constant additions and improvements not only in the geographical data, but also in the technical process of surveying. The first ordnance map came out in 1801, of the county of Kent and covering parts of London and Essex.

Geodetic experiments and terraqueous integration

Enlightenment reason was suffused with *l'espirit geometrique* and had immense faith in the map's rhetoric in depicting with certainty property boundaries, the extent of agricultural holdings, national borders, and the inferential geographies of minerology.[68] The mathematical and scientific scope of cartographic surveys involved three interlinked ideas. Firstly, the Enlightenment philosophers developed an epistemological ideal believing that a correct and certain archive of knowledge can be constructed. Secondly, on a pragmatic level, there was a call for improved scientific technologies such as "triangulation" which would provide verifiability and perfection to geographical knowledge. Lastly, the knowledge thus produced had a larger claim to the ability to clinch geodetic abstractions in concrete terms. The field of operation, be it the immediate space of the nation or the empire, in fact, circumscribed the whole of the earth. The determination of the shape and size of the nation or the empire through surveys, therefore, had an important by-product of gauging the shape and size of the globe as well.

Most of the survey and mapping operations from the late eighteenth century onwards aimed to achieve larger scientific and geodetic goals. The cost and logistics of the endeavours were most often justified in the name of the vital scientific information the operation would accrue on the state concerned. The spirit of competition among European nations arising from steadily consolidating nationalist consciousness, initiated a knowledge race, resulting in knowledge archives growing and flourishing in most European states. Similarly, in imagining Great Britain as the apogee of scientific development, the possession of a constantly updating knowledge data base was a primary urge and also a crucial prerequisite. As spaces, near and far, were harnessed and brought under a scalar ordering, knowledge emerging from the processes was also brought under the fold of a science regime, possessed and used to legitimise power and authority and finally give it its meaning. This emerged out of an egotistical drive, (involving both individualistic and communitarian in nature) and strengthening will to omniscience. In Edney's words, during this age: "geographical practices served to establish and legitimise Enlightenment's ideological self-image as an inquisitive, rational, knowing, and hence empowered state".[69]

The geodetic operation with which the survey and Mudge's name is closely associated is the measurement of the arc of meridian covering a distance of 196 miles across the British Isles. The objectives of this measurement were: firstly to determine the varying curvature of the meridian passing through the greater part of England, for use in computing the latitudes and longitudes of the triangulation; and secondly, and most importantly, to provide additional information in order to arrive at the correct figure and shape of the earth. Another of Ramsden's novel items of equipment was used for this purpose, a "zenith sector". The zenith sector was itself an ensemble of various instruments such as an eight-foot-long telescope and four-inch object glass.

However, Mudge's measurement of a single arc of meridian, less than three degrees long, did not give minutely reliable results, or serve to determine the figure of the earth. But the combination of many arcs, both along meridians and along parallels, and collation of information from geodetic observations spread across Europe, the United States and India ultimately helped to arrive at a conclusion regarding the shape of the earth. On the other hand, latitudes and longitudes derived from the Ordnance Survey were required for nautical charts, especially of the coastline of the islands. At the beginning of the nineteenth century there was considerable ambiguity about the location of islands, especially with regard to their longitudes. The arc of meridian was a crucial by-product with which spaces on earth were determined.

As discussed previously, one of the most essential prerequisites for the trigonometric surveys was the search and construction of the elevated points, crucial for making calculations with the newly introduced scientific apparatus.

If natural elevations were not available, towers were erected. The panoramic vista available at one sweeping glance provided the structural imagination on which the paradigm of the map could be established. As is my ongoing argument throughout the book, the same view could trigger poetic, aesthetic and technological responses. One such epicentre of British cartographic activity was near the Lake District, during an episode recounted by none other than William Wordsworth, who had visited the district when the survey of the region was in full swing. The survey was of the summit called the Black Combe, a view from which not only provided a most sublime experience, but also presented a visual revelation of the entire United Kingdom, England, Scotland, Wales and Ireland in one sweeping gaze.[70] Mudge declared that the sight from "the solitary Mountain Black Combe ... commands a more extensive view than any point in Britain" for the view extended up to the Isle of Man and the Irish coasts.[71] The exposure to the view and Capt. Mudge's activities on site resulted in Wordsworth's poems, "Inscription: Written with a Slate Pencil on a Stone, on the Side of the Mountain of Black Combe" and "View from the Top of Black Combe", both published together in a volume in 1815. In one, Wordsworth exalts:

> From the Summit of Black Combe ... the amplest range
>
> Of unobstructed prospect may be seen
>
> That British ground commands ...
>
> Low dusky tracts,
>
> Where Trent is Nursed
>
> Far southward ...
>
> Cambrian Hills to the south west, a multitudinous show;
>
> And in line with eye sight linked with these,
>
> The hoary peaks of Scotland that gave birth
>
> To Teviot's Stream, to Annan, Tweed, and Clyde[72]

The sight was a revelation and proclamation of "Britain's calm felicity and power", a celebration of the imperial gaze.[73] By now, the rebel-free landscape bore no signs of the bloody uprisings in the last century which had triggered these map making exercises and brought them on the ground. In its place was a seamless:

> grand terraqueous spectacle,
>
> From centre to circumference, unveiled.[74]

Nautical maps and territorial layout

In his "Of The True Greatness of Britain", written for King James in 1608, Francis Bacon writes that the true measurement of greatness of a nation "consisteth in the commandment of the sea".[75] From here on, it was generally believed that the future of Great Britain lay in the control of seas. Sea faring and thus charting of the waters were seen as crucial activities for the consolidation of the nation. Hydrography emerged as an important practice which ultimately controlled the visual imagination of the bound and fortified space of a nation. The newly formed local and national spaces were set against the backdrop of the expanding world of global economy in the face of oceanic travel and trade. As a precondition, this required a consolidation of the notion of the national self, countered against what was considered the "other", based on the notions of sameness and alterity. These questions were addressed and resolved by first demarcating the physical boundaries of the geo-space. The sea was perceived both as a boundary and as a horizon of possibilities to connect and negotiate distances. It was both a bridge and an insulation reflecting a perpetual nationalist contradiction between containment and contact.[76]

In 1808, Thomas Hurd, the then hydrographer to the Board of Admiralty, suggested "the necessity of an application being made to the Board of Ordnance that this office be allowed to have a copy of such parts of Colonel Mudge's Military Survey of England as respects the sea coast thereof together with all remarkable objects in the vicinity, as may be judged useful to navigation".[77] The Hydrographic Office was founded within a few years of the establishment of the Ordnance Survey, in 1795. However, the roots of the formation of the office were nearly the same as the Ordnance Survey's. It pertained to the necessity of naval defence of the island after the experience of the revolt in the Scottish Highlands, and can be traced to General Roy's piecemeal coastal surveys. Cooperation between military engineer and hydrographic surveyor was not frequent in the first half of the eighteenth century as it was in the latter. It was Graeme Spence who, after retiring from undertaking coastal surveys, was employed in the Admiralty to reorient another marine surveyor's (Murdoch Mackenzie) earlier compass surveys of the West Country with the recent results of the Grand Trigonometrical Survey then under way.

Meanwhile, Dalrymple, who was formerly hydrographer with the East India Company, became the hydrographer general to the newly constituted office. With constant prodding, Dalrymple was finally able to garner attention to the need and urgency of marine surveys. "It is a disgrace to this country", he said, "that the hydrography of our coasts is not accurately delineated".[78] He delineated a novel method of marine survey in his *Essay on Nautical Surveying* written while he was hydrographer in 1806. He named it the "Quincunx" from an arrangement of trees. It consisted of a central vessel anchored in the sea surrounded by four smaller anchored vessels arranged in a square. The pattern of the anchored vessels would form a sea triangulation to which soundings and other hydrographic information could be related. The scale of the surveys could be found by measuring the distance between any two of the anchored vessels. Dalrymple proposed using the speed of sound when the distance was considerable. Dalrymple insisted:

> No other method is competant to give an exact chart of the Banks and Soundings out of sight of land.[79]

The quincunx method was never used during Dalrymple's lifetime because the process required too many vessels. Moreover, he quarrelled with the office and was subsequently dismissed.

It was during Hurd's term of office as hydrographer that the Admiralty chart took on a style which collated triangulation on land and extended it to the coastal areas and the sea. During his term, the surveyor George Thomas made use of Dalrymple's method, but with the help of the Dickinson Patent Buoy and Beacon machine instead of vessels, because, as Thomas wrote to Hurd:

> by their height above the water and their perpendicular, as well as fixed position, a series of triangles may be carried out from the shore to any extent within the limits of the soundings.[80]

In 1817, Hurd entrusted Thomas with the task of proceeding with his survey northwards to the Orkneys and, if possible, further to the Shetland Islands "in conjunction with Colonel Mudge, who, ... is very desirous of cooperating with you on these parts of the coast". This association with the Ordnance Survey proved fruitful later, in 1825, when Thomas started his detailed hydrographic survey of the Shetland Isles. The primary triangulation of these northern islands had been measured by Colby in 1821, shortly after his appointment as director in succession to Mudge, but it was left incomplete when, in 1825, the attention and resources were entirely focussed towards the six-inch Trigonometrical Survey of Ireland. Thomas, therefore, had no other option than first carrying out a large scale topographic survey, to which he could then connect his marine examinations. Thomas's greatest contribution to British marine cartography was the survey of the Orkneys and the Shetland Isles. The survey of the Shetland Isles, marked the first step towards accurately charting the whole of the coastal waters of the British Isles.

Close, two-way collaboration was established with the Ordnance Survey which continued to supply triangulation data and, in turn, naval surveyors provided coastal detail for Ordnance maps. All these developments provided a springboard for the ambitious re-survey of the coastal waters of Britain, known as the "Grand Survey of the British Isles", which kept the Hydrographic Office preoccupied for much of the mid-nineteenth century.

However, what was called the Grand Survey of the British Isles saw its completion with Francis Beaufort, who was the hydrographer from 1829 onwards. Under him, different surveyors were simultaneously employed and engaged on various sections of the coast. In this way, long stretches of coastal areas and home waters were charted including the main navigable rivers. Areas which had already undergone the cartographic routine of the Ordnance Survey, were easily surveyed and triangulated by the hydrographer. As the survey progressed, it became clear that the sea bed and the coastal regions are constantly under flux and continuously reshaped by tides and waves. With the limited resources under its command, the Office could not dream of the luxury of updating such information on a regular basis. Later in the century, harbour boards were established to tackle this problem at the local level with important ports being given the duty of conducting surveys at frequent intervals thus adding to the marine cartographic archive. Pilot books were issued with updated and renewed information for smooth navigation.

Initially Beaufort's scheme, constrained by slender resources, was restricted to charting only the home waters. As the work successfully progressed, the seas of the world came under the purview of marine cartography, flanked by the growing mercantile and defence interests of the British state.[81] Soon the systematic triangulation of coastal beds and unknown sea waters curved out and strove to protect land masses under British dominion, located far and near. Most importantly, the Ordnance and the Hydrographic Surveys, jointly, were able to tame the unknown through systematic plotting, which thereby produced a geographically bound spatial entity. This gave birth to a geographic space in visual and thereby comprehensible terms while locating it in geodetic space on the globe. The emergence of Great Britain as a unified nation was the outcome of this joint venture. While the Ordnance Survey brought the innermost recesses of the Isles under surveillance and into public discourse, so the Hydrographic Survey charted the bounds of the dominion, creating a marine frontier. Similar practices were being simultaneously undertaken to carve out the British dominions in its overseas colonies, therefore shaping both the core as well as the periphery. As the trajectory of James Rennell, one of the leading colonial cartographers in India, shows, a training in hydrography proved to be of great advantage in tracking waterways in unknown lands, making inroads into the remote hinterlands of an uncharted territory. The next chapter will deal with the cartographic construction of India as a spatial unit, enmeshed within practices of British political advancement and its scientific ideology.

Notes

1 Leibniz. Gottfried. Dissertatio Exoterica De Statu Praesenti et Incrementis Novissimis Deque Usu Geometriae. Vol. 7. 1676. p. 326.
2 Bacon, Francis. *The Works of Francis Bacon.* James Spedding et al. 7 Vols. London: Longmans, 1862–72. Vol. 7, pp. 48–49, 51; See Elden, Stuart. *The Birth of Territory.*

Chicago: University of Chicago Press, 2013. See chapter "The Extension of the State". pp. 7–8

3 Locke, John. "An essay concerning human understanding". In *Works of John Locke in Nine Volumes*. Vol. 1. London: Rivington, 1824.

4 Hobbes, Thomas. *Leviathan*. (1660) UK: Penguin, 2003.

5 Turner, Henry S. "Literature and Mapping in Early Modern England 1520–1628". In J.B. Harley and David Woodward eds. *The History of Cartography*. Vol. 6. Part 1. Chicago: University of Chicago Press. 1987. p. 412.

6 See Harvey, P.D.A. *The History of Topographical Maps: Symbols, Pictures and Surveys*. London: Thames and Hudson, 1980.

7 Cited in Harvey, David. *The Condition of Postmodernity: An Enquiry into the Origins of Cultural Change*. Oxford: Basil Blackwell, 1989. p. 245.

8 Ibid. p. 245.

9 Klein, Bernard, *Maps and the Writing of Space in Early Modern England and Ireland*. Hampshire: Palgrave, 2001. p. 3.

10 Harvey, David. *The Condition of Postmodernity*. p. 250.

11 Cosgrove, Denis. "Maps, Mapping, Modernity: Art and Cartography in the Twentieth Century". *Imago Mundi*. 57:1, 2005. pp. 35–54.

12 See Edgerton, Samuel. *Renaissance Rediscovery of Linear Perspective*. New York: Basic Books, 1975. Also see Alpers, Svetlana. *The Art of Describing: Dutch Art in the Seventeenth Century*. Chicago: John Murray, 1983 (specifically the essay "The Mapping Impulse in Dutch Art").

13 Alpers, Svetlana. "The Mapping Impulse in Dutch Art". *The Art of Describing: Dutch Art in the Seventeenth Century*. Chicago: John Murray, 1983 pp. 133–4.

14 Harvey, David. *The Condition of Postmodernity*. p. 246

15 Edgerton, Samuel. 1976. Cited in Harvey, David. *The Condition of Postmodernity*. p. 245.

16 Harvey, David. *The Condition of Postmodernity*. p. 244.

17 Ryan, Simon. *The Cartographic Eye: How Explorers Saw Australia*. Melbourne: Cambridge University Press, 1996. p. 102.

18 Cited in Delano-Smith, Catherine and Roger J.P. Kain. *English Maps: A History*. "The British Library Studies in Map History" Vol. II, London: The British Library, 1999. p. 122.

19 Ibid. p. 122.

20 Smith, David. *Antique Maps of the British Isles*. London: BT Batsford Ltd., 1982. pp. 36–7.

21 Edney, Matthew. "Reconsidering Enlightenment Geography and Map Making: Reconnaissance, Mapping, Archive". In David N. Livingstone and Charles W.J. Withers eds. *Geography and Enlightenment*. Chicago: Chicago University Press, 1999. p. 191.

22 Edney, Matthew. "Reconsidering Enlightenment Geography and Map Making". p. 191.

23 Ibid. p. 193.

24 Klein, Bernard. *Maps and the Writing of Space*. p. 43–4.

25 Ibid. p. 44.

26 Harvey, P.D.A. "English Estate Maps: Their Early History and Their Use as Historical Evidence" in David Buisseret ed. *Rural Images, Estate Maps in the Old and New Worlds*. Chicago: University of Chicago Press, 1996. p. 27.

27 Delano-Smith, Catherine and Roger J.P. Kain. *English Maps*. Vol. II, London: The British Library, 1999. p. 122.

28 Ibid. pp. 112–13.

29 Ibid. p. 116.

30 Helgerson, *Representations*. p. 56.
31 Ibid. p. 62.
32 Klein, Bernard. "Constructing the Space of the Nation: Geography, Maps and the Discovery of Britain in the Early Modern Period". *Journal for the Study of British Cultures*. Vol. 4/1–2, 1997. pp. 11–29. p. 16.
33 Ibid. p. 18.
34 Cited in Helgerson, Richard. "The Land Speaks: Cartography, Chorography and Subversion in Renaissance England". *Representations*, Vol. 16, Autumn 1986. p. 66.
35 Smith, David. *Antique Maps of the British Isles*. London: BT Batsford Ltd., 1982. p. 20.
36 Cross refer Chapter 5 for a study of Camden's *Britannia*.
37 Klein, Bernard. "Constructing the Space of the Nation". p. 19.
38 Helgerson, Richard. "The Land Speaks: Cartography, Chorography and Subversion in Renaissance England". *Representations*, Vol. 16, Autumn 1986. pp. 50–85.
39 Klein, Bernard. "Constructing the Space of the Nation: Geography, Maps and the Discovery of Britain in the Early Modern Period". *Journal for the Study of British Cultures*. Vol. 4/1–2, 1997. p. 11–12.
40 Smith, David. *Antique Maps of the British Isles*. London: BT Batsford Ltd., 1982. p. 40.
41 R.S.A. Min. Comm. (Polite Arts), 15th March, 1759. Cited in Harley, J.B. "The Society of Arts and the Surveys of English Counties 1759–1809". *Journal of the Royal Society of Arts*, 112, (1963–4). pp. 44.
42 Cited in Harley, J.B. "The Society of Arts and the Surveys of English Counties 1759–1809". *Journal of the Royal Society of Arts*, 112, (1963–4). p. 44.
43 R.S.A. Min. Comm. (Polite Arts), 10th March, 1760. Cited in Harley, J.B. "The Society of Arts and the Surveys of English Counties 1759–1809". *Journal of the Royal Society of Arts*, 112, (1963–4). pp. 44–5.
44 See Dove, Jane. "Geographical board game: promoting tourism and travel in Georgian England and Wales". *Journal of Tourism History*. Vol. 8, Issue 1. 2016. pp. 1–18.
45 See Victoria and Albert Museum of Childhood Collections http://collections.vam.ac.uk/item/O26289/tour-through-england-and-wales-board-game-wallis-john/, accessed on 27 June 2018. Also see Hewitt, Rachel. *Map of a Nation. A Biography of the Ordnance Survey*. UK: Granta, 2013. p. 249.
46 Tooley, R.V. *Maps and Mapmakers*. London: BT Batsford Ltd., 1952. pp. 51–5.
47 Chakrabarti, Pratik. *Western Science in Modern India: Metropolitan Methods, Colonial Practices*. Delhi: Permanent Black, 2004. p. 70.
48 Hewitt, Rachel. "Mapping and Romanticism". *The Wordsworth Circle*, Vol. 42, No. 2, 2011. pp. 157–65. p. 159.
49 The penchant for survey literature written in a language of science was a parallel phenomenon arising out of the same syndrome. See Edney, Matthew H. *Mapping an Empire: The Geographical Construction of British India*. Chicago: University of Chicago Press, 1997. p. 41.
50 Edney, Matthew H. *Mapping an Empire: The Geographical Construction of British India*. Chicago: University of Chicago Press, 1997. pp. 104–18, 319–25.
51 Charles Edward, the grandson of James II, was the young Scottish pretender to the British throne. In August 1745, he and his troops occupied Edinburgh and Carlisle and marched as far south as Derby. He was compelled to retreat and finally defeated by the Duke of Cumberland, son of George II, at Culloden in April, 1746, finally putting an end to the Anglo-Scot antagonism.
52 Close, Charles F. *The Early Years of the Ordnance Survey*. (1926) Great Britain: David and Charles, 1969. pp. 2–3.

53 Some interesting people were appointed in producing these maps. Though none of the maps are signed, the artist Paul Sandby, who later became an instructor in Royal Military Academy, and about whom I have written in Chapter 4, seems to have been on the list.

54 Elizabethan maps of the British Isles, as discussed earlier, already expressed the idea of incorporating Ireland as one of the components of the national conglomerate.

55 Cited in Close, Charles F. *The Early Years of the Ordnance Survey*. p. 100.

56 Cited in Close, Charles F. *The Early Years of the Ordnance Survey*. p. 101.

57 See Cronin, Nessa. "Writing the 'New Geography': Cartographic Discourse and Colonial Governmentality in William Petty's *The Political Anatomy of Ireland*". *Historical Geography*. Vol. 42, 2014. pp. 58–71.

58 Cited in Close, Charles F. *The Early Years of the Ordnance Survey*. p. 107.

59 Cronin, Nessa. "Lived and Learned Landscapes. Literary Geographies and the Irish Topographical Tradition. In Marie Mianowski ed. *Irish Contemporary Landscapes in Literature and the Arts*. Palgrave Macmillan, 2012. pp. 106–18.

60 Cronin, Nessa. "Lived and Learned Landscapes. Literary Geographies and the Irish Topographical Tradition. In Marie Mianowski ed. *Irish Contemporary Landscapes in Literature and the Arts*. Palgrave Macmillan, 2012. pp. 106–18.

61 Cited in Close, Charles F. *The Early Years of the Ordnance Survey*. p. 107.

62 Close, Charles F. *The Early Years of the Ordnance Survey*. p. 28.

63 Cited in Close, Charles F. *The Early Years of the Ordnance Survey*. p. 12.

64 Mudge, William. *An Account of the Operations Carried Out for Accomplishing a Trigonometrical Survey of England and Wales*. Vol. II Philosophical Transactions. London: 1801. p. 7.

65 For a detailed recounting of the technology and persons involved in the Ordnance and the Trigonometrical Survey in Great Britain, see Hewitt, Rachel. *Map of a Nation: A Biography of the Ordnance Survey*. UK: Granta, 2013.

66 Close, Charles F. *The Early Years of the Ordnance Survey*. p. 23.

67 Sen, Sudipta. *Distant Sovereignty: National Imperialism and the Origin of British India*. London: Routledge, 2002. p. 65.

68 Withers, Charles W.J. "Art, Science, Cartography and the Eye of the Beholder", *Journal of Interdisciplinary History*. Vol. 42, Issue 3. 2012. pp. 429–37.

69 Matthew H. Edney, "Reconsidering Enlightenment Geography and Map Making: Reconnaissance, Mapping, Archive". in Livingstone, David N. and Charles W.J. Withers eds. *Geography and Enlightenment*. Chicago: University of Chicago Press, 1999. p. 173.

70 This episode is talked of in detail in Hewitt, Rachel. *Map of a Nation*. 2013. pp. 248–9.

71 Cited in Hewitt. 2013. p. 248.

72 Wordsworth, William. "View from the Top of Black Comb". *Poems*. Vol. 1. London: Longman. 1807.

73 An alternative interpretation is offered in Ford, Thomas F. *Wordsworth and the Poetics of Air*. Cambridge: Cambridge University Press, 2018. pp. 80–1.

74 Wordsworth, William. "Upon a Stone on the Side of Black Comb". *Poems*. Vol. II. London: Longman, 1815.

75 Bacon, Francis. "Of the True Greatness of Britain". Eds. James Spedding et al. The Works of Francis Bacon. Cambridge: Cambridge University Press, 2011. pp. 37–8.

76 Hogan, Sarah. "Of Islands and Bridges: Figures of Uneven Development in Bacon's *New Atlantis*". *Journal of Early Modern Cultural Studies*. Vol. 12, Issue 3. 2012. pp. 28–59.

77 Public Record Office, Admiralty 1/3523, letter dated 1 Nov. 1808. Cited in Robinson, A.H.W. *Marine Cartography in Britain, A History of the Sea Chart to 1855*. Great Britain: Leicester University Press, 1962. p. 128.

78 Public Record Office, Admiralty I, Hydrographic Office. Letter dated 13 June 1804. Cited in Robinson, A.H.W. *Marine Cartography in Britain, A History of the Sea Chart to 1855*. Great Britain: Leicester University Press, 1962. p. 104.
79 Cited in Robinson, A.H.W. *Marine Cartography in Britain*. p. 129.
80 Ibid.
81 There was constant pressure on the Board of Admiralty to make its charts available to the civilian merchant marine. These therefore served not only the needs of increasing traffic generated in rapidly industrialising England, but also raised much needed revenue to finance further surveys. See Delano-Smith, Catherine and Roger J. P. Kain. *English Maps: A History*. Vol. II, 1999. p. 225.

2

MAPPING INDIA

Rennell and Lambton

First comes a white man for innocent travel, then come two white men to draw a map, next comes an entire contingent to conquer the land.[1]

From a popular indigenous proverb from the nineteenth century Western Frontier

This is at once a reminder of cartography's military rationale as of its symbolic function. The temporal sequence might not have always been the same, but the interconnected nature of the three definitely holds ground. They indicate both the logic and progression of spatial mechanisms and the structural affinities between them. The validation of imperialism's belligerence abroad worked on the assumption of empty space and an oft-imagined threat. Anne Godlewska rephrases the map's ideological underpinning in these words:

This territory is and has long been ours; here is the centre of the universe; what counts is not the people but the state; territorial conquest is a glorious and righteous mission; if we do not claim this land, the enemies you most fear will.[2]

Seemingly disparate cultural practices of travel and cartographic art attained militaristic colouring as a direct fallout of imperial schemes of spatial aggrandisement. By systematically amassing topographical and statistical data, thereby subjecting them to the cartographic grid, cartography converted space into a visible object of knowledge which could be controlled, occupied and managed. As it transpired in England's first internal empire, eventuating into a spatial reorganisation into what came to be known as Great Britain, so the space produced by the cartographic transaction in South Asia enabled a spatial transformation redefining an organic space into a functional space – that of the colony. Just as art gives form to matter, this semantic transfer of the organic space into a functional space through the technology of the map proved to be decisive in defining a geo-political entity, making what came to be known as modern India.

That this was entirely a British device can be established from studying the history of the maps on and in India. Their eventual transition in the latter half of the eighteenth century along with their evolution and trajectory thenceforth, manifest an imperial design to articulate colonial territory. The Mughal Empire itself, on which British Empire was modelled, did not extend beyond the Deccans, the southern peninsula being ruled by numerous petty kings and provincial princes. The consolidation of the culturally disparate regions under a single scopic regime is the consequence of the colonially crafted cartographic image.

The construction of a monolithic or universal space was the imperial endeavour. It constructed space as a mensurable and finite entity. It embraced the Euclidean and Cartesian concept of the measurable space constructed and verified by the mathematical web. The two-dimensional representation of the infinite space helped to reduce it to a comprehensible finitude. Imperial cartography and homogeneous mapping practices encouraged the construction of space in positivist terms, where alternative perceptions of this objective space was never given credence. To think of space as that which could be produced socio-culturally or that varied cultures could produce various spaces, would make colonial constructions impossible. Therefore the native inhabitants were not allowed to have a space different to that of the coloniser. They were merely thought to have under-utilised space which imperialism only took to its natural culmination. The construction of the monolithic space enabled imperialism to annex and hierarchise space, to use it for its own advantages.[3] Colonial cartography involved two stages of acting upon space: firstly, the erasure of pre-existing indigenous spaces to recast it in universal terms and secondly, inscribing universal space with power to outline a material place in the colonial consciousness. In this chapter, I shall concentrate largely on the representational strategies which served to foreground the colonial sphere, but from which emerged the space of the future nation/country with seemingly naturalised boundaries. Drawing upon the theory of socio-political space as developed by Lefebvre, this shall be an analysis of how space is actively produced, in this case by scientific discourse surrounding cartography, necessitated by imperialism. Looking at the historical sequence of map production techniques, and the cultures of mapping in a single region, can lead to crucial insights into the changing understanding of the shape and extent of a given geographical space and unravel a system of cultural meaning which determines social relationships.

Early ventures in mapping India

What was understood as India in the West has an intriguing history. Till a considerable time into the eighteenth century, there was no clear idea about the extent of what was called India. Merely, the location of some of the important sea ports were known through the existing Portolan maps or sea charts due to the flourishing trade that went on between these and the West. India has remained in the European consciousness for a long time and therefore on

European atlases. Most maps show evidence of combining travellers' tales that had so far been published or gathered from hearsay. The early maps which survive, such as that of Ortelius, Mercator, de Jode and Hondius had all omitted certain features and selected portions. In the earlier editions of Ptolemy, the Indian peninsula is shown greatly foreshortened. Ceylon appears sometimes to the right and sometimes to the left of the peninsula. A map of the "Indian Empire" by Bertelli appeared in Venice in 1565. In the atlas of Ortelius, India is not represented as a single entity, but grouped with the East Indies. It is decorated with the arms of Portugal, ships and sea monsters. Mercator's map likewise covers a wide area, and India is shown in an even more attenuated form. Most of these maps did not show the Himalayas. Others showed incorrect sources and courses of the Ganges. In a map of Asia published in 1637 by Mercator's son Rumold, the Ganges is shown flowing east to the China Sea. "Palimbotra" is on this Ganges, while "Delhi" is on the "Guenga" on the peninsula. The "Vindius" mountains are much further north than Bengal, and the lettering of "Indostan als India intra Gangem" meaning "India within the Ganges" starts above the Bay of Bengal and extends China to the ocean in the east.[4]

The first English map of the "Mogol Territories" was published by T. Sterne, globe maker on 1619. It was drawn by William Baffin on information supplied by Thomas Roe after his visit to Jahangir's court as ambassador from James I in 1615. This map was copied by Purchas in 1625 in *Purchas his Pilgrimes* Vol. I, Edward Terry's *A Voyage to the East Indies* (1628), John Ogilby's *Asia* (1673) and many others. It was also copied and published in Paris in 1663. Several small travel books, by Du Val and Mallet in France and Seller and Marden in England, provided reduced copies of Dutch maps for the northern part. In France, other maps appeared by Sanson, Jaillot, Delisle, d'Anville and Bellin. In England maps of India featured in the works of popular cartographers like Herman Moll and Bowen, but these too were not accurate.[5]

Publishers were contented to reprint or copy older maps as long as their public was satisfied, for they were cheaper. Printing new maps were costly projects and involved funding and patronage. Though copying from older maps was cheaper than designing a new map, it resulted in the same mistakes being repeated again and again becoming a dogma unless there was very definite witness, preferably eye witness, to vouch for the maps being incorrect. Therefore, once a name and a rough location had been established, it was repeated long after the kingdom had disappeared or renamed. Thus the kingdom of Narsinga had been reported by the Portuguese in the sixteenth century, when the kings of Vijaynagar were powerful throughout south India. The town was sacked in 1565, yet Narsinga appears on maps as late as 1720. Similarly, the town of Golconda, founded in 1518, was abandoned in 1589 as it was considered unhealthy. A new capital was built at Bhagnagar and the province was annexed by Aurangzeb in 1687. It formed the centre of the Mughal province of the Deccan until Asaf Jah broke away in 1722 and built his own capital at Hyderabad. The kingdom and town of Golconda were still appearing on maps of India as late as 1780.[6]

In London in 1710, Herman Moll (d.1732) drew a small map which showed the whole of India on one page, the first important one to give prominence to the subcontinent as a separate entity. In 1701, he had first drawn "India or the Mogul's Empire" which was extended eastwards to include Cambodia and Cochin-China. His second map was called "The West Part of India, or the Mogul's Empire" and included only the subcontinent, though the name was still incorrect, for the southern part of India was never part of the Moghul Empire. He redrew the map in 1726 and now called it "India Proper, or the Empire of the Great Mogul", which shows the growing awareness in England of the identity of India, and the confusion caused earlier by the use of the name India for the whole of Asia.

The people in the best position to record the geography of India were, in fact, the Christian missionaries. They were often resident in India for many years and many of them acquired the skill necessary for taking accurate astronomical and land measurements. They also travelled to the remotest of locations. The Jesuits had established their first mission in 1542 and they were soon settled in many parts of the country. Father Monserrat had travelled to the court of Akbar in 1579, and then to Kabul, recording his journey in letters and a map which was not made public. From 1702 the Paris Foreign Missions Society arranged the publication of letters from the missions and a few maps drawn by their servants. These are usually clubbed under "Church Cartography". Two maps of India were published in 1722, one a very sketchy outline of a route inland, and the other a more comprehensive map of the peninsula. This had been sent home by Father Bouchet in 1719 and was used by Guillaume de l'Isle (1675–1726) in his map of south India in 1722. Later, in 1734, Father Boudier was invited to Jaipur by Raja Jai Singh to study the astronomical laboratory he had built there. On his way Boudier took many measurements which were consulted by d'Anville for his large map of India in 1752. "Church cartography" caused the excellence in map production to shift to the hands of France.

French geographer d'Anville (1697–1782), compiled the first modern map of India which was published in French, 1751–2, and was based entirely on knowledge derived from the routes of various travellers recorded in history and from some rough charts of the coast. D'Anville may be called the first scientific mapmaker. Shorn of embellishments, this was the first European map which made claims to geographical accuracy and research. For his large two-sheet map of India, published in 1752, he used a number of source books from Ptolemy's lists to a Turkish geography called *Kiatib-Shalebi*, an Indian geographical text in Tamil which he called *Puwanasaccarum*, as well as books by later travellers. He was not a traveller himself for the age of in situ cartography in India was still a decade away, but from his home in Paris he maintained a wide correspondence and received all new charts that were sent back to France, especially by the missionaries. D'Anville was the first to publish all his sources for a new map of India. His *Eclaircissemens* (1753), formed the basis of Rennell's survey and memoirs years later.[7] This set the trend for cartographers to publish their memoirs to accompany their large and "original" maps, narrating in minute detail the many sources which formed the basis of the map and their reconciliation or the criteria by which they were accepted or rejected. Matthew Edney points out that:

> a map's accompaniment by a memoir, was a sign of the cartographer's pretension that the map ought to be considered as a cartographic landmark. Through his memoir, a cartographer assured his public that a map was based on the best available sources and he displayed his own conscientiousness and intellectual virtuosity."[8]

This foregrounded the map's author as the "creator of the map's knowledge".[9]

Enlightenment and the imperial surveyor: the "aura of the scientist"

British nationalism was directly linked to colonialism which, for the British, implied the promise of wealth, investment and remittance. The activities of the East India Company aroused enormous interest and curiosity among the public especially with the Company's annexation of Bengal. The event was all the more glorified as it was a decisive territorial victory over France against which Britain's own nationalism was pitched. As Sudipta Sen points out, the vision of territorial possession in the colonies was directly fastened to the concept of defensible and bounded property as applicable in the home nation. He further points out that the rights of the East India Company were extended not only to exclusive trade but to the legitimate possession of tangible property of a corporate body politic, that is, territory and revenue. While Sen explores texts by numerous theorists and administrators in eighteenth-century Britain to explain this point, Thomas Pownall's statement in his tract called *The Right, Interest, and Duty of the State as Concerned in the Affairs of the East Indies* (1773), which he cites, succinctly validates this argument:

> People now at last begin to view those Indian affairs, not simply as beneficial appendages connected to the empire; but from the participation of their revenues being wrought into the very composition and frame of our finances; from the commerce of that country being indissolubly interwoven with our whole system of commerce; from the intercommunication of funded property between the Company and the state – people in general from these views begin to see such an union of interest, such a co-existence between the two, that they tremble with horror even at the imagination of the downfall of this Indian part of our system; knowing that it must necessarily involve with this fall, the ruin of the whole edifice of the English Empire.[10]

The rule of property in certain ways legitimised colonial extension. Likewise, Adam Anderson, a prominent British historian of political economy in the latter half of the eighteenth century, conflates the logic of property with that of dominion in his An Historical and Chronological Deduction of the Origin of Commerce:

> Public property excludes communion amongst nations; private property is communion amongst persons. For, as particular persons, which they possess privately of other persons: so countries and territories, like greater manors, divided each from the other by limits and borders, are the public properties of nations, which they possess exclusively of one another.[11]

Property and land rights, as seen in the preceding chapter, are the determinants of cadastral and estate cartography in Britain, which led to a complete overhauling of previous modes of map making, implanting the artistic tradition in the domain of science which perceived land as a mensurable entity. As Harvey points out, this new intellectual formulation is related to other rationalising practices which emerged during the same time in other fields such as in commerce, banking, book-keeping, trade and agricultural production under centralised land management.[12] The new age geography gave rise to reconnaissance and field survey that dominated the British domestic scene and were automatically transposed onto Britain's empire which got enmeshed in British order of property and possession. In explicit relation to Great Britain, the homegrown formula exported not just efficient methods of structural surveillance intrinsic to mathematical surveying, but a whole ensemble of views and perceptions to subordinate space. In India, the defining perception of landscape was mixed with that of insubordinate groups who tyrannised and tried to disturb the well-organised administrative space produced by the British in India. As Bernard Klein points out, in the case of British cartography in Ireland:

> the map ... is ... expected to – create the image of a rebel-free landscape, a plane space subject to the imperial gaze ... pre cartography by contrast, has the rebel roaming the wild, unconquered landscape at will, escaping the grip of culture and the fixity of the cartographer's plot alike.[13]

The colonial rhetoric surrounding the representation of India constructed its space as an inherently transgressive realm of savage rebels where renewal of political control was contingent upon the mechanism of systematic cartographic description. The surveyor's view, here, becomes identical to the perspective of power which restores the chaotic space to normality and order.

The professional existence of these surveyors was fashioned upon their superior technological competence. As Harvey points out, as Cartesian principles of rationality became integrated into the Enlightenment project:

> It signalled a break in artistic and architectural practice from artisan and vernacular traditions towards intellectual activity and the aura of the artist, scientist, or entrepreneur as a creative individual.[14]

The cases of both James Rennell and William Lambton, discussed below, shall exemplify the key theme of the discourse of surveying and the tussle over claims on sophistication and technology. It heralded the new age of surveying as a carry-over from General Roy and Mudge's cartographic practices in Britain. This celebrated the survey equipment where the uniqueness of the instruments defined the uniqueness of their user. The apparent novelty of geometrical thought rested finally with a claim to lay open to view, via geometry, truths otherwise unavailable. The surveyors' memoirs, in this case, were merely professional in nature, trying to express the laborious process of extracting data from an immensely chaotic and disorganised mass of unsifted information.

James Rennell's "Map of Hindoostan"

D'Anville had recently collected all existing knowledge of India, derived from routes of solitary travellers and rough charts of the coasts. His map of India appeared five years before the Battle of Plassey. Eight years after the episode, James Rennell was at work in the newly ceded territories of Bengal and Bihar. Initially, in 1760, there was an order passed by the local Council of Bengal which wanted an accurate estimate of its revenue provinces. Once it assumed power to collect revenue on behalf of the Mughal monarch and emerged as a major territorial power, there arose several requirements for a map. As the revenue had to be based on the amount of land under cultivation, it needed to be quantified and demarcated in the Company's new map. Moreover, in view of defence of the territory conquered, exploration of sea, river and land routes emerged as a crucial requirement. The north-west of the territory was constantly threatened by the presence of the Marathas, and at sea, the French were still a considerable power. To counter the two looming powers, strategic information needed to be gathered.

As can be safely inferred from the preceding chapter, geographical knowledge and military expedience are closely intertwined. The military surveys following the Jacobite Rebellion closely parallel cartographic activities in Bengal. In fact, scholars have pointed out the similarity between the characters of Roy and Rennell, both of whom were engineering officers, later elected fellows of the Royal Society. Rennell had entered the navy and distinguished himself as a midshipman at the Siege of Pondicherry. He took up a position with the army under Clive, rose to the rank of major and ultimately became the surveyor-general of India at the pinnacle of his career. Edney, however, suggests that the office was rather limited in its institutional scope and it might not have been as glorified as it appears. He received little aid and had only one or two people under him. He also faced much resistance and hostility from local inhabitants and in one instance, received a crippling injury which physically incapacitated him for life.[15] He was not intended to be the head of a body for organised survey, for anything of the sort did not exist at the time. It was a position intended primarily for gathering geographical information and construction of maps.[16] It was in this sense that the two terms, "geographer" and "map maker" were interchangeable with each other in this age of Enlightenment, in that they provided unambiguous, coherent and comprehensive data from a whole host of disparate sources.

The immediate call for the map came from Robert Orme, the first historiographer of the new age of British conquest ensuing after Plassey, which completely transformed the East India Company's position in the subcontinent. Orme was a major propagator of map use in defence of the Company-controlled territory.[17] In his undated autograph "Essay on the Art of War", probably meant for Lord Clive, Orme writes,

> We have in general very few good charts in India. No wonder. Our Generals have not paid that attention to the subject which it requires … If those in the Administration were sensible of the advantages resulting from it, they would never scruple the expence [sic] I would have a Plan of your whole Frontier, with the Engineer's observations from League to League. And where you have any Defiles, they should be accurately described, surveys having first been made with the most minute exactness … Route surveys … From a compleat Engineer you may go much further. He is not to confine himself to the roads only, but the situation of the country … Embrace therefore every opportunity … to send officers into a Country, where you may soon have occasion to march an Army.[18]

Orme required maps in order to write a book on military achievements in India. In a letter to Lord Clive, dated 21 November 1764, Orme suggests, "Make me a vast map of Bengal, in which not only the outlines of the provinces, but also the different subdivisions of Burdwan, Beerboom etc., may be justly marked … Take astronomical observations of longitude".[19] To this, Clive responds on 29 September 1765, "You shall have very exact charts of Bengal, Bahar, and Orissa, and of the Mogul Empire as far as Delhi at least. A map of the Ganges likewise, and all the other rivers of consequence".[20] Soon, the first baseline for the survey was laid near Calcutta. (For a similar measurement of the Calcutta baseline at a later date, see Figure 2.1) Rennell's survey covered an area 900 miles long by 360 to 240 miles wide, from the eastern confines of Bengal to Agra, and from the foot of the Himalayas to Calpi. Rennell's training as a maritime surveyor under Alexander Dalrymple proved to be a boon for his survey work in Bengal. In view of the East India Company's spatial movement in the subcontinent from the littoral to the interior, Rennell's basic training in hydrography, helped him chart the river coasts with ease by basing his techniques on maritime science. "Modelling a current as an ocean's river", Rennell could successfully conjecture the gradient of the rivers gradually while his survey slowly moved towards the source of the rivers. This movement is significant as it also gave direction to the colony's expansionist gaze. Also significant in this respect, is the synergy in itineraries of the colonial men of science and the arts. The Company travellers, surveyors and artists subsequently took up the same route for their individual journeys in their multi-faceted endeavours to enrich the colonial archive. These ambulatory machinations in turn successively framed the core and the periphery of the colony in chalking out the full possible extent of their rule through the latter half of the eighteenth and early part of the nineteenth century. With Rennell, began a new age wherein the figure of the cartographer became an imperial role model and a vanguard. The functional role of the cartographer was to determine the coordinates within which the colonial knowledge archive could be systematically raised and reared.[21]

Figure 2.1 Measuring of Calcutta baseline with trig point in the background. This was the Great Trigonometrical Survey measurement of the Calcutta baseline by George Everest. It shows a Ramsden chain being set with tents to prevent expansion from heat

Image courtesy: Wikimedia Commons[22]

The *Bengal Atlas*, published in 1781, was compiled from around 500 surveys conducted by Rennell and ten other officers and engineers appointed under him. Alongside the surveys of the main rivers in the region, the Ganges and the Brahmaputra and several of their tributories, these officers also amassed a great wealth of information about existing trade routes and roads making way for later revenue surveys. The *Bengal Atlas* was dedicated to Robert Clive, Warren Hastings, Hector Munro, Verelst and the Company.[23] Rennell's was the first systematic survey conducted in the region and can be interpreted as the inception of the Survey of India. Rennell's task was "to form one general chart from those already made"[24] and therefore by compiling a final map out of the geographical knowledge already in circulation, he gave coherence to the seemingly disjointed and varied information. The drawing and control of the map thus signified both the control of the knowledge archive and the region it depicted. The construction of India as a place was imbricated with other colonial cultural practices. Appearing at a time when colonial historiography was at its formative stages in India, the preliminary maps and surveys were used by the historian Robert Orme for his *A History of the Military Transactions of the British Nation in Indostan (1745–1760),* which appeared in 1763. Not only did the constitution of geographical knowledge of India entail a retrieval of history, it also served to write new histories as well.

The Map of Hindoostan was compiled from the evidence of various route surveys and was first published in 1782. It subsequently went through three editions with its accompanying *Memoir of a Map of Hindoostan; or the Moghul Empire* published in 1788, dedicated to the president of The Royal Society as an "attempt to improve the geography of India and the neighbouring countries". For the 1788 edition, Rennell claimed that he had collected far more information from various sources, which enabled him to produce a map which was more accurate and therefore decided to draw it on a larger scale. The quantity of land represented in it was claimed by Rennell as equal to half of Europe. As more and more facts were gathered to fill in the void inner space, the possibility of representation of "truth" in the map increased:

> Considering the vast extent of India, and how little its interior parts have been visited by Europeans, till the latter part of the last century, it ought rather to surprise us, that so much geographical matter should be collected during so short a period; especially, where so little has been contributed towards it by the natives themselves as in the present case.[25]

Earlier route surveys by Huddart, Pearse and Fullarton effected an outline of the so-far unknown shape of the southern peninsular region, which complemented what was then known of the shape and size of the Mughal Empire in the north. Thus, for the first time, India could be viewed as a whole from four detached sheets.

His *Memoir* also carried "sketches" of the history of the Mughal Empire and the Mahrattas, the narrative of the former to indicate the possibility of a similar British Empire, and the latter narrative was to describe the formidable foe who needed to be vanquished. India was not envisioned as a fixed entity since the days of Alexander. The turbulence of history had made it into a malleable entity and the purpose of selectively reconstructed narratives of history was meant to impress the idea of the non-fixity of political spaces within a geographically united space. Rennell notes in his memoir:

> I shall not attempt to trace the various fluctuations of boundary that took place in this empire … It is sufficient for my purpose that I have already impressed on the mind of the reader, an idea that the provinces of Hindoostan proper have seldom continued under one head, during a period of 20 successive years from the earliest history, down to the reign of Akbar in the 16th Century … and that sometimes the empire of Delhi was confined within the proper limits of the province of that name.[26]

Rennell was the first colonial cartographer to establish the fashion of writing a memoir alongside the construction of the map. As pointed out before, geographical memoirs were produced by geographers in order to legitimise their intellectual pretensions. By the late eighteenth century, these memoirs had become regular features accompanying a geographer's major work. They formed the trademark of geographers who were often supported and patronised by the state or those who had claims to social status. The purpose of the memoirs was to explain the process of collecting and collating their data, crucial for validating the objectivity of the map, thus certifying its claims to factuality and certainty.[27] Rennell was once accused of being "adept" in an "occult" science in his maps of western Africa, for failing to provide a document about how he teased geographical knowledge from a mass of conflicting data.[28] His *Memoir of the Map of Hindoostan* was the outcome of the lesson learnt from that mistake. Thus, Rennell heralded the age of scientific accuracy and precision in cartography which eventually became a methodological requirement.

Rennell's map thus was a conceptual unifier of geographical knowledge. It was eclectic, gathering as much information from indigenous sources as from European. Apart from the *Ain-i-Akbari* which gave a fair idea of the reaches and extent of the Mughal Empire, he made use of several local charts and maps. He mentions four of these in his *Memoir*. The first was a large map of Punjab with the names in Persian. The roads in the north-west had not yet been surveyed and Rennell found the map particularly helpful for the information the map offered about Lahore, Multan and Attock and the exact placing of the five rivers. He wrote in his *Memoir*:

> I consider the MS as a valuable acquisition, for't not only conveys a distinct idea of the courses and the names of the five rivers, which we

never had before: but with the aid of Ayin Acbaree, sets us right as to the identity of the rivers crossed by Alexander.[29]

Another map used by Rennell was by a "native of Guzerat" which "gave the form of Guzerat with more accuracy than most of the European maps can boast". Rennell also mentioned a "Hindoo map of Bundela or Bundeland, including, generally, the tract between the Betwah and Soane rivers, and from the Ganges to the Nerbudda", with the names in Persian. The fourth map was a Malabar map; or rather a map drawn by a native of the Carnatic ... the map alluded to, is not constructed, by a scale, but rudely sketched out without much proportion being observed either in the bearings or distances of places, from each other: and the names, and the distances between the stages; are written in the Malabar language.

What is striking in Rennell's statements is his complete erasure of authorship of native maps. It is the European master who is to judge the worth of these maps in order to assimilate these small parts into his immense project. It was the task of compilation, of reconciling different geographical sources and combining various surveys was the high art of the eighteenth century. Though indigenous sources were vital to the British surveys these were barely mentioned and subordinated beneath a "visual and graphic rhetoric of cartographic knowledge".[30] They all fed into the grand design of a "single cartographic archive" where information and materials were standardised.

Rennell's map was a means and metaphor for a reordering of space according to imperialist designs. Finally, he might claim to have expunged all emotive and mythic elements from the map's spatial imagery, but these popular elements continued in the historical narratives of the elaborate cartouches. The depiction of the brahmin handing over the Hindoo Shastras to Britannia overtly refers to not only a military and geographical conquest, but also the conquest of the native cultural and epistemological space, making the British the intellectual masters of the native landscape.[31] What Rennell only initiated, later British cartographers completed. The colonial project of governance entailed first defining the territory and fixing the boundary, before an understanding of the realm contained therein in terms of people and things. Later surveys like the Great Trigonometrical Survey initiated by William Lambton in the early nineteenth century and the topographic surveys of Colin Mackenzie served to stabilise the space politically and geographically.

Geodetic goals

The story of Lambton's cartographic pursuit in India encode different perspectives on the territorial, political and conceptual struggles attendant on the emergence of the colony. His cartographic fantasy removed land from a pre-existing social sphere, placing it firmly in a constructed universal scheme. The cartographic synthesis thus encoded the familiar in an unfamiliar language, converting the local and the regional into a meta framework of the scientific universal coordinates, effacing the deep links between people and their lived space.

Lambton himself regarded the general survey and the general map as a very important part of his labours, though geodesy was "the higher branch" which remained entirely in his hands. The objective was to make higher science serve the utilitarian purpose of colonialism. "I shall offer this plan as a specimen", he writes in his memoir:

> of what the higher branches of my survey may be applied to, and how far practical science may be combined with publick utility, and it will be gratifying to me, after having extended my operations from Cape Comorin to the banks of the Kistna, to see them become the foundation of various useful works ... I shall feel peculiar satisfaction if, while my labours are directed to the advancement of science in general, they may at the same time contribute to the more immediate benefit of my country.[32]

Early in December 1799, while Mackenzie was making preparations for his topographical survey of Mysore, Lambton put forward his first proposal for a trigonometrical survey to fix prominent points over the whole south peninsula. This followed on the heels of the successful colonial suppression of another armed rebellion in peninsular India in 1792, led by Tipu Sultan, and the "Siege of Seringapatam" at the hands of an army led by Cornwallis. Writing about the singularity and efficacy of his proposed idea, Lambton writes in Asiatic Research:

> Having long reflected on the great advantage to general Geography that would be derived from extending a survey across the Peninsula of India for ... determining the positions of the principal geographical points; and seeing that by the success of the British arms ... country is acquired ... which not only opens a free communication with the Malabar Coast, but ... affords a more admirable means of connecting that with the Coast of Coromandel by an uninterrupted series of triangles, and of continuing that series to an almost unlimited extent in every other direction ...
>
> It is scarcely necessary to say what the advantage will be of ascertaining the great geographical features ... upon correct mathematical principles; for then, after surveys of different districts have been made in the usual mode, they can be combined into one general Map.[33]

Lambton stressed to the government the practical benefits which would emanate from his geodetic survey. According to him, it would provide a veritable extendable lattice into which results from more detailed but less accurate local or regional surveys could be incorporated.

In 1781, Lambton was appointed ensign in Lord Fauconberg's Regiment of Foot, a provincial or home service regiment in Britain. In 1782 he transferred with the rank of ensign to the 33rd Regiment with which he stayed for a long while. He was initially deputed to North America and then Canada to join the regiment. He was here appointed as a land surveyor both by the War Office as well as the Board of Ordnance.[34] He read deeply while stationed in America, taking a special interest in geodesy and following closely the work of General Roy and of the Ordnance Survey of Great Britain. When, in 1796, the 33rd Regiment was ordered to the East Indies, Lambton was obliged to choose between his civil and military positions. He chose to go to India with his regiment, which was then commanded by Arthur Wesley, later Wellesley. Lambton arrived in Bengal with his regiment in 1797. He later moved to Madras in the September of 1798. Through a letter of introduction from his Canadian patron, Brooke Watson, to Sir Alured Clark, commander-in-chief in India, he soon secured the staff position of brigade major to the king's troops under Fort St. George in 1799. It was during this time that he actively wrote papers for the *Asiatic Researches* dabbling in various scientific and mathematical issues.[35] These fetched attention from Wellesley, who proved to be a helpful patron for Lambton's later endeavours in India. Lambton sprung into prominence during the expedition against Seringapatam in the fourth Anglo-Mysore War against Tipu Sultan. The war was supposedly saved by Lambton's apt intervention in pointing out the correct direction by the location of stars when General Baird was leading his army at night towards the enemy camp rather than to safety.

It was in 1799 that Lambton presented a memorial to the governor of Madras proposing a trigonometric survey connecting the Malabar and Coromandel coasts. With a seconding of the proposal by both Mackenzie and Wellesley, and after a sanction by Close from Bangalore on 6 February 1800, formal orders were issued for the start of the survey. The instruments were purchased from Dr Dinwiddie in Calcutta, where Lambton had initially seen them. Lambton's instruments were a theodolite, a zenith sector and steel chains.[36] In this, his demands for instruments were modelled on those used by William Roy. He was given further detailed orders by Webbe to orient his survey with Mackenzie's topographical survey which had also commenced in the peninsula:

> You have been already made acquainted with the intention ... to employ you in an Astronomical Survey in the Peninsula but chiefly in the territories lately subdued ...
>
> A considerable establishment under the direction of Capt. Mackenzie having already commenced a detailed Survey of the provinces of Mysoor and the Southern part of the Peninsula, his Lordship is desirous that, without departing from the purposes of general geography which your labours will have principally in view, they may ... be made to coincide with those of Capt. Mackenzie, so as to enable him with

greater facility to combine the details of his Survey, and to verify the positions of the most remarkable Stations.[37]

He was thereupon asked to submit the full details of his proposal.

The essential features of Lambton's proposals were that his survey would be based on "correct mathematical principles", that it would extend right across the peninsula, that it would be capable of extension in every direction, and that it would be a reliable basis for all other surveys, "that this survey should precede all others, that data may be readily prepared, and the work become the legitimate foundation of every other survey, whether geographical, military or statistical".[38] However, it had an added geodetic goal. His closing words in his memoir of 1801, expresses an objective to "accomplish a desideratum still more sublime, viz, to determine by actual measurement the magnitude and figure of the earth, an object of the utmost importance in the higher branches of mechanics and physical astronomy."[39] It was this scientific end which finally secured approval for Lambton's scheme. In his proposal, he discusses in detail the special features of his proposed work and the precautions needed to ensure correct mathematical principles:

> It has been the usual practice ... to work upon a series of plane triangles ... thinking the curvature of the Earth of too little consequence to be taken into consideration; and the only mode of correcting was by observing Jupiter's satellites, occultation of stars etc., for determining the longitude. ... It is easy to see the errors that must result from extending the survey over a portion of the globe comprehending a number of degrees both in Latitude and Longitude. ...
>
> the first operation for obtaining a datum ... is by the measurement of a *base line*, which being reduced to the level becomes a part of a great circle on the surface of the Earth. ... From thence is derived new data to proceed in all directions, recollecting that ... the observed ... angle is to be corrected again to the angle made by the chords.[40]

Copies of Lambton's proposal and of Mackenzie's *Plan of the Mysore Survey* were passed to Rennell who had by then returned back to London. However, Rennell completely misunderstood Lambton's proposals. As Phillimore assumes, he might have been baffled by the government order appointing Lambton in charge of an "Astronomical Survey", and that for Mackenzie's survey of Mysore "principal points ought ... to be corrected by *Astronomical* observations connected by a series of triangles". Rennell presumed that while Mackenzie carried out a topographical survey of Mysore, Lambton was to conduct a completely independent series of astronomical observations, on which Mackenzie's survey would be subsequently adjusted. This misconception compelled him to express contempt towards the proposals as "one of the most extraordinary that has been heard of". As for accuracy, Rennell opined that his own method of deriving maps from route surveys was scientifically sound. Copies of his letters of protest were sent both to Lambton and Mackenzie. Lambton, much disturbed by Rennell's objections, wrote down a detailed refutation, which ultimately forced Rennell to withdraw his protest. Nevil Maskelyne, the Geographer Royal and uncle to Edward Clive, the governor of Madras, being persuaded by Clive, also took Lambton's side. His intervention ultimately broke the deadlock and the misunderstanding between the two cartographers belonging to two generations.

The Great Trigonometrical Survey of India

The term "Astronomical Survey" was often used by the government in its documents, while Lambton himself generally described himself as being "on Geographical Survey", or "on General Survey" right up to 1815. The expression "Trigonometrical Survey" appears on Lambton's charts and memoirs:

> The trigonometrical part of this survey is the foundation from which all distances and situations of places are deduced; a true delineation of the river vallies, ranges of mountains, with some noted points near the ghauts and passes, will also be a foundation for more minute topographical surveys such as are immediately wanted for military purposes.[41]

Lambton's survey proposals were approved on 6 February 1800 and began between 1800 and 1802, with the preliminary survey of Mysore and the measurement of the baseline at St Thomas's Mount in Madras in 1802.

The Observatory in Madras, where observations and records stretched back many decades and where the dependable measurements could be taken from, was the starting point of the sequence of peninsular surveys. Its value in relation to the Greenwich meridian, the zero degree longitude, and the Equator, the zero degree latitude, was the sheet anchor to Lambton's survey. Madras was the Indian counterpart of Greenwich.[42] As Markham commented:

> The longitude of Madras is important as that of the secondary meridian, or substitute for the prime meridian of Greenwich Observatory, from which observations for longitude in the Indian Survey are reckoned. Every station and place in that Survey will be erroneous if the longitude of Madras is in error. In other words, the accuracy with which the entire map of India, as a whole will be placed on a globe, will correspond precisely to the accuracy with which the geographical position of the Madras observatory has been determined.[43]

It was a crucial means to translate and rearrange the geographical space in relation to English coordinates. Observations from this point had been taken from 1787, but the building of the Madras Observatory was erected in 1792. Goldingham, who had succeeded Michael Topping, was the astronomer at the observatory and Lambton's contemporary.[44]

The Trigonometrical Survey inaugurated itself with the measurement of a baseline near Madras. The next leg of the survey was the measurement of the arc of meridian. His programme was to measure an arc of more than three degrees in length astride meridian 78°, which was to stretch from Cape Comorin to the Himalayas. The measurement of the meridional arc was crucial in view of Lambton's geodesic pursuit. Maskelyne wrote, after discussions with Rennell in 1806:

> Among the subjects which are purely scientific, the measurement of an extensive arc on the meridian will doubtless (attract) the first attention, being ... grand desideratum to compare with what it is doing in England and France, and with what was recently been done at the polar circle.[45]

The arc was a device to conjoin and combine disparate experiments and measurements in different regions and then spatially unite them. The regions colonised by European nations used them as extensions of their laboratory spaces originally located in the home country. As triangles formed parts of an imagined whole (i.e. the earth, which then would be reduced from infinite to finite), similarly, the arc would gradually progress till it encircled the entire earth reducing it to a comprehensible and manageable structure. The arc and triangulation enforced the idea of a continuum which was a useful mechanism of expanding "invasive prospects".[46] Once the Arc Series was completed from Cape Comorin to Bangalore, in 1811, Lambton and his team turned their attention to extend northwards to the Himalayas, which gradually emerged as the frontier to the British colony.

The site chosen was a stretch of level ground between St Thomas's Mount and another hill seven and a half miles to the south. A series of triangles, in two degrees of latitude was then carried across the peninsula. The distance across the peninsula, at this point, was found to be 360 miles, while the best maps till that date had given the distance as 400 miles. Thus, the absolute need for a trigonometrical survey based on scientific precision, in contrast to other methods, was justified. The report on his meridional arc in the Carnatic was submitted in October 1803, one copy being passed to the Asiatic Society at Calcutta, and published in the Asiatic Researches VIII. This report forms the first volume of the manuscript reports of the Trigonometrical Survey; the second part, covering operations across the peninsula, 1803–6, was submitted in 1807. The General Map of the South Peninsula was submitted in 1810 along with a memoir.

Lambton became captain in the 33rd Foot in 1806 and became a major in 1808. Though the 33rd Foot returned to England from Madras in 1811, Lambton stayed back as superintendent of the Indian survey at the Company's expense. In 1818, his survey was transferred from Madras to be under the control of the supreme government under Fort William. It was now that the survey received the name with which it would be recognised henceforth. It was named the Grand or Great Trigonometrical Survey or GTS of India. His importance therefore surpassed that of Mackenzie who was surveyor general of India. With Lambton's survey being officially transferred, he was directly answerable only to the government and not Mackenzie under whose authority all survey activities in all the three presidencies for so long fell.[47]

In 1818, once his third report was complete, he proposed extending his arc further north. Agra was fixed as a suitable point by the government. He initiated moving his headquarters from Hyderabad to Nagpur, but died before the transfer was complete. Lambton's arc was 10° or 700 miles at the time of his death. It was already longer than the European arc and much admired as a scientific triumph in the European world. It was left to Everest ultimately to complete Lambton's scheme. After Lambton's death in 1823, the Trigonometrical Survey was carried forward by George Everest, who was previously Lambton's subaltern (see Figures 2.1 and 2.2). He succeeded in measuring the length of the meridional arc, from southern India to Dehra Dun in the north.

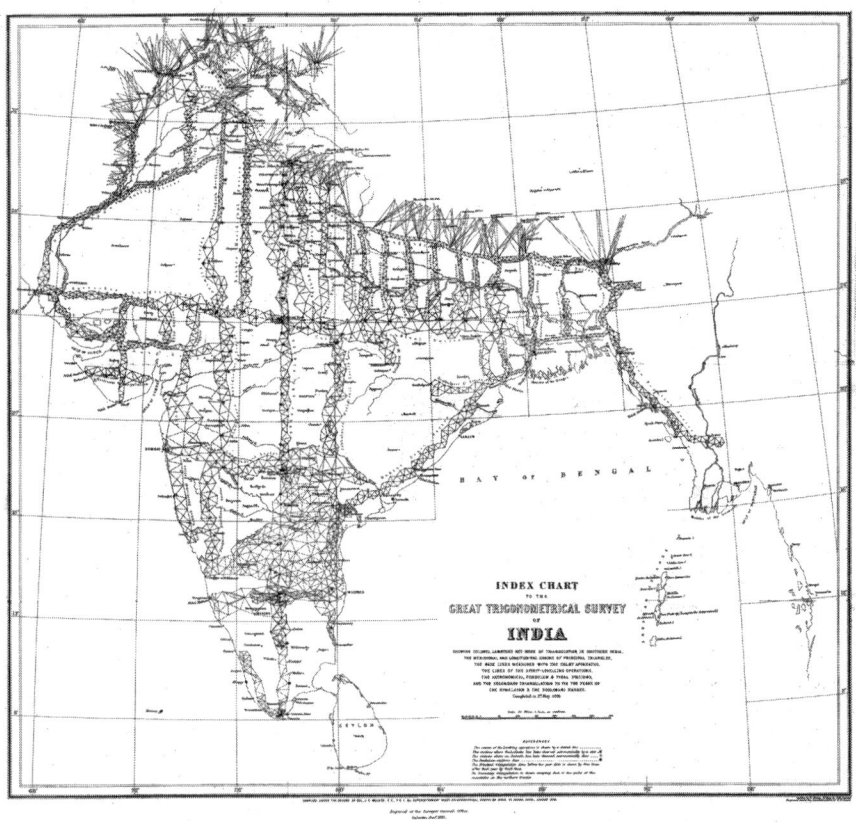

Figure 2.2 Index Chart to the Great Trigonometrical Survey of India, showing Lambton's network of triangulation from southern India. George Everest later extended the GTS across northern India and the Gangetic Valley up to the Himalayas

Source: Survey of India

Mixed appraisal of the survey

Lambton's enterprise had met with considerable criticism towards the beginning, though consequently his feats won him membership in the French Academy and fellowship of the Royal Society as well as the Asiatic Society. It was of course essential for him to write his memoir to demonstrate the value of his experiments. Though Rennell was appeased by Maskelyne's interventions and Lambton's refutations, there were several more criticisms to his work and the amount spent on the activity. The members of the Finance Committee of Madras "appear to have had great difficulty in comprehending the object of Col. Lambton's survey". One of the committee's leading members having voiced the criticisms reflects the general idea that the survey was an entire waste of resources and time:

> If any traveller wished to proceed to Seringapatam, need only say to head palankeen bearer, and he vouched, that he would find his way to that place without having recourse to Col. Lambton's map.[48]

The committee plagued Lambton with numerous questions and comments. Lambton was forced to react passionately. Lord Bentinck requested Lambton to resist representing his feelings in public correspondence. The survey was almost at the verge of being called off.

On the other hand, there were also people who supported the survey from its very start. The then quartermaster general, Lt Col. John Munro, on having heard that the government contemplated the abolition of the survey, waited on the governor for the purpose of representing the utility of the operations from a military point of view. He argued that the Topographical Survey, which was crucial in ascertaining British territory and strategising defence, was itself dependent on the Trigonometrical Survey and its triangulation for its accuracy. On his assertion that the survey was under the discretion of the Military Institution, and that it was ultimately its work, the intention to abandon the survey was finally annihilated.

Much support was garnered in the name of science and scientific achievement and it was especially taken up as a matter of national pride. William Petrie, who acted as governor after Bentinck's departure, and who was also an amateur astronomer remarked:

> I have repeatedly submitted to the Hon'ble Company my sentiments of this splendid work. Its merits ... require no proofs of my testimony, and when the Fame of the Conquest and Extensive Dominion has passed away, a page may remain on the records of Science to shew that under the fostering and liberal protection of the East India Company, a Survey has been carried on in a part of their Eastern Empire, verified and determined by a Series of Astronomical and Mathematical

Measures, not inferior in Science and Accuracy to the Brilliant Labors of the English and Fresh Astronomers.

The value of Major Lambton's work has been justly appreciated, not only by Mathematicians in our own Country, but by that Dept. there can be no national warfare.[49]

Another of Lambton's supporters was Scott of the Madras Civil Service. He was the first judge in the Court of Appeal. He was a known scholar of science and was generally consulted by the Madras Government on various issues pertaining to science. Scott, too, in the same vein as Petrie, talks of Lambton's achievements as of pride to the nation:

The ... very great importance of Major Lambton's Survey, is ... but little understood. I fear there are but few among us who consider the ascertaining the lengths of three or four degrees of the meridian, and as many of Longitudes, as of any importance or who conceive that much scientific knowledge, or much labour, is necessary for accomplishing it.

He is full of reproach for the unenterprising British nation as against the French which, according to him, achieved enormous feats in science. Lambton's contribution to science, if allowed to continue, would surmount and supersede these other scientifically advanced European nations:

The opinions of the Learned in Europe however, are very different; witness the expensive expeditions sent by the French to the Polar Circle and Equator ... Major Lambton will, if not prematurely interrupted, in a short time have ascertained the length of a greater arc of the Meridian than was done either in Lapland or Peru

Combative nationalism supported and embraced the cause of science to assert greatness. In India, French power, though by now having nearly disappeared, was in fact the reason behind major wars which saw the primary consolidation of the Company estate both in Bengal and Seringapatam. The contours of the Company's territory needed to be redrawn also to assert increasing British authority as against the French receding one. Scott further pleads for Lambton's endeavours both in the name of a greater pursuit of scientific knowledge as also practical purposes of administration:

It is only by having the correct length of degrees of the Meridian and Longitude in different Latitudes that the great desideratum can be obtained, of establishing what the true figure of the Earth ... really is; some may consider this a matter of mere curiosity, without considering its real importance in Navigation, Geography, and Astronomy, and, where France has done so much and they are still going on in England, do not let us be so stupidly ignorant as not to set a proper value on what Major Lambton is doing.[50]

George Everest, who took over the charge from Lambton, described the GTS as "perhaps of itself the most Herculean undertaking on which any Government ever embarked".[51]

Inscribing power

The specular method of the survey completely overwrote as well as overrode the lived space which already existed in the experiential reality of the natives. Lambton and his men were moving over the full extent of the peninsula, from one province to another, as though in a sweeping glance. They had no chance of ground level interaction with either the colonial officials posted at various places or the resident natives. Pre-existing provincial boundaries did not matter, nor cultural, religious or linguistic divides. The spectacle of the surveyor on field work, parading his measuring apparatus and performing his measurements surrounded by suspicious natives had a distinct air of theatricality about it. The mobility of the survey team assumed, as it were, an appurtenance of property, a right of way which came incidentally with the possession of property. Their only concern was the ready access to vantage points, such as mountain or hill tops, or temple and fort steeples. At a later stage, where natural vantage points were not available, trig stations and towers were erected, many of which still lie as forsaken monuments all across the subcontinent.[52] This outlook is instrumental in discursively constructing a gaze, an authoritative view point. The entire activity was based on this construction of the gaze which proclaimed a mastery of space. Space, thus fixed, was mathematically proscribed by the central observer, as Michel de Certeau writes of the Cartesian system's positioning of the observer/surveyor:

> A Cartesian attitude ... is an effort to delimit one's own place in a world bewitched by the invisible powers of the Other ... It is also a mastery of places through sight. The division of space makes possible a panoptic practice proceeding from a place whence the eye can transform foreign forces into objects that can be observed and measured, and thus control and "include" them within its scope of vision.[53]

The search for elevated viewpoints (which marked other modes of spatial representation like the landscape prospects or panorama paintings) is itself loaded with imperial rhetoric and symbolism which sought to format the earth's surface into a system of spatial hierarchy of differential geographies of that of the colonial core and periphery. Figuratively, height takes over as a way in which value is attributed. The vertical height stands for the imagined "chain of being": the European, and especially the British, being stationed higher than other spaces in the world. To quote de Certeau once again in explaining the construction of the point of view of the colonial surveyor:

> His elevation transforms him into a voyeur. It puts him at a distance. It transforms the bewitching world by which one was "possessed" into a text that lies before one's eyes. It allows one to read it, to be Solar Eye, looking down like a God. The exaltation of a scopic and gnostic drive; the fiction of knowledge is related to this lust to be a viewpoint and nothing more.[54]

However, the occupation of elevated vantage points in the localities surveyed often had its share of problems and conflicts, but Lambton remained unsubdued by them. One of the officers involved in the Trigonometrical Survey, Warren wrote to the collector in Chittoor in 1803 about the outrage his activities created among the local Poligars when the Narnicul Droog was taken as a station for triangulation:

> Neither myself, nor the delegate which you sent me, were aware of any Poligar retaining still any authority, Civil or Military, in your Districts and ... Narnicul Droog being one of my points, without any further ceremony, I directed one of my flags to be placed on that hill and the morning followed, intending to observe at the station.
>
> No obstacles was offered me as I entered the bound hedge and Jungle which surrounds the Fort, but I noticed a number of men hurrying from the village ... with matchlocks, swords, and daggers, who entering the jungle at various places met ... and opposed with great clamour my proceeding any further. I thought at first that they only wanted to see my passport, or that they questioned how far I was authorised by you to visit the Fort of Narnicul Droog, but in this I was mistaken; they answered to all that I urged that I had no business there without the Poligar's leave, and that I must return to the village ... until it was obtained, and meanwhile that I would meet with due attention there.
>
> As it would have been vain to resist, I directed my bearers to return, and resolved on acquainting you with what had happened.

To this, the collector replied:

> Had I been aware of your intention to observe from Narnicul Droog I could have warned you of the reception you were likely to experience from the Poligar there, who has been for some months back in a state of disobedience and refractoriness ... I therefore think it would be improper to hazard an opportunity for the repetition of similar insult by insisting on accomplishing the object of your public functions in Naracul Polliam, and that it would be preferable to desist from that attempt.[55]

This was not the last time that Warren would face such a situation. Shortly after, Warren met with similar treatment at another hill in the same area. Here is what he writes about it in his letter:

> Having had occasion to send a Flag to be placed on Bungarry Droog Hill near Munglee, I gave directions to my Lascars to that effect and, as you are so good as to assist me with a letter to the Poligar of that place ... I concluded ... that no possible objection could be made to its admission. To my no small surprise, however, the people I sent ... informed me that ... they were stopped by some Tanna Peons, who signified to them that they could not pass without the Poligar's leave. On this my Lascar delivered your letter which was conveyed to him by one of his own people. The Poligar returned for answer that he could not allow the Flag to be placed in the Droog, by the reason that as it commanded a view of his habitation his women might be exposed to view.[56]

Wishing to avoid an unpleasant interface, Warren directed his men to another adjacent hill, which would serve his purpose. He asked his "lascars" to plant the flag on the other hill. However, to Warren's dismay, the same objection was raised for this hill as well "on account of its commanding a view of the Pettah". Instead, the Poligar's men pointed out a "small hill" in the plain at some distance, and said that the surveyors could place their flag on that if they wanted. Since Warren needed to place the flag on the highest of hills, and "preceding and succeeding points" this was hardly a solution. His every move frustrated, he writes to the Collector about his desperation: "These in a hilly tract like this are generally the highest, and almost everywhere the stronghold of a Poligar". The collector suggested that the team abstain from further survey in the Chittoor Polliams for the time being and Lambton was forced to inform the government that he had abandoned the attempt to carry triangles through Chittoor.

A few years later, Lambton faced other incidents of native intolerance in north-west Mysore. This time it was towards a British officer, De Penning, for having cut down a peepul tree in order to clear and create the view. The case was sent to the Resident and then to the government, but Lambton took offence to the government sending an officer "as far as Shikarpur" to make an enquiry into the incident when he had stated that De Penning was not at fault.[57] Lambton failed to realise why his survey faced such obstacles and blamed it partly on British administrative inefficiency and failure to communicate with the natives. In reply to a letter from the government, Lambton replied:

> With respect to placing flags upon pagodas, mounde in forts, etc, I have only to say that when I crossed the Peninsula in 1804–5, there was scarcely a pagoda or Droog in the Mysoor country that was not a station, and I never met with the smallest objection to placing flags, either one or the other.

> Even in the bigotted country of Tanjore, I ascended no less than 12 Coverams [gopuram], and without those lofty buildings I never could have got through the country. At Ramisseram [Rameshwaram] I was permitted to place the Instrument directly over the cell which contained the Sawmy, and all that too when there was a general apprehension of the Christian Religion being propagated.[58]

That Lambton could a few years back station himself atop Gopurams and temple peaks, was, according to him, a sign of able governance of the British. The complete indifference and unawareness of sacred spaces of the natives was flaunted in the name of the rational operation. In fact, those sacred spaces were transformed into ratified spaces of British power and surveillance from which science could be implemented. Before extending his triangulation to the Nizam's province southward, Lambton implored the local officials and the resident Henry Russell to make sure there was cooperation by the natives:

> In order to state to you my particular objects and wants, that you may give full explanation to His Highness the Nizam, or the different Vakeels residing at his Court; for unless there be a readiness everywhere to aid and accommodate, it will be impossible for me to carry on a work of this nature, especially if any obstacles be thrown in my way.
>
> I am aware of the jealousy of all the native powers, as well as that of their subordinate chiefs, on seeing any description of survey carried on within their districts; but, mine being of a more general and extensive nature than those which they have been accustomed to notice, and not embracing statistical objects, or such as excite their suspicion, I am in hopes that by a little address they may be induced to view it without alarm.[59]

It is clear that the surveyor is generally deemed an outsider by the local inhabitants and is seen as a socially disruptive force. It is ironic that the survey which was supposed to usher in science and rationality, actually alienated the natives further and kept them away from the results. In fact there are instances whereby the surveyor is often perceived by the natives in terms quite contrary to rational behaviour. In an anecdote in his travelogue, W.H. Sleeman noted that in central India Everest's practice of surveying at night, when the atmosphere was clear, led

> the peasantry to believe that men who were required to do their work by the aid of fires lighted in the dead of night upon *high places*, and work which none but themselves could comprehend, must hold communion with supernatural beings.[60]

On the other hand, nothing like Rennell's matter of fact acknowledgment of native sources ever figure in any of Lambton's survey records. The advent of such mathematical emplotments and geodetic geometry as that of Lambton's, marks a point of departure from the earlier methods which relied heavily on local knowledge. Lambton's Trigonometric Survey is indicative of a process that removed land from its location in popular memory and upset the tradition of a limited localised setting, where space was lived and experienced. By this means, a completely "unruly" social space that existed was underwritten by a scheme of self-empowerment which was most closely associated with the visualising of space through elaborate, geometry-based survey instruments. Such conceptualisation of space produced by the mechanical eye, the theodolite, is the "dominated – and hence passively experienced – space which the imagination seeks to change and appropriate".[61] It subjected cultural and geographic variety of an extensive region to the levelling and flattening sameness of the mathematical order, and from this ground plan constructed the globe.[62] It is as Harvey puts it:

> The application of mathematical principles produces "a formal ensemble of abstract places" and "collates on the same plane heterogeneous places, some received from tradition and others produced by observation". The map is, in effect, a homogenization and deification of the rich diversity of spatial itineraries and spatial stories. It "eliminates little by little" all traces of "the practices that produce it.[63]

Naturalised frontiers: spaces of knowledge

According to de Certeau, the "establishment of a break between a place appropriated as one's own and its other" entails a number of processes. Among these is the "ability to transform the uncertainties of history into readable spaces" which is legitimised by "power of knowledge". An autonomous space is the product of knowledge which in turn depends on a "certain power (which) is the precondition of this knowledge and not merely its effect or its attribute ... It produces itself in and through this knowledge".[64]

The concept of boundary or frontier is one of the most essential as well as the most interesting with respect to construction of space. As land came to be redefined as territory or area, it automatically entailed a boundary to make it a mensurable entity with fixed shape and bounds. According to Sudipta Sen:

> The mapping of India permitted a certain play of visual imagination that belied the episodic nature of the Indian conquest and the indefinite frontiers of the new state. In this sense it was an indispensable exercise of authority, re-ordering the country on paper.[65]

The high-altitude superficial view converts the physical limits to visibility into material margins resulting into imperial frontiers or fixing territorial closures. The cartographic methods and ontology subscribed to by the surveyors presupposed India as a geographical expression with natural boundaries. Notions of protective and natural boundaries surrounding India were taken as axiomatic. The colonial techniques of enumeration and calibration stabilised it into the unquestioned boundary of the modern nation state. The British Raj turned out to be the most data intensive "paper empire".[66] The East India Company's acquisition of territory and simultaneously the movement of the survey, up the Gangetic and doab plains to Punjab, and then to the mountain ranges beyond the Indus, and in the north-east, towards Tibet and Burma, can be seen as a search for a permanent and viable frontier. India could be conceived of as a geographical entity only when its bounds were defined, and only when conceptual boundaries were secure enough. Colonial cartography, "with one stroke of the pen disabled medieval notions of a fractured, incomplete India". Ian J. Barrow, in the context of the idea of colonial frontiers points out the imperialist perception of land, which sought to transform space into a place. Under these circumstances land was understood only as area:

> If land came to be thought of primarily as area, capable of being mathematically compared, boundaries may then be considered "natural" since an area must logically have its parameters if it is to constitute a place.[67]

In a separate study, Barrow shows how the order in which the popular India maps of the Society for Dissemination for Useful Knowledge (SDUK), published 1831–35 in London, modelled after the East India Company's *Atlas of India*, followed the time–space chronology of cartographic expansion of India. The first to be engraved for its political significance was the Bengal map, which, by now, had a long history of survey and mapping, since Rennell's operation in the area. Following the GTS, the maps of the southern parts came next and preceded the later maps of central and northern regions, the Punjab maps being in great demand for following news of military campaigns in the region. The last ones to come out were the ones showing north India, containing large blank chunks in the Himalayas which clearly had not yet been properly surveyed and therefore could not match required standards of accuracy and verifiable knowledge.[68] These and other such maps invited the reader to traverse the region as birthed through the cartographic expeditions. It functioned like a reverse denouement, journeying from the known to the unknown.

The construction of the boundary of India was the outcome of three successive phases of British cartography in India. Interestingly, what was generally known as the North West Frontier continually shifted further west and northwards over the two centuries of British rule in India. During Rennell's era of cartography in late eighteenth century, frontiers were largely undelimited. The route survey served as a penetrative tool and as a facilitator to expansion beyond a frontier. The trigonometric survey of the first half of the nineteenth century extended the baseline through a frontier, and information regarding the interior locations was scientifically obtained and consolidated. In the latter half of the nineteenth century and the first two decades of the twentieth century, the cartographers' bureaucracy, the Survey of India and the Boundary Commission, were established which discursively settled the notions of protective and natural boundaries surrounding India. According to Edney too, the logic behind the colonial expansion was the securing of safe and stable frontiers. Edney remarks:

> Whatever the opinion of the Directors and politicians in London, the Company expanded in India almost by necessity; territorial growth was the imperial equivalent of commercial expansion dictated by the political economies of the day. Few, if any administrators, argued that a state was an organism that must grow to survive, but all understood the logic, if not the need, to subdue the peripheral areas so that the core domains might be made more secure.[69]

In this regard, as Clements Markham puts it:

> The structure of the great Himalayan mass which bounds India to the north is the branch of the subject to which attention is naturally drawn in the first place, and it is that to which both travellers and systematic geographers have devoted the largest share of their labours.[70]

Also, developing from the island nation back in Britain, there was a linear conception of space: that the further one went inland from the shores, the more impenetrable the landscape became and the more uncivilised the inhabitants. Therefore, the reports of a long line of travellers preceded any attempt to map it. Markham talks of early European travellers such as the Missionaries Desideri, Freyre and Antonio di Andrada who were simply appalled by the snowy mass and the eternal winter. The Himalayas drew the attention of the British as a natural barrier of the plains they colonised, which were watered by the rivers which had their source hidden somewhere in the snowy ranges. Also, the actual height of the peaks were a mystery and the desire to ascertain their height was overwhelming. At the first glance, there seemed to be nothing to lend a clue to the development of the mountain masses, and there appeared an assemblage of elevated peaks confusedly heaped together. A profusion of geographical narratives thus sprang out of the need to subject these great ranges to analysis (a few of which are discussed in Chapter 6). Capt. Herbert was one of the first British officers to explore the Himalayan region. He primarily came up with the idea that, by tracing the courses of the rivers and their tributary streams, a clue would be found to lead an observer out of this labyrinth. Following his theory there were a number of expeditions which ramified into various directions subjecting these lofty highlands to rigorous topographical, geological and botanical mapping. It was also generally believed that the resolution of the mystery of the Himalayas would unfold numerous unanswered questions about the plains and of physical geography in general too.

The rivers flowing from the Himalayas and forming the two great systems of the Indus and Ganges, which were deemed to be the nerve centre of the Indian civilisation were studied with minute attention. The physical laws regulating the direction, oscillation and volume of the rivers were explored by Fergusson who also studied the Gangetic deltas. His extensive diorama of the Ganges is discussed in Chapter 4. Fergusson, through his study points out an important hydrographic law, namely, that the mouths of tributaries shift upwards along the main stream in consequence of the decrease of slope caused by the rise of the delta, which causes the tributaries to increase the angle at which they fall into the Ganges.[71] Similarly, in what came to be called the North East Frontier, the tracking of the course of the Brahmaputra posed a continual interest from Rennell to Lt Richard Wilcox. Whether the Tsangpo flowed into the Brahmaputra or the Irawaddy was a classic conundrum which preoccupied geographical societies for a long time. Zou and Kumar have elaborated on how the categorical intelligence gathering in the Brahmaputra and the Barak Valley, and the surrounding hills, were initiated after the Anglo-Burmese war of 1824–6. The war triggered an "information panic" in the East India Company's administration which added impetus and led to fervent topographic surveys and mapping in the eastern and greater Himalayas for the sake of defence of its colony.[72]

Various geological studies functioned to support the inherent distinctness of the cultural space of the Himalayas from the plains. Falconer and Cautley's joint study of fossil remains in the Himalayas and the Siwaliks led them to conject that at an early tertiary epoch, the subcontinent was a large island situated in a bight between the Himalaya and the Hindu Kush ranges.[73] Several upheavals led to the island getting connected to the ancient continent. The collision led to the lower foothills or the Siwaliks being formed and the Himalayas attaining greater height. The geological disparity with the plains posed it as a natural boundary between the island-like subcontinent and what lay on the other side of the mountains.[74] The frontier, therefore, was a "metaphor for knowledge ... an area to be reached and eventually traversed, but a zone, nevertheless, demarcating what was and was not known".[75] It determined the extent of known territory.

The pilots in marine cartography

Where in the north the Himalayas proved to be a formidable barrier, especially against an impending threat of Russian invasion later on in the nineteenth century, the sea in the south was deemed an equally invincible natural fortification. In fact, as sea-faring traders, with the British in India marine cartography preceded any survey on land.

From early times, armed vessels were employed at Bombay to protect the East India Company's ships from pirates. It was as Bombay Marine that the British consolidated their naval forces from 1742 onwards. Later on this would be renamed the Indian Navy and existed from 1832 to 1862. Though there was opportunity to train surveyors here, it was not until the days of Rennell and Dalrymple that marine survey received its fair share of encouragement. John Ritchie, who was hydrographical surveyor from 1770 to 1785, surveyed the coasts of the Bay of Bengal and the outlets of the Ganges. Many of his sea charts were engraved by Dalrymple and were used by Rennell extensively as his material for the survey of Bengal. In the late eighteenth century, a number of ships like the *Endeavour,* and the *Panther,* surveyed the west coast from Daman and Diu to Cape Comorin. Some of these charts too were engraved by Dalrymple. Another of the East India Company's packets *Antelope* sailed from Macao, but was wrecked near the Pelew Islands. All these ships were captained by McCluer, who surveyed the coasts of West Indies and other Caribbean islands too. The physical world, in this case, is the result of the view of the observer moving through space, in this case the sea. Space thus constructed was the outcome of the mobile eye and therefore had to employ fit apparatus for stabilising it.

Daniel Ross, who was called the "Father of the Indian Surveys", introduced scientific accuracy to marine surveys through triangulation. He measured bases on shore by running a ten-foot rod along a cord stretched tight between the extreme points, and kept in position by stakes, the direction being verified by a telescope. Exactly in the manner as British waters were surveyed, when work on shore was impracticable, recourse was taken to measurement by sound. The vessels were anchored when the weather was calm, and the time was taken between the flash and report of the gun, on the assumption that sound travels 1140 feet per second. Angles were taken with the help of a sextant, and the triangulation was verified by frequent astronomical observation.[76]

Many of the surveys of islands adjacent to the Indian peninsula were surveyed on boats fully manned and armed in apprehension of outbreaks from the native inhabitants who were thought to be savage. Before taking observations on shore it was generally necessary to station outposts in the jungle to prevent surprises. Most of these surveys were conducted over a huge expanse of sea area often encompassing the entire sea route from England to the Far East. Many of them were written in the manner of detailed topographical accounts and were aimed at helping navigation. Such hydrographic accounts were generally called Pilots and a number of them were published in the nineteenth century, like the *West Coast of Hindostan Pilot*, the *Persian Gulf Pilot* and the *Bay of Bengal Pilot*. For example, the *Bay of Bengal Pilot or the Sailing Directions for the Coasts of Ceylon, India and Siam, from Colombo to Salang Island; Nicobar and Andaman Islands* was published by the Hydrographic Department; it contained not only instructions for navigation, but comprised of a detailed outline and description of the shore. It also contained translation charts of English nautical terms and directions into several South Asian tongues like "Hindustanee", "Bengalee" and "Malay", along with "Hindoo Astronomical table", an Indian timetable as well as a Bengalee timetable. The computation of

93

time is usually against Madras time from the location of the Madras Observatory. These usually charted distance according to the time required to reach a specific point and frequently incorporated first person narratives. For example, the *Bay of Bengal Pilot* contained the remarks of Capt. Blanchard of the *Riviere d'Abord*:

> I sailed from Calcutta river for Reunion ... the wind being tolerably strong from the south-west and the sea rough ... I experienced, without intermission, hard gales with strong squalls from W.S.W. and S. W., which did not admit of my making the coast of Orissa, but drove me over to the coast of Pegu.
>
> I determined to pass to leeward of the Andamans ... I passed half a mile northward of the Little Coco, which at that end is quite steep; the bank of coral fringing the south and west sides terminates near the north point and approaches nearer the coast. From that point the dangers at the South end of Great Coco were visible to the eastward, though the horizon was indistinct."[77]

The genre, though determined by the institutional strictures which ensured they remained within specific tenets related to objectivity, yet, explorative epistemology was ultimately dependent on the explorer/writer as the producer of truth. The construction of vision too is dependent entirely on the figure of the explorer (and his is the central voice which narrates it too). Most importantly, seeing, here, can be understood as a possessive force, a mode of appropriation. The horizon (being) indistinct, space is thus shown as continuous and unified, and this integrity of space is confirmed by being arranged around the pivotal human subject. Space thus becomes a product of the human eye which bestows on it a sovereignty and implicates it into property ownership.[78]

The marine survey was occasionally linked up with the survey on land by way of rivers, and thereby came about a complete circumscription of the company's terrain. For example in the west, the hydrographic surveys of Pottinger and John Wood of the River Indus, from the sea upwards, were linked with Del Hoste and Alexander Burnes's survey of the area.[79] John Wood's work in the Indus area commenced in 1835. He accompanied Burnes afterwards in his mission to Kabul. Wood made a survey of the Indus from its mouth to Attock. At Kalabagh, where the Indus descended from the Salt Range, Wood found it difficult to move against the current. However, undeterred, Wood took the land route and reached Attock completing the survey amidst falls and rapids. After reaching Kabul with Burnes, he crossed the mountains to Kunduz. He is generally deemed the first European after Marco Polo to reach what was called the "Bum-i-Dunya" or the roof of the world.[80] He received the prestigious gold medal from the Royal Geographical Society for this stupendous feat. In future years, of course, there were many more such expeditions in the area which posed as an useful buffer between British India and Russia.

The surveys and the cartographic representations thus gave a political stability to a dynamic and volatile space marking it at the same time as an imperial property. The circulation, integration and collation of data collected from ergonomic spaces of action on the ground, were meticulously and painstakingly sent back to the science hubs in the metropolis to be assessed and corroborated by various scholarly societies, academies, associations and their prominent doyens and patrons who reviewed, processed and corroborated the information to ultimately make holistic claims on the basis of its system of orientation.[81] The boundary fortified abstract entity which figured on the map became the first and most authoritative referent to reality, making the formal representation more real than ever before. The artificially constructed space which derived consolidation through the cartographic image was, in effect, a discursive formation of Enlightenment science and rationality. As cartography made visible, so surveys made accessible the remotest spaces on earth. Vision entailed control; knowledge led to power.

Notes

1 Gore, David. *Soldiers, Saints and Scallywags: Stirring Tales from Family History*. UK: David Gore, 2009. p. 42. The translation of the Afghan/Pushtun adage given here is: "First comes one Englishman, as a traveller or for [shikar] hunting; then come two and make a map; then comes an army and takes the country. Therefore it is better to kill the first Englishman."

2 Godlewska, Anne. "The Idea of the Map". In S. Hanson ed. *Ten Geographic Ideas that changed the World*. New Brunswick: Rutgers University Press, 1997. pp. 15–39. p. 36.

3 Ryan, Simon. *The Cartographic Eye: How Explorers Saw Australia*. Cambridge: Cambridge University Press. 1996. pp. 3–4.

4 See early maps in Gole, Susan. *India Within the Ganges*. New Delhi: Jay Prints, 1983.

5 Tooley, R.V. *Maps and Map Makers*. London: B.T. Batsford, 1952. p. 104.

6 Gole, Susan. *India Within the Ganges*. New Delhi: Jay Prints, 1983. p. 67. For more recent reprints of early maps also see, Lahiri, Manosi. *Mapping India*. New Delhi: Niyogi Books, 2012.

7 *E'clairecissemens Ge'ographiques sur la Carte de L'inde*. Rennell mentions this many times in his memoirs. See Rennell, James: *Memoir of a Map of Hindoostan, or the Mogul Empire*. London: C. Nicol, 1793. pp. viii, 36, 44, 154, 187, 274, 295; also see Markham, Clements R. *Major James Rennell and the Rise of Modern English Geography*. London: Cassell and Co., 1895.

8 Edney, Matthew H. *Mapping an Empire*. Chicago: University of Chicago Press, 1997. p.98.

9 Edney, Matthew H. *Mapping an Empire*. p.100.

10 Cited in Sen, Sudipta. *Distant Sovereignty: National Imperialism and the Origins of British India*. New York: Routledge, 2002. p. 7.

11 Cited in Sen, Sudipta. *Distant Sovereignty*. New York: Routledge, 2002. p. 21.

12 Harvey, David. *Condition of Postmodernity: An Enquiry into the Origins of Cultural Change*. Oxford: Basil Blackwell, 1989. p. 245.

13 Klein, Bernard. *Maps and the Writing of Space in Early Modern England and Ireland*. Basingstoke: Palgrave, 2001. p. 63.

14 Harvey, David. *Condition of Postmodernity*. p. 245.
15 Cited in Bravo, Michael. Bravo, Michael T. "Precision and Curiosity in Scientific Travel: James Rennell and the Orientalist Geography of the New Imperial Age (1760–1830)". In Jas Elsner and Joan Pau Rubies eds. Voyages and Visions: Towards a Cultural History of Travel. London: Reaktion Books, 1999. p. 173.
16 Edney, Matthew H. *Mapping an Empire: The Geographical Construction of British India 1765–1843*. Chicago: University of Chicago Press, 1997. p. 134.
17 Robert Orme (1728–1801), the author of two significant texts of the early colonial period: *A History of Military Transactions: British Nation in Hindostan 1745–1760* (1763), and *Of the Government and People of Indostan* (1805).
18 Orme MSS. 303 (109) quoted in Phillimore, R.H. ed. *Historical Records of the Survey of India*. Vol. 1 (18th Century). Dehradun, India: Survey of India, 1945. p. 28–9.
19 Quoted in Phillimore, R.H. ed. *Historical Records of the Survey of India*. Vol. 1 (18th Century). p. 22.
20 Ibid. There were other impetuses as well for maps, other than Orme. For example, Phillimore mentions Lt Ferguson whose journal of his Western expedition was not considered accurate enough as his compass had gone wrong.
21 Bravo, Michael. "Precision and Curiosity in Scientific Travel". p. 178.
22 https://commons.wiki media.org/wiki/File: Calcutta_ Baseline_18 32.jpg
23 Kalpagam, U. "Cartography in Colonial India". *Economic and Political Weekly*. Vol. 30, No. 30.29, July 1995. pp. 87–98.
24 Clive's letter to the Court of Directors in London, 30 March 1767. La Touche E. ed. *The Journals of Major James Rennell, first surveyor general of India, written for the information of governors of Bengal during his surveys of the Ganges and the Brahmaputra rivers, 1764–1767*. Calcutta: The Asiatic Society, 1910; also see Heaney, G. F. "Rennell and the Surveyors of India". *The Geographical Journal*. Vol. 134, No. 3, 1968. pp. 318–325.
25 Rennell, James. "Preface to the First Edition". *Memoir of a map of Hindoostan*. (rpt.1976). p. ix.
26 Rennell, James. *Memoir of a map of Hindoostan*. (rpt.1976). p. 109
27 Bravo, Michael T. "Precision and Curiosity in Scientific Travel". p. 172.
28 Edney, Matthew H. "Reconsidering Enlightenment Geography and Map Making: Reconnaissance, Mapping, Archive". Livingstone, David N. and Charles W.J. Withers eds. *Geography and Entertainment*. Chicago: Chicago University Press, 1999. p. 187.
29 Rennell, James. *Memoir of a Map of Hindoostan*. (1793). New Delhi: Editions Indian, 1976. p. 212.
30 Edney. (1997). p.69.
31 The symbolic content of the title cartouche in Rennell's *Map of Hindoostan* has been dealt with in detail by Matthew Edney and Sudipta Sen.
32 Lambton, William. *Memoir*. Ddn. 85, 14–12–10. Cited in Phillimore, R.H. ed. *Historical Records of the Survey of India*. Vol. 2 (1800–1815), India: Office of Survey of India. 1950. p. 245.
33 Lambton, William. *Asiatic Research* VII, 1801. p. 312. Cited in Phillimore, R.H. 1950. p. 233.
34 During the period Lambton was surveying in New Brunswick, a mountain was named Lambton's Mountain after or by him. The name appears on some early maps. The peak was later known as Big Bald Mountain.
35 Lambton's "Theory of Walls" and "Maximum of mechanical power and the effects of machines in motion" were the two papers communicated to the Asiatic Society.
36 The theodolite used by Lambton has a story of its own. It was modelled on the theodolite constructed by Ramsden and used by Roy. The three-foot theodolite, by

William Cary, once apprenticed to Jesse Ramsden, was captured on the passage to India by a French frigate and landed at Mauritius, but it was ultimately returned to its destination by a chivalric French governor Caen with a complimentary letter to the governor of Madras. The chain was modelled on the one used by Roy. The zenith sector was of five-foot radius and was constructed by Ramsden. The chain was meant as a present for the emperor of China from Lord Macartney's embassy but was declined. It was ultimately handed over to Mr Dinwiddie, the astronomer to the mission, as part payment for his services. See Markham, Clements R. *A Memoir of the Indian Surveys*. London: W.H. Allen & Co., 1878. p. 60–61.

37 Cited in Phillimore, R.H. ed. *Historical Records of the Survey of India*. Vol. II (18th century), India: Office of Survey of India, 1950. p. 234.

38 Lambton, William in a letter dated 28 January 1811. Cited in Phillimore, R.H. Vol. I. 1950. p. 233.

39 Cited in Phillimore, R.H. Vol. II. 1950. p. 250.

40 *Asiatic Research* VII, 1801. p. 318. Cited in Phillimore, R.H. Vol. II. 1950. p. 250.

41 Cited in Phillimore, R.H. Vol. II. 1950. p. 237.

42 Keay, John. *The Great Arc: The Dramatic Tale of how India was Mapped and Everest was Named*. London: HarperCollins, 2000. p. 70.

43 Markham, Clements R. *A Memoir of the Indian Surveys*. p. 65.

44 See Goldingham, J. Madras Observatory Papers. Madras: College Press of St George, 1826.

45 Cited in Phillimore, R.H., Vol. II. 1950. p. 265.

46 The phrase is borrowed from Sen, Sudipta. *Distant Sovereignty*. p. 57.

47 Baigent, Elizabeth. "Lambton, William". *Oxford Dictionary of National Biography*. Oxford University Press, Sept 2004; online edn. Jan 2008. [http://www.oxforddnb.com/view/article/15948, accessed 24 Oct 2009]

48 *Calcutta Review* IV (80). Cited in Phillimore, R.H., Vol.II. 1950. p. 264.

49 Cited in Phillimore, R.H. Vol. II. 1950. p. 265.

50 Cited in Phillimore, R.H. Vol. II. 1950. p. 265.

51 Cited in Phillimore, R.H. Vol. IV. 1950. p. 18.

52 See the website http://www.surveyingempires.org/ for details of a recently completed project on the existing trig stations erected as GTS survey towers nearly two hundred years ago. The interviews conducted reveal some interesting, amusing and alternative local anecdotes on these colonial monuments interspersed with local mythology and folk recounts. Also see Singh, Shiv Sahay. "Of Trigonometry and towers – and two centuries of history". *The Hindu*. Kolkata, 13 December 2017.

53 Certeau, Michel de. *Practice of Everyday Life*. Vol. 1. Berkeley: University of California Press. 1984. p. 36.

54 Ibid.

55 Letters of 25 and 27 September 1803; MPC 14–10–03. Cited in Phillimore, R.H., Vol. II. 1950. p. 369.

56 Cited in Phillimore, R.H., Vol. II. 1950. p. 369.

57 Government later stated that the complaints against De Penning "were much exaggerated, but his conduct in striking the public servants at Chundergooty is considered to be highly reprehensible". See Phillimore, R.H. Historical Records of the Survey of India. Vol. 2. 1950. p. 372.

58 Ddn. 63 (337), 18–12–4. Cited in Phillimore, R.H. Vol. II. 1950. p. 372.

59 Ddn. 146 (3), 1–3–13. Cited in Phillimore, R.H. Vol. II. 1950. p. 372.

60 Sleeman, Sir William Henry. *Rambles and Recollections of an Indian Official*. Vol. 1. London: J. Hatchard, 1844. p. 258.

61 Lefebvre, Henri. *The Production of Space*. Oxford: Blackwell, 1991. p. 41.

62 See Klein, Bernard. *Maps and the Writing of Space in Early Modern England and Ireland*. Basingstoke: Palgrave, 2001. p. 41–9.

63 Harvey, David. *Condition of Postmodernism*. pp. 252–3.

64 de Certeau, Michel. *Practice of Everyday Life*. p. 36.

65 Sen, Sudipta. *Distant Sovereignty*. 2002. p. 57.

66 Zou, David Zumlallian and M. Satish Kumar. "Mapping a Colonial Borderland: Objectifying the Geo-Body of India's Northeast". *The Journal of Asian Studies*. Vol. 70. No. 1. 2011. pp. 141–70. p. 150.

67 Barrow, Ian J. "Moving Frontiers: Changing Colonial Notions of the Indian Frontiers". *South Asia Graduate Research Journal*. Vol. 1 No. 2. Fall 1994. pp. 3–28.

68 Barrow, Ian J. "India for the Working Classes: The Maps of the Society for the Diffusion of Useful Knowledge". *Modern Asian Studies*, 38: 3, 2004. pp. 677–702. pp. 686–7.

69 Edney, Matthew. "Mapping and Empire: British Trigonometrical Surveys in India and the European Concept of Systematic Survey, 1799–1843". (Ph.D. Dissertation: University of Wisconsin-Madison, 1990). p. 394.

70 Markham, Clements R.A. *Memoir on the Indian Surveys*. 1871. p. 247.

71 Fergusson, James. "On Recent Changes in the Delta of the Ganges". *Quarterly Journal of the Geological Society*, 19, 1 Feb. 1863. pp. 321–54.

72 Zou, David Zumlallian and M. Satish Kumar. "Mapping a Colonial Borderland: Objectifying the Geo-Body of India's Northeast". *The Journal of Asian Studies*. Vol. 70. No. 1. 2011. pp. 141–70. p. 151–2.

73 Falconer, H. and P.T. Cautley. "*Sivatherium Gigantium*, a new fossil ruminant genus from the valley of the Markanda in the Siwalik branch of the Sub Himalayan Mountains." *Asiatic Research*, 19, 1836. pp. 1–24; Falconer, H. and P.T. Cautley. "Note on the fossil Hippopotamus of the Siwalik Hills". *Asiatic Research*, 19, 1836. pp. 39–53; Falconer, H. and P.T. Cautley. "On some fossil remains of *Anoplotherium* and *Giraffe* from the Siwalik Hills, in the north of India. *Proceedings of Geological Society*, London. 4, 1843–44. pp. 235–349.

74 See Chakrabarti, Pratik. *Western Science in Modern India: Metropolitan Methods Colonial Practices*. Delhi: Permanent Black, 2004. p. 72.

75 Barrow. "Moving Frontiers".

76 Markham, Clements R. *Memoir of the Survey of India*. p. 11.

77 Imray, James F. *The Bay of Bengal Pilot. A Nautical Directory for the Principal Rivers, Harbours, And Anchorages, Contained within the Bay of Bengal; also for Ceylon, Andaman and Nicobar Islands, and the North Coast of Sumatra*. London: James Imray and Son, Chart and Nautical Book Publishers, 1879. pp. 306–7.

78 Osborne, Peter D. *Travelling light: photography, travel and visual culture*. Manchester: Manchester University Press, 2000. pp. 5–6.

79 Del Hoste co-authored the *Route Book of the Mission to Sind in 1833, with Sketch Maps* with Lt Patterson. A. Burnes constructed a map of the Indus and Punjab rivers along with a paper which elaborated the process of its construction.

80 Wood, John. *Personal Narrative of a Journey to the Source of the River Oxus, by the Route of the Indus, Kabool, and Badakshan*. 1845.

81 For example, James Rennell, after his retirement, associated with other prominent figures of science like Joseph Banks in London. They formed an important centre and hub of scientific and cartographic discourse emanating from London which processed data and information collected from all over the world.

Part II

LANDSCAPES OF CONTROL

This section invites a debate on the constructed nature of spaces through a discussion on landscape paintings. What is seen as a naturalised representation of landscape in fact involves a great deal of artifice arising out of specific socio-historical circumstances. As a specifically European genre and born of a cartographic impulse, it was both a product and contributor to processes of spatialisation. When the fervour reached Britain, it gathered the artistic convention and socio-political currents to construct not only the view but also impacted on the appearance of the space. A landscape painting is therefore a whole ensemble of economic and political relations translated into and invested onto a representation which designed views and viewership in defining ways through the eighteenth century to the nineteenth century. The landscape painting, therefore, can be seen as a figurative device to visually control space. The construction of space in such terms is not outside the pale of relations of power. This section will see the cultural production of landscapes in both Great Britain and India in relation to each other and as products of the imperial eye of power.

3

ESTATES, GARDENS AND ENCLOSURES

Aesthetic framing of British landscapes

In painting it is more than plain that the interpretation of space is a function of matter, which sometimes limits space and sometimes destroys its limits. ... It has been like some observatory whence both sight and study might embrace within one and the same perspective the greatest possible number of objects and their greatest possible diversity.

The Life of Forms, Henri Focillon[1]

Denis Cosgrove has famously stated that landscape as a form in the West has its roots in Renaissance humanism and is by nature, bourgeois and individualistic. It cannot be seen free from its ideological overlays in history, its visual power closely duplicating real power humans exercise over land as property. As a "way of seeing" it employs the basic technique of perspective which is homologous to an originary geometry employed also in merchant trading, accounting, navigation, land survey, mapping and use of artillery. Closely linked with the concept of the "prospect" (a word which has the twin meaning of view as well as a possibility of future advancement), perspective landscape is urban in nature: its basic principles are first deployed to the city space before travelling outwards to subjugate and bring under domination all land which lies external to it.[2] In the course of this chapter, we shall see how the expansive vision of the landscape moves from the city to the countryside and subsequently to the peripheries, both inside home and in the overseas colonies, in an apparent synchronicity with the movement of the cartographic gaze.

Both paintings and maps were born from an identical impulse to represent the seen environment. Wild nature, forests, skies, mountains and seas got associated with the word "landscape" as late as the eighteenth century. And it was around this time too, that images of landscape scenery and nature became identifiable with its specific locale. Olwig, in his fascinating book reveals through a historical study of the etymology of the word "landscape", how the idea of landscape got defined as the scenery of a geographical body or "country" while simultaneously enmeshing itself with the idea of country also as a polity defined by political, legal and ethnic criteria.[3] This apparent unity, created by the identification of a political community with the physical bounds of a geographical body and its scenic surface, can therefore mask a contested terrain or conceal rifts among differing ethnicities and polities. The scenic splendour of landscape paintings diverts attention from and disguises these very political schisms in pictorial terms.

The common assumption is that the term "landscape" emerged in sixteenth-century Europe to mean a painting which dealt with natural scenery. However, the term had a life even before this. "Landscape" in various early Germanic languages as also in old English usage commonly designated an area or region. Paintings or scenes from such a landscape were also termed "landscape". However, the first paintings from the sixteenth century which came to be called landscapes depicted not merely the scenery but also the life and culture of a region or an area of activity. In order to understand the transition of meaning and reference of the word, it is important to look further into the relationship between the form of representation and the content or that which is represented; for to talk about landscape is, after all, to talk about representation.

Landscape as a scheme of representation, no less than the cartographic scheme of map making, is an artifice which is entangled in a host of codes. Historically, however, the two have differed as landscapes are thought not to have served pragmatic purpose as have maps. Landscapes have existed in an entirely different aesthetic realm altogether, conveying a complex web of desires while carrying references to a whole set of notions and traditions in art. However, in moving forward with the central idea of spatiality, which underlies my study, it is necessary to engage with questions of how pictorial meanings (in landscape paintings) are manufactured and manipulated "in a milieu of enacted or invented painterly marks".[4] For this, of course, what is required is a recounting of the genesis and historical evolution of this genre while locating it in the discourse of contemporary aesthetics. This would entail a discussion on the evolution of the larger meaning of landscape and the ways in which these have been used to define country and the place of the body politic. Olwig discusses the construction of a unique geographical identity through texts, both written and pictorial, with respect to Anglophone cultural geographies, or of Britain as conveyor of geographical and material changes occurring elsewhere such as in America. By this means, place is imagined and conceived through a level of abstraction which a quantitative cartographic and geometrical logic can only supplement and concretise.

Landscape of the body politic

Though the common perception of landscape is associated with painting, Olwig shows how landscape derives from the "ideas of custom, law and community expressed in "Sachsenspiegel" and Scandinavian bodies of "landscape law".[5]

Therefore the paintings of early Dutch landscape painters like Pieter Brueghel (c. 1525–69) often depict the customs of the time although frequently portrayed in imaginary settings, in the format of "proverbs'", a contemporary artistic genre. Brueghel's paintings are a combination of both "landscape" and "genre", concerned with the everyday existence of common people. This everyday world of common customs and traditions got inscribed into the material world in landscape paintings therefore emphasising lived space; for art at the time was seen as a medium of preserving and transmitting social and geographical knowledge. The validation of this idea comes from the writing of Henry Peacham, a sixteenth-century English landscape painter who admired the Dutch landscape paintings of the time. Art, according to Peacham:

> bringeth home with us from the farthest part of the world in our bosoms whatsoever is rare and worth of observance, as the general map of the country, the rivers, harbours, havens, promontories, etc, within the landscape; of fair hills, fruitful valleys … it shows us the rites of their religion, their houses, their weapons, and manner of war.[6]

According to Olwig, landscape, in Renaissance Europe referred to a particular notion of polity rather than a territory, which steadily underwent a transition to become markable in a world of scalar fixation. Olwig, hereby, employs the metaphor of performance, spectacle and staging in order to understand the process of place-making. Other scholars, like Roger Benjamin, also endorse this idea of landscape as staging a place:

> What is needed is a view of landscape conceiving place as achieved through a process of staging rather than transcription. The term "staging" suggests land-scape as a play of artifice more than an engagement with brute fact.[7]

Whereas what is quoted above refers to the idea of staging as an interpretive tool in order to look at landscape art, Olwig gives the idea a historical legitimacy by relating landscape art to theatrical devices and activities of stage design. It won't be a deviation here to point out that the popular imagery of the world as a theatre used in Renaissance drama, was reciprocated in the titles of atlases and maps produced at the time, for example, John Speed's *Theatre of the Empire of Great Britain*. The construction of landscapes according to principles of theatrical design and stage conventions soon followed suit.[8] In going back to the early seventeenth century, Olwig discusses about the theatrical innovations in the masques performed in the court of the newly crowned King James I of England, (James VI of Scotland). This was a crucial juncture in the history of Great Britain undergoing transitions at several levels. England and Scotland were not yet politically united as one country and the throne of England had just passed from the house of the Tudors to the house of the Stuarts of

Scotland. It was also the time of tussle regarding who represented the land – the parliament or the king. The English parliament constantly resisted James's attempts to unite Scotland and England although he was the king of both. According to the English parliament, the country of England was manifested as a polity through its representation by parliament, as legitimated by age-old conventions and customs. The parliament would not have the same legitimacy when it came to Scotland and therefore it resisted the union. At this time, the conflict between court and country intensified, operating at political levels as two distinct ways of life. This conflict found its way into literature of the times in the pastorals and plays which celebrated rustic life as one of authentic natural life and condemned the court as corrupt and artificial. Duke Senior says in Shakespeare's *As You Like It*:

Hath not old custom made this life more sweet

Than that of painted pomp? Are not these woods

More free from peril than the envious court?

Here feel we but the penalty of Adam,

The seasons' difference; as, the icy fang

And churlish chiding of the winter's wind,

Which, when it bites and blows upon my body,

Even till I shrink with cold, I smile and say,

"This is no flattery." (2.1.2)

The idea reiterates itself not only in the dramatic texts but also in the conceptualisation of the theatre space itself.[9] As early as in the fourteenth and fifteenth centuries Sebastiano Serlio's book on stage architecture showed illustrations of two out of the three types of static scenery that Serlio based on Roman theatre. The two are the *scene tragica* or tragic scene and the *scene satirica* or the satiric or comic scene. The *scene tragica* had an illustration of an inside of a castle whereas the *scene satirica* was illustrated with an outdoor scene which consisted of trees and gardens.[10] The power of the idea of "country" lay in the way it embodied the memory of the rights that constituted people's place and community identity. This was related to the idea of landscape as the word "landschaft" was translated as country, region or province.[11] On the other hand, the monarch himself or herself ought to have been the very embodiment of the polity or body politic. This dichotomy could only be resolved through spectacle. The Stuart court, for this purpose, made use of the masque which already held political and social power as a customary tradition. The masque operated at an elite level in the court in the same manner as rustic rituals and festivities operated amongst the commoners in order to forge a sense of community solidarity. Various stage mechanisms, like the stage scenery, constructed an illusion of natural habitus of a community of people, literally the stage upon which the court and the king could enact and perform power:

> In the masque, however, custom was replaced with costume, a matter of design and style, just as, through the representation of landscape as perspective scenery, unchanging geometrical principles replaced the evolving laws of custom. In the masque the landscape ceased to be the "habitus" of a people, an environment shaped through customary practices and bodily activity; instead, it became the scene upon which the personages of the state performed their roles under the authorial gaze of the ruler.[12]

Staging the nation: landscapes in stage scenery

The first traces of a premature idea of Britain as a nation emerged out of such practices. For example, in Ben Jonson's *The Masque of Darkness*, first performed in the Stuart court in 1605, the symbolism of the nation state was all pervasive. The masque was famously stage designed by Inigo Jones:

> First, for the scene was drawn a Landschap ... which falling, an artificial scene was seen to shoot forth, as if flowed to the land raised with waves which seemed to move, and in some places the billow to break, as imitating that orderly disorder which is common in nature ... The scene behind seemed a vast sea, and united with this that flowed forth, from the termination or horizon of which (being on the level of the state, which was placed in the upper end of the hall) were drawn by the lines of perspective, the whole work shooting downwards from the eye ... So much for the bodily part, which was of Master Inigo Jones his design and act.[13]

The focal point or convergence of the lines of perspective was the eye of the monarch, in this case King James I, seated on an elevated throne above the general public. The geographical body of Britain was defined by the circumferential seas which, while separating it from the rest of Europe, gave it a unique identity. This was obviously possible only when England, Scotland, Wales and Ireland came together as a consolidated territory. Here, the very abstraction of a unified geographical body was embodied, represented by James I as it was he who brought about this unification by virtue of his blood and his marriage to the princess of Denmark, through which part of Scotland's disputed territory, the Orkneys, was settled as dowry. This was the first time that the British Isles was addressed as a single geographical unit. In the masque the geo-body of Britain is female and is united through marriage to the king. Numerous cartographic images in Renaissance England and the national iconography in general, depicted the British territory as female.[14] The tradition gathered even more currency during the reign of Elizabeth I when it was only but natural that her feminine body would be identified with the British Empire. Her figure was visualised as Astrea, the goddess of justice and natural law and the Stuarts inherited the same tradition. Also, it is essential to mention at this point the predominant belief of cyclical renaissance of empires corresponding to the cyclical rotation in space. This spatial movement followed the solar cycle, by the logic of which, empires would at various historical moments be translated to a point west from its previous location. The Romans had believed that their empire had been transferred to the west from its previous Greek seat, and during the Renaissance, Britain was thought to be the location of the golden empire translated further west. According to Olwig:

> The unified geometric space of the map, as well as the landscape scene, facilitated the ability to imagine that these historically and geographically diverse "countries" made up a single country at the grand scale of Britain – masking the divergences between them by reducing qualitative differences to a question of quantitative scale. Distinctive places, each with its own history and customs, were thereby reduced to locations within the spatial coordinates of the map.[15]

For a people, the very act of inhabiting a characteristic and historically allusive native landscape evoked a sense of belonging and therefore resolved the problem of legitimacy without altering allegiances and loyalties to local provinces. The landscape evoked also an earlier Britain and tried to obliterate the memory of contemporary disunity.

The seventeenth and the eighteenth centuries witnessed the rise of the theatre as a vehicle for grand spectacle in the whole of Europe along with Britain. Princes and noble families directed vast funds into the staging of courtly productions, well aware of their propaganda value. It had long been conventional for European theatrical settings to include symbols of heavenly figures, or a monarch in an elevated position, or both. In the Stuart court, this symbolic position of a monarch controlling space from above was not only exploited to the full but in some cases also supplemented by that of a monarch controlling the illusory horizontal space of landscape scenery. The staged perspective, like the one in *The Masque of Blackness*, could create through the depiction of the scenery of Britain's landscape, an illusion of a spatial entity suggesting the state which the monarch controlled. What is identifiable here, is a Ptolemaic chorographic impulse for, while laying down the geographic space of Britain, it simultaneously marks out the boundary of the state (for "choros" technically means the boundary of the extension of some thing or the container or receptacle of a body).[16] The great upsurge of cartographic activity in Britain at this time, strove to give it a clear demarcation and identity as rivalries with other European kingdoms/ nation states increased.

The science of perspective played a very crucial role in the history and development of paintings, especially topographical and landscape paintings, in fashioning and ordering a sight or vision. Perspective is also the dominant technique with the help of which a European-style cartographic culture took off. According to Erwin Panofsky, perspective is the central component of Western "will to form". It is an expression of a schema linking the social, cognitive, psychological and technical practices of the Western culture.[17] An early fifteenth-century Florentine painter, Filippo Brunelleschi, who used standard surveying techniques, calculations and triangulation for his drawings, is said to be its first proponent. Numerous experiments to create illusion and depth of space were undertaken from the time of Brunelleschi and perhaps even before him. Brunelleschi was the first to use the scheme of linear perspective to construct a form of peep show to heighten illusion. His techniques of creation of an imagined space which could serve the artist's need were later exhausted by future artists. Once rationality became the maxim which defined processes of representation, art slowly moved to the realm of certainties. Epistemological principles took over and it became irrefutable that the "real" could be plotted exactly in space. This entailed also, the formulae for representing, limiting and controlling space. Mathematics, for the first time defined fine art, a nexus which would reach frenzied heights by the eighteenth and nineteenth centuries, especially in Britain. The new European academic tradition embraced an expressive dimension which was aimed at inducing a sense of high philosophical principles and a formal dimension, which required scientific exactitude in execution. Nothing could be arbitrary with respect to these formal requirements; for example, the appearance of an object would depend on the tangents of the planes of the forms in relation to the eye of the spectator, the angle of incident light and its distance.[18]

During the European Renaissance, representations of architecture and buildings in drawings and paintings were first initiated by exploiting these new innovations in the science of perspective used in their projection. This in turn, not only provided a cue for painting townscapes but also led to conceptualising the inner space of buildings. Initial innovations in theatrical scenery attempted to reconstruct the inside of these buildings as settings for action on stage. Such spatial architecture, then, slowly travelled from indoors to portraying outdoor scenes to depict townscapes/landscapes as the backdrop of theatre action.[19] Inigo Jones was one of the first to have perfected the art of scenery making in the context of the seventeenth-century English proscenium. Interestingly, this transition in the depiction in the backdrop followed an opposite trajectory of theatre performances, moving from outdoors to indoors. As theatre moved from open spaces to enclosed space, usually the court, the perspective too changed from multi-point perspectives of common people to a single point perspective of the head of the state whom the play was designed to entertain. Therefore, during the seventeenth and eighteenth centuries, the theatre scene was a topos that catered to the taste and was devoted to the cause of the monarch. This, along with other similar kinds of innovations, like the peep box structure, facilitated and bore upon topographical and geographical representation in several ways.

Land, enclosures, estates and the cartographic gaze

The unspoken relationship between maps and landscape became still more eminent in the eighteenth century when land underwent large scale surveys and remapping.[20] As Stephen Daniels points out:

> The rise in landscape painting in Britain during the eighteenth and early nineteenth century coincides with a renaissance in British cartography. ... Cartography emerged as a specialist, but not entirely separate, discipline. ... the remapping of Britain was part of a broader revisioning of the country by travel writers, antiquarians, landscape gardeners, and landscape painters.[21]

The phenomenon can be read as an all-pervasive cultural movement during this time. This requires an interrogation of the social history of land ownership during the time. The Glorious Revolution of 1688, also called a "bloodless revolution", was a bourgeois revolution which did not overthrow the monarchy and create a republic, it merely overthrew James II and installed monarchs who would accept the authority of the parliament and function as per parliamentary jurisdiction. This was the triumph of a Whig-dominated parliament – the landed gentry making land the centre of attention as country seats of an oligarchy of great estate owners, and landscape the dominant art form from this time onwards to the eighteenth century. This caused obvious repercussions in the way landscape was perceived. For the first time, also, the concept of enclosure evolved as a means to demarcate property. In the Middle Ages there was no large-scale manipulation of the landscape in the display of personal status. It was only with the evolution of private property that the landscape began to be deliberately and extensively shaped for social and aesthetic purposes. The social elite, in this manner, displayed power over land and inscribed their property with their own preferences. The dissolution of the monasteries in the 1530s under Henry VIII and the Anglican Reformation, ushered in a period of massive landholding. Previously, there existed little or no trace of privately-owned land as land was generally owned either by church or the monarch. After the dissolution of church, vast acreages were suddenly more or less up for grabs. Although the Crown initially took over the estates of the monastic institutions, most of these lands had passed into private hands in order to raise revenue to fund continental military campaigns. This resulted in a hyperactive land market where property was sold and resold rapidly. It was the local gentry which benefited the most, though great magnates of national importance also took over huge portions of the land.[22]

Whereas the former Renaissance garden, a century back, appeared to be an extension of the architecture of the mansion, the Arcadian landscape gardens, *le jardin anglais,* of the following century were oriented towards the surrounding countryside and were consciously blended with it as "natural". The boundary between the garden and countryside was deliberately blurred by erecting a fence as a ditch below eye level so that there seemed to be no barrier between the garden and the outside world.[23] The subterfuge obliterated boundaries and helped create the illusion of a pastoral golden age. Though the image of the land estates disguised their dependence on non-agricultural forms of income, it indicates the tremendous cultural importance of landed property in the eighteenth century. As English society became increasingly dependent on commercial and industrial wealth, the possession of land became an ever-more powerful symbol of social status.[24] The upwardly mobile class who profited from either trade or other sectors aspired to own land which symbolised social status. These were cordoned off from the labourers and field workers with enclosures and hedges which limited their access to the estates.[25] The great estates flourished in the areas where land could be acquired easily and relatively cheaply. By the nineteenth century, the great land owners had taken control of vast portions of upland moorland areas which was relatively of poorer quality and meant usually for grazing rather than cultivation. Only some of this was improved and converted to farmland upon acquisition, while most of these formed elaborate landscapes. Extensive planned landscapes such as these form a striking contrast to the landscapes that were shaped and moulded by innumerable hands over the centuries. The boundary provided the opportunity to shape the landscape and display aesthetic taste.[26] The grandly formal, geometric landscaping was based on the principles of Le Notre imported from the court of Louis XIV.[27] Both in England and France these landscapes were seen as symbolic of the power of the landed elite. The avenues and alignments running out into the landscape showed the areas under the owner's control and emphasised the country house as a pivotal point in the landscape. Most of these landscapes were inspired by continental landscape art and paintings. It had been a part of the fashionable and elite cultural and educational practices of the time to despatch young Englishmen on what was called the "Grand Tour" to places like France and Italy. The places were seen to be rich in culture and these gentlemen, exposed to the highly influential landscape paintings of Salvator Rosa, Claude Lorrain and Nicholas Poussin, brought home ideas of the "sublime" and romantic scenery. Freshly back from the European tour and deeply impressed by the paintings of the grand masters, the youngsters sought to inscribe similar designs on the land of their own estates.[28]

In the eighteenth and the nineteenth centuries, there came about the advent of another movement directly related to landscape in paintings and architecture called the "picturesque", its main proponents in architecture being Uvedale Price (1747–1829) and Richard Payne Knight (1751–1824).[29] They believed that landscapes and gardens should imitate nature in the raw and parks should attempt to capture the true spirit of the Italian artists, especially Salvator Rosa, like whose paintings, the landscape gardens would have drama and ruggedness that would serve to inspire awe. Accordingly, on many of the estates, conifers and other exotic trees were planted to make the scene suitably wild. On some estates rocky slopes, waterfalls and fountains were artificially created. Within these rugged landscapes, a host of ornaments in the gothic and rural style were created. The consciously archaic style of these features romanticised rural communities of the past and their ability to survive in a hostile untamed landscape.

Social changes materially constructed landscapes as private property, even as art, specifically continental landscape paintings, influenced the aesthetic response to it. Both of these have a bearing on the construction of space/place of the nation and ownership. Whereas on the one hand these created space for the elite landlord and marked the region out as his property, simultaneously they constructed and altered entire geographies to be identified with a name. I shall turn to this point later on, when studying perspectives on colonial space and its representation in pictographic terms to see how similar the processes of spatial construction were, that were deployed in two different parts of the world near about the same time.

The correlation between maps and paintings as spatial productions can further be explored when looking at the cartographic and pictographic representations the period triggered. The map itself was admired as a work of art as late as the eighteenth century before finally surrendering itself irrevocably to the realm of absolute physical science. And as works of art, the maps commissioned by estate owners, often served as ornamental show pieces proudly displayed in country houses. Concurrently, the eighteenth and the nineteenth centuries saw an unprecedented upsurge in landscape paintings. These too found their place on the walls of manors of wealthy landlords.[30] Humphry Repton (1752–1818), a noted architect and landscape theorist in the primary decades of nineteenth century, observed that paintings and sculptures had lost their original didactic purpose and had come to be valued as wealth, furniture and ornament:

> for whatever might be the original uses of pictures or statues, they are now only considered as ornaments, which, by their number and excellence distinguish the taste, the wealth, and dignity of their possessors.[31]

Thomas Gainsborough's famous genre painting *Mr and Mrs Andrews* (1749), probably commissioned for the same reason, contains in it, a demonstration of the socio-economic values of the time.[32] As many wealthy landlords liked to get their estates painted along with themselves as proud owners of these, there emerged distinct by-products of landscapes: the country house portrait and the garden conversation piece. Such paintings, often commissioned by the sitter or the landlord himself, would be displayed in the manor as a pictorial miniature of all that he controlled. In *Mr and Mrs Andrews*, a newly-wed couple pose before the trunk of a mighty oak. The depiction of Auberies, their estate, with its fields, meadows and trees, takes up more space than the double portrait itself. Andrews represents the upwardly mobile class as he poses as a member of the landed gentry, the free-holding class of squires and recent peers, who not only owned most of the country but had parliament in their hands, too. Without their consent, George II could neither impose taxes nor raise an army. The view from the garden bench gives the appearance of a boundless idyll in tune with the ideal parkland proposed by the famous propagators of landscape gardens discussed below. It even includes an occasional cluster of trees aesthetically indispensable for a natural interruption of vision, of vistas that would otherwise seem too wide, or too symmetrical. Andrews's real source of income from his financing business in London is carefully concealed, as land and agriculture seem to be his source of livelihood and position. The fields, the agricultural produce (symbolised by the piled-up haystacks), the dogs, the seated wife, all of these appear to be subordinated by Andrews's erect figure, gun in hand, master of all that is seen in the portrait. The painting, meant as a wedding portrait, also epitomised the correct match or wise marriage of the times as marriage happened to be the most significant way of merging and con-solidating land and estates.[33]

Spaces of art and science: travel, maps and landscape paintings

The eighteenth and nineteenth centuries saw an unforeseen profusion of land-scape and topographical paintings from within Britain. Many of these paintings were the direct outcome of the socio-historical background related to the rise of an increasingly prosperous class of landed gentry. The gentry sustained the demand for depiction of land in the English countryside they owned. Much of the English self-fashioning was modelled on the tastes of the European cultural elite. Following on from this, patterns of representation of landscapes were inspired by works of Continental artists and masters, such as Claude Lorrain, Nicholas Poussin and Salvator Rosa, whose paintings were either viewed abroad or acquired in the course of the Grand Tour.[34] The continental tours, popularly called the Grand Tour and undertaken by young gentlemen, acted as an important reason behind the production of landscape paintings which were "sublime" and "beautiful". These travels also served as the basis for creating pictographic records of memorable views, more or less the same purpose that photography would serve from the second half of the nineteenth century onwards. Therefore these paintings had strong connections with travel and recounting. A popular periodical acknowledging the bond between travel and landscape paintings states:

> The great encouragement which has been manifested of late years for the cultivation of landscape drawing, has originated principally in the love which has been evinced for making tours, to explore the beautiful scenery of our island ... the merit of having created so general a love for travelling.[35]

Here, the magazine also discusses the touring books written and popularised by Reverend William Gilpin (1724–1804), the chief proponent of the "picturesque", which had become a cult by the nineteenth century in Britain.[36] During the 1790s, three works provided an aesthetic and instigated this movement: Richard Payne Knight's didactic poem *The Landscape* (1794), Uvedale Price's *An Essay on the Picturesque* (1794) and Humphry Repton's *Sketches and hints on picturesque gardening* (1795). The "picturesque", of course, had a host of jargon and techniques of its own. Where the Burkean notions of beautiful and sublime celebrated smoothness, the "picturesque" heralded a taste for the wild and the irregular which managed to incorporate a vast and diverse range of subjects, such as old cottages, gnarled trees, ruins, crags, beggars and gypsies (which could branch off and feed antiquarian, historical, scientific, botanical and anthropological interests).[37] This encouraged sketching expeditions in and around the British countryside. Just as rural scenes were modelled on Rosa and Lorrain, urban scenery drew upon Canaletto's (1697–1768) clear, sharp, well drawn, brightly coloured views.[38] Almost certainly, tour guides and water-colour instruction books trained the public on the conventions of understanding and appreciating nature *in situ* which increased the value of the painting manifold in comparison with those done purely from imagination. For the first time, therefore, first hand factual observation became more important than imagination. All kinds of technical and optical devices surfaced in the market which aided the view-making and factual observation. These were called "artists' viewers" and comprised of small lenses like pocket magnifying glasses, the "Claude Glass", a type of convex mirror, the camera obscura and the camera lucida.[39] These instruments helped the untrained and amateurs to first of all recognise a scene as characteristically aesthetic and, thereby, reproduce the same in drawings according to the rules and prescriptions laid down by the professors of the "picturesque". The art of seeing nature, according to John Constable (1776–1837), one of the most famous English landscape artists in the naturalist tradition, is "as much to be acquired as the art of reading the Egyptian hieroglyphics."[40] According to Gombrich:

> stimulated by the rise of science and the new interest in factual observation, questions of vision were much debated by artists of the nineteenth century.[41]

Moreover, Gombrich draws our attention to a new edge to this utterance which was targeted towards the general public rather than trained artists. He identifies a slow but steady journey which unfolds over centuries culminating in the nineteenth century. This, Gombrich calls a progress towards visual truth, motivated to disentangle what is seen from what is already known. Hence, such guided observation would be tantamount to enquiry into the hitherto unknown laws of nature, invoking a scientific objectivity. When Carolus Linnaeus published his *Systema Naturae* in 1735, which elaborated a classification of vegetation from over the world, Joseph Banks went on his Endeavour Tour with Captain Cook 1768–71, following which, Thornton's *A New Illustration of the Sexual System of Carolus von Linnaeus* (in which *The Temple of Flora: Or Garden of Nature* was the third part) came out in English between 1797 and 1807. These caught the aesthetic imagination and the scientific temper of the nation, which was tasting the fruits of foreign travel and colonial expansion. The popularity of publications of botanical treatises with intricately hand-painted specimens of flora spread over to the landscape form; for most of the painters of such productions were not professional botanical draughtsmen, but mainly practitioners of landscapes and portraits. The knowledge and skill honed from these productions, which often involved exotic specimens gathered from overseas colonies, were transferred onto the more lucrative and dominant forms of the day by adding details and texture in landscapes and backgrounds to portraits. Together, they went on to create a quintessentially eighteenth-century "naturalism" and spatiality even within these forms which, most of the time, were a curious mix of scientific observation and the fantastical and dramatic. Similarly, new academic discourses in the field of geology made way for a more phenomenalist mode of representation, since both geologists and landscape artists were concerned with the appearance of the natural world, marrying science and art in a direct way.[42] Many eighteenth-century traveller-naturalists had a well-developed visual awareness of the topographical phenomena that they studied. They were initially equipped to communicate those details in verbal terms, but gradually felt the need to communicate their observations in visual terms. Many of them collaborated with painters or tried to sketch and paint themselves to create visual records and aids of their observations. This also marked the introduction of a disciplinary academic discourse in art.[43]

This brings us to representative spatiality in another form. Stephen Daniels remarks on the symbiotic influence cartography and landscape paintings had on one another. The socio-cultural reasons for the large scale production of the two forms in England have just been discussed. While the map and topographical paintings complemented each other, on the other hand, they also competed against each other. While surveyors gazed through theodolites, artists and painters pondered on elements of correct perspectives and representations according to principles of the "picturesque". Theorists of the picturesque however considered cartography, or for that matter any topographic or panoramic view, as inferior in nature and a brazen display of power (over property). They lobbied for a cultivation of taste and advocated representation of short-focus views which would evoke the sense of drama of the great landscape painters. In 1805, Henry Fuseli, as professor of the Royal Academy, denounced maps, saying:

that kind of landscape which is entirely occupied with the tame delineation of a given spot; an enumeration of hill and dale, clumps of trees, shrubs, water, meadows, cottages and houses, what is commonly called Views ... The landscape of Titian, of Mola, of Salvator, of the Poussins, Claude, Rubens, Elzheimer, Rembrandt and Wilson, spurns all relation to this kind of mapwork.[44]

However, even with these strictures, landscape artists continued to get influenced by map work and often their work was recast according to survey and activities. Stephen Daniel speaks of how two of England's most important landscape artists, John Constable and J.M.W. Turner, were influenced by the visual model of the map in expanding their horizon to be able to accommodate a gamut of geographical information, knowledge, associations and references. Constable was influenced by other scientific and meteorological texts and diagrams, for example, in his famous series of studies of cloud formations, many of which he incorporated into his landscape scenes. His paintings of the Stour Valley were influenced by the large scale Ordnance Survey that went on at that time including a survey by his father. As Daniels says:

If we are seeking parallels for Constable's documentary style – its elevated views of sites from various angles, its use of landmarks for orientation, its even focus, its detailed differentiation and integration of features, its specification of these features' form and function – we should look beyond painting to maps and manuals of surveying and military drawing.[45]

Constable's paintings were so noteworthy for their accuracy and details that he was offered the post of drawing master to cadets learning to depict terrain at the new Royal Military College at Great Marlow. Though he ultimately desisted joining the service, the fact bears an important reference to the nexus between survey and art, between science and aesthetics. Another acclaimed topographical painter of the eighteenth century, Paul Sandby (1731–1809), credited with popularising scenes of Scottish Highlands, began his career as a draftsman of the military survey of Scotland. J.M.W. Turner, too, made a two-month sketching tour of the West Country on a commission to produce watercolours for W.B. Cooke's *Picturesque Views of the Southern Coast*. In this series of paintings, Turner embraces the prospect or elevated view which is a conventional feature of topography and maps because of its capability of scalar transformation of a huge space into a small space while accommodating a great deal of detail. For these paintings, Turner closely observed the physical and human geography of places and also consulted a detailed gazetteer on the region, from which he made notes on its history and antiquities, geology and manufactures. Arguably, these paintings were influenced by the maps of the area by the Ordnance Survey which were published in the first decade of the nineteenth century. The final engravings of these maps, according to Daniels:

do convey the substance of the landscape. Through graduated hachuring and fine delineation of rock formations (of coastal cliffs and moorland tors), the West Country emerges as a three dimensional, physiographic image, as a landscape with a structure as well as a surface.[46]

Another painter, Robert Dawson, who was a contemporary of Turner and Constable, is said to have influenced these drawings and perfected the genre of topographical paintings. Dawson worked as an instructor at the Royal Military College before joining the Ordnance Survey.[47] He developed a characteristic style of drawing relief that used precise elevation points called "spot heights", brush and watercolour and hachuring. The method was to select a sight from a raised altitude and convert it into a view or a prospect. It worked in the same manner as trig points worked in the case of triangulation. It soon was upheld as the "British National Style" originally borrowed from the French manual, *Memorial topographique et militaire*. Dawson talked of his paintings as applications of "natural-history-principle of drawing". In these sorts of paintings, the physical features and geological structure, along with the foundation of the land, needed to be first understood before representation.

Gombrich argues that in the Western tradition, painting had been pursued as a science through a process of "ceaseless experimentation".[48] As if to match those very words, Constable asserts in one of his lectures:

> Painting is a science, and should be pursued as an enquiry into the laws of nature. Why, then, may not landscapes be considered as a branch of natural philosophy, of which pictures are but the experiments?[49]

In an age where representation was an epistemological exercise, it was bound to be incorporated within the paradigm of science: for paintings are assumed to deal with visual facts which are reported by the eyes and recorded by the hands.[50] Martin Kemp discusses the concept of imitation of nature, based on scientific principles which found inspiration in two branches of optics – the geometrical science of perspective, discussed earlier, and the physical science of colour.[51] From the eighteenth century onwards, the act of reproduction of nature was implicated within an existing arena of technical jargon. It was, after all, an entire action not only of picture-making but also of view-making.

Seeing the nation: aesthetic construction of Great Britain

There were, however, certain challenges to view-making. It required a knowledge and training of how to see which also entailed a systematic collection of data and an analytical gaze. What was even more compelling was the recognition of that data as a visual record. This difficulty could be overcome by the use of taxonomic series binding individual images together within an overall cognitive system which guided and informed it.[52] For this purpose, an unstinting adherence to realism in the name of objectivity was invoked. Just as the idea of "family" was gaining importance in science, so could variety be strung together as different aspects of the same thing. Being an age of compendia, in this sense, "illustrative topography could be grafted onto a parent stock of eighteenth century science", and to antiquarian/historical studies as well.[53] According to the members of the Society of Antiquaries, the historical imagination of the nation in the eighteenth century could be mitigated through and better understood by combining historical research and accurate illustration. Richard Grough, in his *Anecdotes of British Topography* (1786) and *Sepulchral Monuments in Great Britain* (1786) pioneers this movement. Soon, John Carter's (1748–1817) work such as his *Views of Ancient Buildings in England*, published serially from 1786 to 1793, followed suit. A deluge of topographic views and illustrations of gothic architectural ruins from across Great Britain soon followed. Taken together, these sought to curate British history and what medieval Britain had achieved. By 1821, the culture of topographical and architectural illustrations was so popular that the *Quarterly Review* proclaimed:

> Every nook in our island has now been completely ransacked, and described by our tourists and topographers. If we call over the Counties one by one, their historians will be seen marshalling their ranks in quarto and in folio ... Nor has the pencil been employed with less diligence than the pen. It would be difficult to name any structure of the "olden time" which has not been transmitted into the portfolio and the library.[54]

The act of reading each of these images relates to an outward trajectory: of piecing them together in a system as a whole. Images from various parts of the British Isles could be stitched together to reproduce Britain's past, working as visual knowledge of a territory. This, of course, had an impact on public taste and generated, along with an academic desire for re-evaluation of medieval Britain, a possibility for picturesque topography, in line with illustrative accuracy and a systematic ordering of knowledge.[55]

With the advent of cheaper and amenable print technology, the historic past was made accessible to the learned audience through illustrated compendia of topographies and costumes and natural history. The romantic genre of the "ruin" poetry developed hand in hand with this. Abandoned architecture such as abbeys, chapels and castles in the hinterlands of the island, rather than classical European remains, became the objects of serious academic interest and contemplative imagination. Therefore, just as the sight of the Tintern Abbey and the adjacent Wye Valley in Wales could instil in William Wordsworth a contemplative rumination, inspiring him to write the "Lines written a few miles above Tintern Abbey" (1798), the same structure was the subject of a number of watercolour compositions by J.M.W. Turner, such as *Ruins of West Front* and *The Chancel and crossing of Tintern Abbey* in 1794, a dramatic, melancholic scene by Philip James de Loutherbourg titled *Visitor to a Moonlit Churchyard* in 1790, a print by Thomas Hearne of Th*e Iron Forge at Tintern* (1795) and a host of other paintings by William Havell, Edward Dayes and William Henry Bartlet.[56] There evolved a tension between aesthetic and scholarly approaches which coincided with the clash between the artist's personal experience of the place and a purely academic interest in exact recording. Caught between the two, this duality was not easy to deal with but often crossed paths as the national history and geography had to be made aesthetically pleasing and therefore had to adhere to certain picturesque ideals as well. Hence, certain registers got enmeshed in the recording of the ancient relics which managed to lift these ruins out of their mundane context and mixed them with nostalgia for a bygone majestic past. This was most often the engravers' doing. Landscape scenes were fraught with being considered as a form of evidence to be used by professors of antiquaries and architecture and, on the other hand, symbolic of inspired contemplation by poets and artists rather than being a mere "kind of mapwork".[57]

The appropriation of peripheries into a quintessentially British national aesthetic of the picturesque is strikingly visible in Paul Sandby's work, *The Virtuosi's Museum* (1778–81), one of the most important commercial print ventures in the late eighteenth century, covering the whole of Britain.[58] In the late 1730s, when the Buck Brothers published their engravings of ancient castles and cathedrals, popularly called *Buck's Antiquities*, they could only cover England and Wales. In Sandby's' *Virtuosi*, there was a series of 26 plates depicting scenes from Scotland, Wales and Ireland. In the landscapes of the Scottish Highlands, Sandby uses his sketches made as a military draughtsman right after the suppression of the Jacobite uprising at the Battle of Culloden. (See Fig. 3.1 & 3.2) This was also the time that these terrains were being mapped fervently and General Roy was assigned the duty of producing the first map of this region called the Great Map. Sandby uses his sketches made during his four-year stay in Scotland, beginning in 1747, a good 30 years later, and transforms the military topographic sketches undertaken as on-the-spot views according to evolving conventions of aesthetic appreciation.[59] Around this time, through his works like the *Observations on the Highlands of Scotland,* Gilpin too was successfully engaging distant landscapes within an antiquarian rhetoric of the picturesque:

> Among the picturesque appendages of this wild country, we may consider the flocks, and herds, which frequent them. Here we have stronger ideas, than any other part of the highland presents, of that primeval state, when man and beast were joint tenants of the plain. The highlander, and his cattle seem entirely to have this social connection. They lead their whole lives together, and in their diet, beverage and habitation discover less difference, than is found between the higher and the lower members of any luxurious state.[60]

117

Likewise in Sandby's plates, herds and flocks of cattle on mountains and glens, and foliage on ruins, are carefully introduced along with an ambient *chiaroscuro*. The views were packaged in ornate oval frames to cater to the new customers of these productions, namely the gentry rather than the army. The prospect views which were earlier plain and formal, largely sticking to cartographic stylistics, hues and brush strokes, were invigorated by adding Highland figures, both civil and military, often interacting with each other generating an impression of peace and harmony.[61] Gradually, a new territory was being opened up to touristic gaze, which encompassed the Celtic and Gaelic fringes and which reconstituted the nation afresh.

The popularity of Sandby's depictions contributed to the spread of the picturesque convention which was initially limited to the English countryside. Gilpin, in one of his art expeditions to Scotland in search of picturesque, in 1776 found the mountains suitable for picturesque compositions:

> Their broken lines and surfaces mix variety enough with their simplicity to make them often noble subjects in painting; tho as we have observed, they less accommodated to drawing. Indeed these wild scenes of sublimity, unadorned by a single tree, form in themselves a very grand species of landscape.[62]

Figure 3.1 Surveying Scotland: *View Near Loch Rannoch* by Paul Sandby, 1749
Reproduced with the permission of British Library

Figure 3.2 Surveying Scotland: *Party of Six Surveyors, Highlands in Distance* by Paul
      Sandby, 1750
Reproduced with the permission of the British Library

Gilpin's drawings and their aquatints highlight the unevenness of the mountains
through a *chiaroscuro* effect. The drawings were mostly nondescript; had not
letterpress explained and named the places, the castles and the mountains, they
would not be recognised or identified. Just a few years after Gilpin's tour,
Joseph Farington (1747–1821), another painter and an important socialite of the
Royal Academy, made trips to Scotland between 1778 and 1790, when perhaps
Gilpin's *Observations* was not yet published. Though he made many sketches
of the mainland, the castles near and around Edinburgh, by the time he arrived
in the Highlands, he had no great commendation to offer in his *Sketchbook*):

119

> I recollect in no part of the world such a length witht [sic] objects to engage notice ... a continuation of barrenness and drearyness.[63]

Farington made a few more journeys and by his third journey in 1801 he was able to appreciate the Highland scenery to the same tune as Gilpin: "my attention was engaged in watching the effects of light and shade on the Mountains".[64] By now the Celtic fringes were appropriated by the picturesque gaze and were granted the pictorial status that was given to the Lake District or the Welsh countryside.

Another eminent painter who travelled to Scotland and sketched Scottish scenery was William Daniell (1769–1837). Daniell, a friend and a close acquaintance of Farington, had already earned a name and a place for himself after his expedition to India with his uncle, Thomas Daniell (1749–1840) with whom he produced a series of Indian landscapes and topographies, later published between 1795 and 1808 as the *Oriental Scenery*. By virtue of this work, he acquired the enviable position of a royal academician, having been elected in preference to even Constable. Between 1807 and 1812, he compiled a series titled *Interesting Selections from Animated Nature with Illustrative Scenery* which incorporated scenes from his travels in England, Scotland and Wales. His next project, and one of national interest, was *A Voyage Round Great Britain*, which was completed after undertaking six trips across the island from 1813 to 1823. Specialising as a marine painter, he had plans to circumnavigate the island in its entirety by sea, painting the coasts, shorelines, beaches and harbours. But, ultimately, he was forced to drop the coastal tour and take up land journey for the most of it. A lot of the plates were meant for a specifically geological audience as he produced some of them to illustrate the volcanic geology in the Inner Hebrides.[65] Charlotte Klonk analyses how Daniell's depictions, sometimes of the same spot, change over time. For example, in his paintings of Fingal's Cave on the Isle of Staffa, (See Fig. 3.3) a transformation of his style is most noteworthy. This is an iconic site of volcanic origin, made famous through Romantic associations. The isle must have derived its name from an old Norse word meaning "save", (for the interlocking basalt columnar formations found there abundantly) when these parts were regularly invaded by the Vikings. MacPherson's (1736–96) Ossianic poetry did much to popularise the spot in 1762 when he adapted Gaelic and Celtic legends in order to rewrite a fanciful history of the newly acquired regions for a primarily English readership with an unforeseen antiquarian interest and enthusiasm for new information and an insatiable desire for new and stranger sights.[66] To weld with this imagination, Daniell had made an earlier painting of the place in his *Interesting Selections*, which took many artistic liberties, making the cave look like a vaulted cathedral-like structure. (See Fig. 3.4) Subsequent visits from geologists and accompanying draughtsmen annulled this view, toning it down from its former fantastic grandeur to one of a great natural geological curiosity. None other than Joseph Banks equated the cave with other geological wonders such as the Stonehenge declaring it as far surpassing the splendour of any man made structure:

Figure 3.3 In Fingal's Cave, Staffa, Hebrides, Scotland by William Daniell, part of
Voyage Round Great Britain (1813–23)
Image courtesy: Creative Commons. Reproduced with the permission of National
Library of Scotland

Figure 3.4 An earlier reproduction of Fingal's Cave, Staffa, Hebrides, Scotland by William Daniell, part of *Interesting Selections* (1807–12)
Reproduced with the permission of National Library of Scotland

each hill, which hung over the columns below, forming an ample pediment ... almost into the shape of those used in architecture .... Compared to this what are the cathedrals or palaces built by men! Mere models or playthings, imitations as diminutive as his works will always be when compared to those of nature. Where is now the boast of the architect! Regularity the only part in which he fancied himself to exceed, his Mistress Nature, is here found in her possession, and here it has been for ages *undescribed*.[67] (My italics)

Since it fell to the artist to illuminate what had hitherto remained, "undescribed", the artist's task was to remain faithful to reality. Influenced by Macculloch's geological surveys and mapping of Scotland published in 1817, Daniell's later paintings of the cave, in his *A Voyage Round Great Britain Undertaken in the Summer of 1813*, strips his earlier depiction of its exaggeration and hyperbole, embracing a factual geological description "in plain and direct language" by placing the subject in the habitat as it stands, giving it "a more just and reasonable standard":

for by placing the different subjects apparently in situations and under circumstances where they are usually seen in nature, a new interest is communicated even to familiar objects, and an air of truth given to all, much more impressive than without such local accompaniments.[68]

Art had definitely gained a new idiom of accuracy, and by the middle of the nineteenth century, topographical paintings had become as much a corrective and reformist medium of spatial representation as any in science.

Gazing out: the empire as spectacle

The public consumption of sceneries and landscape paintings had by now acquired a theatrical status. And like the seventeenth-century masques, they reappeared once again on stage to entertain audiences, only this time equipped with much-advanced technologies of stage craft. By the late seventeenth century the stage hands usually had a painted canvas dropped from above, perhaps by rollers, similar to a curtain. Numerous examples attest to the attention given to "scenes" and the means by which they were made impressive. When Pepys visited Drury Lane while it was closed due to the plague, 1664–5, he talked of the paintings as being "very pretty", and Prince Cosmo III in his visit to the London theatres when travelling in England said, "The scenery is very light, capable of a great many changes, and embellished with beautiful landscapes."[69] In the early eighteenth century what had emerged were techniques of painting angular asymmetrical perspective. Bibiena's innovation of a flexible scheme was introduced on the London stage, resulting in the impression of diagonal placement of scenic architecture, opening the stage up to new loftiness and vastness of space.

There was a growing tendency toward elaborate scenes and props depicting specific places. On 22 May 1736, Francis Hayman, the designer and painter at Drury Lane, brought out "A new Entertainment after the Manner of Spring Garden, Vauxhall with a new scene representing the Place."[70] According to Sconten, during this time:

> in addition to spectacular eye-appeal was added the pleasure of recognition. The theatre has always been the home of illusion, so the scene painter's fancy continued to give local habitation on canvas to imagined scenes from the poet's descriptions.[71]

By the mid-eighteenth century, the nature of the places depicted in theatre scenes changed. In place of imaginary or universal landscapes these now demanded precision and accuracy. A vogue developed of sending stage painters to specific locations to sketch materials for stage scenery. For example, Colman sent scenery painters Richards and Dahl to Stratford-upon-Avon to sketch the amphitheatre there in 1769 for using in the plays *Jubilee* and *Harlequin's Jubilee*, and Dahl was sent on a trip to Windsor in 1771 to sketch scenes to be used in the *Fairy Prince*. Sheridan allowed de Loutherbourg £35 to travel to Kent and Derbyshire to make sketches for scenes for *The Camp* (1778) and *The Wonders of Derbyshire* (Drury Lane, 8 January 1779), which was a pantomime entirely built round scenic views of that county. This also initiated the culture of employing a number of royal academician landscape artists for the painting of stage scenes.[72] In the eighteenth century, actors were said to perform in front of the scenery and not within it. The box set, complete with three walls and a ceiling, did not come into use until the early 1830s. The wings were placed in the groove that, in each of the sets, lay furthest downstage. The number of the sets of grooves corresponded to the number of the wings which, in Drury Lane and Covent Garden, were usually ten, i.e. five on each side.[73] Wing after wing were laid in diminishing perspective until the painted scene was seen at the very back of the stage.

Attempts at rendering the scenery more realistic, at heightening the illusion, were constantly being made, notably by the most eminent scene designer of the late eighteenth century, Philippe James de Loutherbourg. He was the first artist to effectively use cut-out scenes. In *Omai* (1785), a play whose stage craft will be dealt with in detail later, a single view of the frozen ocean was said to contain 42 separate scenes. He painted a great many transparencies by means of which he was able to achieve a three-dimensional picture. Here he broke away from the standard format of Restoration stage. The illusion was also maintained by the fact that the floor of the stage in all the theatres was raked, the slope upwards from the footlights towards the rear, sometimes being fairly steep. John Kemble undertook the actual construction of buildings, with towers, battlements, drawbridges and archways on stage. These scenes were placed not frontally but at an angle on the stage creating a neat perspective. William Capon, working on improvement in the same principle, insisted upon exact archaeological detail.[74]

As the nineteenth century progressed there evolved a very different conception of the stage: the stage space was not simply decorated but also shaped.
Specially painted scenery, upholstery, built-out scenery and extensive use of
objects, combined with improved lighting techniques, radically transformed the
stage. Acts, which were increasingly set in foreign, usually eastern, lands cashed
in on exoticism while the staged spaces celebrated their new-found materiality
under the criteria of "visual truth":

> The physical stage was now asked to create place – or, rather, to
> recreate place – to a degree that was never before expected. The stage
> became the *in situ* display, the perfect reproduction that stood in
> metonymic relation to an Eastern totality. The theatre's re(created)
> geography developed in tandem with its new sense of the physical
> possibilities of the stage space.[75]

Certain geographies developed as imagined space existing outside the common
field of vision. These were the geographies which, though not seen, were being
fervently mapped at this time. Though this process transpired at various points
in the society, for a sizeable number of Britain's population, these new spaces
existed on the stage and other forms of spectacles. Britain's developing conception of the exotic "new world" was inseparable from technological changes
in the processes of reproduction which were employed in theatres and other
places of popular entertainment.

Sybil Rosenfeld refers to the same taste, which was catered for in the
panoramas, being mirrored on the stage.[76] It was de Loutherbourg who really
established and popularised topographical scenery through theatrical practice.
He exploited the emerging excitement with the "new world" and for the first
time introduced it on a grand scale on the stage. De Loutherbourg's and John
O'Keefe's enormously successful pantomime *Omai: or A Trip round the World*,
performed in 1785, provides an example of representation of colonial space and
colonised peoples. (Fig. 3.5) *Omai* has been read by scholars as:

> a translation into entertainment of ethnographic moments in which the
> European strangers confronted the otherness of the Pacific island
> natives, tried to describe that otherness and in that description possess
> them.[77]

It can be said that de Loutherbourg's experiments with exotic settings reached
their culmination in *Omai*. Long before *Omai*, faced with the challenge of
representing an Egyptian setting for *Sethona* (Drury Lane, 1774), de Loutherbourg consulted Montfaucon's voyages and F.L. Norden's *Travels in Egypt and
Nubia* (1757).[78] *Omai*, which was first performed in Drury Lane, was described
in one opening-night review "as the stage edition of Captain Cook's voyage to
Otaheite (Tahiti), Kamchatka, the Friendly Islands [Tonga] &c, &c."[79] The

lavish costume and stage design were scrupulously overseen by John Webber, Captain Cook's chief illustrator on his third voyage. Webber's involvement was required to produce an effect of ethnographic authenticity. According to O'Keefe's *Recollections*,[80] de Loutherbourg also took his designs from the prints and drawings by William Hodges, who had also accompanied Captain Cook on his explorations.[81] Loutherbourg's costumes, props and stage design were also influenced by Ashton Lever, who owned an extensive collection of souvenirs and antiquities from Cook's voyage.[82] Speaking of the pantomime's closing spectacle, a reviewer comments:

> A procession of the natives of different islands and other places visited by Captain Cooke is here introduced. The music preserves the characteristic airs of different people in the procession as much as science can approach barbarity. The APOTHEOSIS of Captain Cooke closes this most admirable assemblage of curious views.[83]

However, reviewers have commented about the superiority of the paintings themselves in comparison to the pantomime, itself, in which they were presented:

> Such a picture – in point of all that constitutes the sublime of the art – the drawing and disposition of the figure – the well expressed countenance … such a picture will immortalise the author as the subject of it – and were there no other merit in the pantomime would hold forth the attractions of an EXHIBITION in itself.[84]

As the century progressed, more localised scenery was shown and was accompanied by the demand for greater accuracy of representation. However, it was not always necessary and neither was it feasible to send scene painters to the spot in faraway places, specially to the East which was increasingly gaining prominence and popularity over the banal English castles and forts or European scenes such as the Pantheon, Marseilles or Venice. In such cases pictures made for other purposes were copied for stage scenery. As in the case of *Omai*, such imitations were the only feasible method when performances were based on the Orient, India or the South Seas. This expansion of the appeal of landscape imagery forms part of an ongoing process of the imperial agenda in which, as Captain Cook's first biographer, Andrew Kippis, put it in 1788, "new spaces (are opened up) for a poetical fancy to range in".[85] According to Roskill, this process persisted until the territory in question (in the above case, Tahiti) was finally occupied and its aesthetic and "poetical fancy" subordinated under commercial and military interests.[86] This trajectory is, of course, recharted and duplicated in multiple places across the globe. There were numerous productions charting the planet with exotic sceneries and performances straddling the Mughal Darbar, Chinese lanterns, Siamese ballet, grand harbours or "romantic views on the

Figure 3.5 Stage scenery from *Omai* (1785) by Philippe de Loutherbourg, inspired by
Captain Cook's journeys to the South Seas
Theatre and Performance Collection, Victoria & Albert Museum

borders of Chinese Tartary" or "an extensive view of the River Ganges" in the
same breath.[87] Theatre's relationship to colonialism emerged over time due to
theatre's inherent locality and spatiality. By such means, the oriental and the
exotic were presented to an audience whose tastes were becoming more and more
antiquarian and romanticised. Progressively, in the nineteenth century the world
existed as a picture or an exhibition to see and possess. This triggered a nuanced

126

and dual process of representing not only the ever-expanding British Empire but also the effects of it through theatre's propagandist claims that it projected a real place and that the representation itself, was authentic and real.

The preceding exposition prepares a groundwork through which we can now view Indian landscapes in British art. The study of the development of social, cultural and intellectual (scientific and mathematical) climate which conditioned the response to nature and landscape in Britain itself will now help understanding of the distinctive way of "seeing" Indian landscape, and thereby of possessing it and shaping it. Seeing is transformed into an active intervention, for, through seeing, land is "discovered" and won. We have already explored the ways in which representations of landscapes in visual culture, along with cartographic activities, constructed the idea of a unified Great Britain as a nation commanded by a single head and later the Commonwealth. It was premised on visual mastery over a given space. The same paradigm could in the eighteenth and nineteenth centuries be re-implemented to express a unified territory, (which Edward Ziter calls "theatrical geography")[88] which existed primarily in the colonial consciousness as part and property of the British Empire.

I shall look at the reception of Indian-Oriental landscapes in the sphere of public entertainment and popular culture in the following chapter which I also reserve for the study of landscape art as practised by the British in India. Indic themes, motifs and ornamentalism have existed in European visual culture ever since the two regions came into contact. Travelogues of visitors to India already existed in Elizabethan and Jacobean England. However, India in landscapes was a much later phenomenon. It emerged at around the same time as continental landscapes thrived in the eighteenth and nineteenth centuries with a rampant touristic impulse propelling it. From the late eighteenth century onwards, India existed in the imperial metropolis not "in" but increasingly "as" spectacle.

Notes

1 Focillon, Henri. *The Life of Forms*. New York: Zone Books, 1992. p. 102.
2 Cosgrove, Denis. "Prospect, Perspective and the Evolution of the Landscape Idea". *Transactions of the Institute of British Geographers*, vol. 10, no. 1, 1985. pp. 45–62.
3 Olwig, Kenneth Robert. *Landscape, Nature and the Body Politic*. Wisconsin, University of Wisconsin Press, 2002.
4 Benjamin, Roger Harold. "The Decorative Landscape, Fauvism, and the Arabesque of Observation". *The Art Bulletin*, Vol. 75, No. 2. (June, 1993). p. 296.
5 Olwig. p. 23. *Sachsenspiegel*, or *Saxon Mirror* was the most famous written body of Germanic customary law. Spiegel is the Germanic word for mirror and is also related to the English word spectacle, showing the interlinkage of the concepts of landscape and spectacle.
6 Peacham, Henry. *The Complete Gentleman*. (1622). Quoted in Olwig. *Sachsenspiegel*. p. 25.
7 Benjamin, Roger. "The Decorative Landscape: The Fauvism and Arabesque in Observation". p. 296.

8 Harvey. David. *The Condition of Postmodernity: An Enquiry into the Origins of Cultural Change*. Oxford: Basil Blackwell. 1989. p. 246.

9 The romantic idealism of the country and nature later resurface in the philosophy of the Romanticists such as Jacques Rene Rousseau whose idea of the "noble savage" reiterated this harmonious union with nature to regain human beings' innate goodness.

10 Serlio, Sebastio. *Tutte l'Opera d' Architettura et Prospectivadi Sebastatediano Serlio Bolognese* (Sebastio Serlio on Architecture) Books 1–5, Venice, 1555.

11 Cosgrove, Denis. "Landscape and landschaft." *German Historical Institute Bulletin* 35. Fall (2004): 57–71.

12 Olwig. *Sachsenspiegel*. p. 61.

13 Jonson, Ben. *The Complete Masques*, ed. Stephen Orgel. New Haven: Yale University Press, 1969. p. 510.

14 Traub, Valerie. "Mapping the Global Body". In Peter Erickson and Clarke Hulse, eds. *Early Modern Visual Culture: Representation, Race, Empire in Renaissance England*. Philadelphia: University of Pennsylvania Press. 2000. pp. 44–97. Ramaswamy, Sumathi. "Maps and Mother Goddesses in Modern India. *Imago Mundi*. Vol. 53. 2001. pp. 97–114.

15 Olwig. *Sachsenspiegel*. p. 85.

16 Olwig. *Sachsenspiegel*. p. 85.

17 Panofsky, Erwin. *Perspective as Symbolic Form*. New York: Zone Books, 1991.

18 For an extensive exposition of the philosophical and intellectual traditions in the evolution of perspective and perspectival science see Kemp, Martin. *The Science of Art: Optical themes in western art from Brunelleschi to Seurat*. New Haven: Yale University Press, 1990.

19 De Bolla, Peter. The *Education of the Eye: Painting, Landscape, and Architecture in Eighteenth-Century Britain*. Stanford University Press, 2003.

20 Harley, J.B. "The re-mapping of England, 1750–1800". *Imago Mundi*. 19. 1965. pp. 56–123.

21 Daniels, Stephen. "Re-visioning Britain: Mapping and Landscape Painting, 1750–1820". Katherine Baetjer ed. *Glorious Nature: British Landscape Painting 1750–1820*. London: Zwemmer, 1994. p. 61.

22 Williamson, Tom and Liz Bellamy. *Property and Landscape: A Social History of Land Ownership and the English Countryside*. London: George Philips. 1987. p. 116; Daniels, Stephen. "The Culture of Cartography". In Geoffrey Cubbit ed. *Imagining Nations*. Manchester: Manchester University Press, 1998. p. 120.

23 Bermingham, Ann. *Landscape and ideology: The English rustic tradition, 1740–1860*. California: University of California Press, 1989. Hunt, John Dixon and Peter Willis, eds. *The genius of the place: The English landscape garden, 1620–1820*. "Introduction". Cambridge, Massachusetts: MIT Press, 1988. pp. 1–45.

24 Barker, Hugh. *Hedge Britannia: a curious history of a British obsession*. A&C Black, 2012.

25 Blomley, Nicholas. "Making private property: enclosure, common right and the work of hedges." *Rural History* 18.1 (2007): 1–21.

26 Harvey, David C. "Ambiguities of the hedge: an exercise in creative pleaching–of moments, memories and meanings." *Landscape History* 38.2 (2017): 109–27.

27 Williamson, Tom and Liz Bellamy. *Property and Landscape*. p. 143.

28 Ibid. p. 144–5.

29 See Rosenthal, Michael, Christiana Payne and Scott Wilcox, eds. *Prospects for the nation: recent essays in British landscape, 1750–1880*. Yale University Press, 1997; Klonk, Charlotte. *Science and the perception of nature: British landscape art in the late eighteenth and early nineteenth centuries*. New Haven: Yale University Press,

1996. Broglio, Ron. *Technologies of the Picturesque: British Art, Poetry, and Instruments, 1750–1830*. Lewisburg: Bucknell University Press, 2008.

30 Whyte, Ian D. *Landscape and History since 1500*. Reaktion Books, 2004. p. 51, 53, 79. Rees, Ronald. "Historical Links between Cartography and Art." *Geographical Review*, Vol. 70, no. 1, 1980, pp. 61–78.

31 Repton, H. "Sketches and Hints on Landscape Gardening"(1795). In *The Landscape Gardening and Landscape Architecture of the late Humphry Repton, Esq.* ed. Quoted in Hemingway, Andrew. *Landscape Imagery and Urban Culture in early 19th. Century Britain*. Cambridge: Cambridge University Press, 1992. p. 40.

32 See https://en.wikipedia.org/wiki/Mr_and_Mrs_Andrews#/media/File:Thomas_Gainsborough_-_Mr_and_Mrs_Andrews.jpg

33 Barrell, John. *The dark side of the landscape: the rural poor in English painting, 1730–1840*. Cambridge: Cambridge University Press, 1980; Egerton, Judy. "National Gallery Catalogues: The British School". 1998: 218. pp. 80–6; Berger, John. *Ways of seeing*. UK: Penguin, 2008. p. 84.

34 Many "Northern" or Dutch painters were also invited to draw the country houses and estates of their English patrons according to conventions of accurate detailing.

35 Ackermann, Rudolph. (?) *Repository of Arts, Literature, Commerce, Manufactures, Fashions and Politics* 9, 1813. Quoted in Standring, Timothy J. "Watercolor Landscape Sketching during the Popular Picturesque era in Britain" in Baetjer, Katherine ed. *Glorious Nature; British Landscape Painting 1750–1850*. London: Zwemmer, 1994. p.73.

36 The idea of the picturesque was introduced later to the Burkean notions of the beautiful and sublime in landscape as an Anglicisation of the French *pittoresque* or Italian *pittoresco*.

37 Klonk, Charlotte. *Science and the perception of nature: British landscape art in the late eighteenth and early nineteenth centuries*. Vol. 57. New Haven: Yale University Press, 1996. p. 27.

38 Interestingly, Canaletto also visited England in the 1740s.

39 The lenses when held up to the eye showed a reduced image of the entire landscape which provided the artist with a ready, suitably composed scene. The Claude Glass is a round or oblong convex mirror, four inches broad, backed with dark foil and bound in leather like a little pocket book. This eighteenth- and nineteenth-century popular device could be held up to reflect the view and the dark foil gave greater depth to the shades and lowered the tone. The resulting diminished picture looked like one by Claude, as Gilpin noted: "It gives the object of nature a soft mellow tinge, like the colouring of that great master." The camera obscura is a box-like device open on one side. On the top were bellows with a convex lens and an adjustable plain mirror in which an image of the landscape could be seen as in the view finder of the camera, today. A sheet of paper was laid on the base, and by optical refraction the image was thrown onto it. The camera lucida was a long stalk fitted at the end with a prism and a group of lenses fixed at eye level, reflecting the landscape below. The artist could see both the image and the pencil point at the same time when drawing complicated subjects.

40 Gombrich, E.H. *Art and Illusion: A Study in the Psychology of Pictorial Representation*. London: Phaidon, 1962. p. 12.

41 Gombrich. *Art and Illusion*. p. 12.

42 Scottish Enlightenment philosophers such as Archibald Alison's (1757–1839) *Essays on Nature and Principles of Taste* (1790) and Dugald Stewart's (1753–1828) various works on the arts and sciences had a marked influence on the perception of natural phenomena.

43 Klonk, Charlotte. *Science and the perception of nature: British landscape art in the late eighteenth and early nineteenth centuries*. Vol. 57. New Haven: Yale University Press, 1996. p. 37.

44 Fuseli, Henry. "Lecture IV – On Inventio, Part II", 1804, in John Knowles ed. *The Life and Writings of J.H. Fuseli Esq., MA, RA*, Vol. II, London: Colburn and Bentley, 1831, p. 217.

45 Daniels, Stephen. "Revisioning Britain". p. 62.

46 Daniels, Stephen. "Revisioning Britain". p. 67–9.

47 Interestingly, Robert Dawson also taught drawing and "surveying in the field" to cadets in the Royal Military College at Woolwich, among whom was George Everest, who later became the surveyor general of India. He also taught and inspired other cartographers posted in different English colonies over the world, such as Robert Hoddle, posted in Plymouth in Australia. See Colville, Berres and Hoddle. "Robert Hoddle: Pioneer Surveyor, 1794–1881." *The Globe*. 57, 2005. pp. 17–26. p. 17.

48 Gombrich, E.H. *Art and Illusion: A Study in the Psychology of Pictorial Representation*. London: Phaidon, 1962. p. 29.

49 Beckett. R.B. ed. *John Constable's Discourses*. Ipswich: Suffolk Records Society, 1970, p. 69. Quoted in Hemingway, Andrew. *Landscape Imagery and Urban Culture in Early Nineteenth Century Britain*. Cambridge: Cambridge University Press, 1992. p. 15.

50 Paris, H.J. "English Water-colour Painters". In W.J. Turner ed. *Aspects of British Art*. London: Collins, 1947.

51 Kemp, Martin. *The Science of Art*. New Haven: Yale University Press, 1990.

52 Smiles, Sam. *Eye Witness: Artists and Visual Documentation in Britain 1770–1830*. Great Britain: Ashgate, 2000. p. 15.

53 Smiles, Sam. p. 59.

54 *Quarterly Review*, Vol XXV, April and July 1821, p. 113. Quoted in Smiles, Sam. *Eye Witness*. p. 57.

55 Smiles, Sam. *Eye Witness*. p. 59.

56 One of the earliest engravings of Tintern was by Samuel and Nathaniel Buck in their series on antiquities in England and Wales, popularly called *Buck's Antiquities* in 1732. Following Gilpin's 1770 tour and the publication of his *Observations on the River Wye*, Tintern became a popular tourist site with visitors from spas in Bristol and Bath taking boats upriver from Chepstow and later with Wye tourists following in Gilpin's trail on a two-day trip from Ross. See https://www.walesonline.co.uk/whats-on/arts-culture-news/famous-paintings-inspired-tintern-abbey-7147126, sourced on 13.07.2018.

57 See Fuseli, Henry. "Lecture IV – On Invention, Part II", 1804, in John Knowles ed. *The Life and Writings of J.H. Fuseli Esq., MA, RA*, Vol. II of 3, London: Colburn and Bentley, 1831, p. 217.

58 Sandby, Paul. *Virtuosi's Museum Containing Select Views in England, Scotland and Ireland*. London: George Kearsley, 1778–81.

59 See Wyld, Helen. "Re-Framing Britain's Past: Paul Sandby and the picturesque tour of Scotland". *The British Art Journal*, Vol. XII, No. 1. pp. 29–36.

60 Gilpin, William. Observations on the Highlands of Scotland. London: R. Blamire, 1789. p. 135.

61 After the successful subordination of the Highland forces at the Battle of Culloden, the Highland regiments which were inducted into the British military forces went on to earn repute for their distinguished service and being remarkably loyal. The figures were not present earlier when Sandby produced the draughts, but were added later when they came out in *The Virtuosi's Museum*. See Sam Smiles for an elaborate discussion on this.

62 Gilpin, William. *Observations relative chiefly to picturesque beauty, made in the year 1776: on several parts of Great Britain; particularly the Highlands of Scotland*. Vol. 2. London: R. Blamire. 1789. p. 122.

63 Quoted in Klonk. *Science and the perception of nature.* p. 73.

64 Ibid. p. 74.

65 Gilpin, William. *Observations relative chiefly to picturesque beauty, made in the year 1776: on several parts of Great Britain; particularly the Highlands of Scotland.* Vol. 2. London: R. Blamire. 1789. p. 156.

66 According to ancient Gaelic and Celtic legends and folklore, Fionn MacCumhail, or Fingal in James MacPherson's Ossianic poetry, was an Irish warrior giant who went to war with a Scottish giant. Fionn built a causeway between Antrim in Northern Ireland and Staffa in Scotland, as a war strategy, the two ends of which still remain.

67 Anon. "Curious Account of the Island of Staffa (one of the Hebrides) communicated to Mr. Pennant, by Joseph Banks, Esq." *The Annual Register or a View of the History, Politics and Literature for the year 1774.* London: J. Dodsley, 1775. p. 89; See Klonk. *Science and the perception of nature.* p. 74. Also see Gordon, John E. "Rediscovering a sense of wonder: geoheritage, geotourism and cultural landscape experiences." *Geoheritage* 4.1–2 (2012): 65–77. p. 69.

68 Ayton, Richard and William Daniell. *A Voyage Round Great Britain undertaken between the years 1813 and 1823,* and *Commencing from the Land' End, Cornwall, with a series of views.* Vol. 3 (8 Vols), London: The Tate Gallery, 1978. pp. 37–8. See also Klonk. *Science and the perception of nature.* p. 69.

69 Avery, I. Emmett and Arthur H. Scouten. *The London Stage 1600–1700: A Critical Introduction.* Carbondale: Southern Illinois University Press, 1968. p. lxxxvii.

70 *London Daily Post and General Advertiser.* Quoted in Scouten, Arthur H. *The London Stage 1729–1749: A Critical Introduction.* Carbondale: Southern Illinois University Press, 1968. p. cxxii.

71 Sconten, Arthur H. *The London Stage 1729–1749.* p. cxxii.

72 Stone Jr, George Winchester. *The London Stage 1747–1776; A Critical Introduction.* Carbondale; Southern Illinois University Press, 1968. p. cxix.

73 Hogan, Charles Beecher. *The London Stage 1776–1800: A Critical Introduction.* Carbondale: Southern Illinois University Press, 1968. p. lix-lxiii.

74 Hogan, Charles Beecher. *The London Stage.* p. lxiv.

75 Ziter, Edward. *The Invention of the Middle East in British Scene Painting and Mise en Scene: 1798–1853.* (Ph.D. Thesis, University of California, 1997).

76 Rosenfeld, Sybil. *Georgian Scene Painters and Scene Painting.* Cambridge: Cambridge University Press, 1981. p.33.

77 Quoted in O'Quinn, Daniel. *Staging Governance: Theatrical Imperialism in London, 1770–1800.* Baltimore: Johns Hopkins University, 2005. p. 74.

78 This is an early example of the taste for Egyptian backgrounds which were rampant later. De Loutherbourg's scene of the legendary Ninus's tomb was inaccurate and faulty as it depicted a pyramid for the tomb. The same scenery, however, came in handy later for the performance of *Semiramis* (Drury Lane, 1776).

79 Quoted in O'Quinn. *Staging Governance.* p. 74.

80 O'Keeffe, J. *Recollections of the life of John O'Keeffe, written by himself.* Vol. 2. London. p. 14.

81 Rosenfeld, Sybil. *Georgian Scene Painters and Scene Painting.* Cambridge: Cambridge University Press, 1981. p. 34.

82 Sir Ashton Lever's Museum, which was otherwise known as the Holophusicon, had an extensive collection of South Sea materials from the Cook expeditions. These were on display next door to Loutherbourg's workshop. Loutherbourg almost certainly used various objects, vestments and headdresses from the collection as models for his design.

83 *Town and Country* (December 1785). Quoted in O'Quinn. *Staging Governance.* p. 75.

84 *Public Advertiser* (24 December 1785). Quoted in O'Quinn. *Staging Governance* p. 83–4.

85 Kippis, Andrew. *A Narrative of the voyages around the world, performed by capt. James Cook*. Ch. VII. (1788) London: Bickers and Son, 1878. p. 510.

86 Roskill, Mark. *The Language of Landscape*. Pennsylvania: Pennsylvania State University Press, 1997. p. 104.

87 A number of playbills in the theatre holdings of the Victoria and Albert Museum advertise spectacular sceneries and tableaux as part of the main performance. Some examples of these are Elizabeth Inchbald's *A Mogul Tale* performed in 1784 at the Theatre Royal, Haymarket; *Chinese Sorcerer, Or The Emperor and His Three Sons* performed in 1823 at the Theatre Royal, Drury Lane, advertising "a Grand Chinese Ballet"; *Burmese War, Or Our Victories in the East* performed in 1826 at the Royal Amphitheatre, Astley's, Westminster Bridge which included "a grand Siamese Ballet and an extensive view of the River Ganges and a funeral ceremony of a burning Hindoo widow".

88 Ziter, Edward. Ziter, Edward, *The Invention of the Middle East in British Scene Painting and Mise en Scene: 1798–1853*. (Ph.D. Thesis, University of California, 1997).

4

FRAMING INDIA
Chinnery and D'Oyly

Whether the Painter – fashioning a work
To Nature's circumambient scenery,
And with his greedy pencil taking in
A whole horizon on all sides, with power,
The Prelude: Or Growth of a Poet's Mind
William Wordsworth[1]

The spatial rationality which emerged in Britain over centuries since the Renaissance, framed not only the British nation as a coherent geographical unit (as seen in the last chapter), but was significantly connected with the extension of its boundary overseas. It is no mere coincidence that the great age of spatial activities in the European continent was also the great age of exploration and imperialism merging with images of imperialist celebration. As these new yet unexplored spaces opened up to European vision they were also incorporated into the idiom of the picturesque which included temples, ruins, rivers and mountains, finding their expression in varied media. A stage, a theatre as it were, was created for unfolding the colonial trajectory and visual appropriation of spaces in the great saga of imperialist victory. The geographies of the British Empire found expression in the geographical nature of the discourse of the public sphere. Historians of art have illustrated how important certain new spaces were for the display of visual culture and the constitution of the viewing public. Significantly, these were the public and private spaces of reception of a scientific geographical consciousness successfully transmitted from their actual sites of production – the colonies.[2] Many public buildings in Britain were inscribed with designs and souvenirs of the East and a common recurring figure in all these was that of Britannia receiving gifts from various continents as tokens of submission. On the other hand, pleasure gardens and places of entertainment, such as the Vauxhall dinner boxes, came to be painted with fantastic orientalist images. In the eighteenth century, Orientalism was a key component in the context of the spaces to which were attached significations of play and pleasure, often clubbed into a schema which sported Chinese, Turkish and Persian scenes and objects. And what better way to look at this theatricality of colonial performance if not the theatrical space itself which contained within it all the undertones of the illicit, profane and the pleasure principle!

In this chapter, I shall first chalk out the visual practices pertaining to the East as played out in Great Britain and thus the resultant cumulative visualisation of India which such practices gave rise to. In the latter segment, I will talk about the *in situ* production by two lesser-known landscape artists and try to locate their exercise within the context of creation of a visual compendium on India. Therefore, I shall progress from the urban appropriation, fascination and popularity of Eastern landscape scenes in Great Britain to their ergological space of actual production in the suburbs of a colony.

Theatrical empire: Indian landscapes in stage scenery

Continuing from where I left off in the preceding chapter, the theatrical space in the late eighteenth and nineteenth centuries, as already seen, was not merely metaphoric but a metonymic expression of a real place. From the late eighteenth century onwards, India became a favourite in the arena of spectacle. As Natasha Eaton points out:

> Although a certain cohesion is said to exist between governmental techniques [in] the colony and the metropole ... the colonial state became far more spectacular than its contemporary.[3]

Eaton refers to the Georgian state in the first half of the eighteenth century being obsessed with sensationalism, voyeurism and vulgarity, where public hangings, masquerades and exhibitions shared equal exposure and popularity. Within this spectrum, the colonial state soon overtook them in "intimidating theatricality", with its Hindu rites and sensational practices, characterising a superstitious and primitive India.[4]

J.S. Bratton mentions the hundreds of spectacular melodramas on military or imperialistic themes that were abundant at this time, with soldiers, sailors or adventurers as their heroes, ending on a patriotic note in the tune of "Rule Britannia".[5] The simulacrum of the stage began to be used for colonialist panegyric.[6] To match the abundance of colonialist themes, theatre scenes bustled with exotic paintings in order to depict the place where the action was to be set. With the popularity of "on the spot" paintings, the works of traveller-painters were frequently consulted. To match the prolific upsurge of paintings of exotic locales, there also emerged special props, pageants and performances recreated from the eye-witness accounts of travellers.

A typical example is James Cobb's *Rama Droog or Wine does Wonders*, first performed in the Covent Garden on 12 November 1798. Comprising both European and many Indian characters, the locale of the drama is India, the scenes being based on drawings made in India by Thomas Daniell.[7] The *Morning Herald* of the following day reviewed the scenes of the play in these words:

A view in the fortress of Ramah Droog, the British captives on one side, the walls of the Palace gardens on the other – a distant view of the hill fort of Ramah Droog – an apartment in the Rajah's Palace, the women of the Zenana dancing and singing – the battlements on the rock – an apartment in the palace – a private apartment belonging to the Vizier – a wood near the Pattah, or town, at the foot of the rock – Zelma's prison – and the outside of a fort.[8]

The playbill of the play refers to a grand procession at the end of Act II of the main piece:

A Return from a Tiger Hunt, to the Rajah's Palace, representing the Rajah on an Elephant, returning from Hunting the Tiger, preceded by his Hircarrahs, or military messengers, and his State Palanquin – the Vizier on another Elephant – the Princess in a Gaurie, drawn by Buffaloes – the Rajah is attended by his Fakeer, or Soothsayer, his Officers of State, and by an Ambassador from Tippoo Sultaun in a Palanquin; also by Nairs (or Soldiers from the South of India), Poligars (or Inhabitants of the Hilly Districts), with their Hunting-dogs, other Indians carrying a dead Tiger, and young Tigers in a Cage; a number of Sepoys – Musicians on Camels and on Foot – Dancing Girls etc.

The procession was not an integral part of the narrative or the plot of the play, but it was of course a necessary measure for creating the spectacle which functioned as a visual register for the alien land, people and culture. A few years earlier, in 1788, another of Cobb's plays, *Love in the East; or Adventures of Twelve Hours* boasted of a variety of new scenery, one of them being, "a view of Calcutta, from a painting done on the spot by Hodges" which opened the scene. While the painting in question could not be identified, probably it is Hodges' famous painting, "View of Calcutta", a boldly painted sketch depicting the Calcutta river bank at sunset (Figure 4.1). In this picture, a version of which now hangs in the Victoria Memorial Hall, Hodges' palette ranges from warm rose tones to yellow and orange in the sunset, clouds and reflections in the water.[9] On the other hand, a likely candidate for the same could also be another of the popular paintings by Hodges titled *A View of Calcutta Taken from Fort William* which was exhibited in the Royal Academy in 1787. In his *Travels*, Hodges claims that the engraved *View of Calcutta from Fort William* was taken from his picture which was painted on the spot. When the painting was exhibited at the Royal Academy, it received the following review:

We learn from those who know the situation, that this is a faithful delineation. Mr. Hodges possesses much of Canaleti's [sic] stile, and though he may fail in his exactness, he has infinitely more variety.[10]

Figure 4.1 View of Calcutta by William Hodges. Many replications of this image are to be found with little variations. One such was probably used as a scenery for Cobb's play, *Love in the East; Or Adventures of Twelve Hours* (1788) performed at the Theatre Royal, Drury Lane

Note: This oil on canvas, exhibited in the Victoria Memorial (R2856), is captioned *View of the Esplanade, Calcutta, from Garden Reach*, circa 1785, and is attributed to Thomas Daniell.

By kind permission of the Trustees of Victoria Memorial Hall, Kolkata

Similar, at least from the spectacle point of view, was Mariana Starke's *The Widow of Malabar*, based on *La Veuve du Malabar* by Antoine Marin Le Mierre. It was first performed at Covent Garden on 5 May 1790. The playbill advertised "a Procession representing the Ceremonies attending the Sacrifice of an Indian Woman on the Funeral Pile of her deceased Husband."[11] The above, which was repeated in all subsequent performances, was an enactment of a set piece. It was so popular that the scenery and pageant were included in other plays as well. With reference to the spectacle of the procession one can possibly draw connection with one of William Hodges' paintings titled *Procession of a Hindoo Woman to the Funeral Pile of Her Husband* which was engraved in England by W. Skelton. In his *Travels in India*, Hodges stated that he was so moved by the ceremonial death of a Hindu woman in Benares that it was some time before he was able to make a drawing of the subject.[12] (Of course, Sati was a familiar and identifiable motif within the iconography related with India.) Daniel O'Quinn discusses about the political problematics in many of the plays related to the British Empire of the age, such as the aforementioned one (in this case Sati) which, instead of being sensitively handled as matters of serious import are under subterfuge behind the spectacle of the drama, was overshadowed by dramatic excesses.[13] In the imperial allegory, tangible economic, political and social problems in the colonies received distanced fantastic solutions. Nevertheless, they could not hide the imperial anxiety to control, regulate and rationalise colonial public and private spaces.

Just as Sati formed a pageantry for entertainment in Starke's *Widow of Malabar*, William Thomas Moncrieff's *The Cataract of the Ganges* (1823) shows a political problematic of female infanticide but manages to subsume it under the artifice of spectacular grandeur and imperial fantasy. *Bell's Life in London*, 2 November 1823, stated that *The Cataract* "is altogether a dashing, splashing, kicking, prancing, raree show". Quinn talks of its real objectives being "in the realm of scene painting, hippodramatic spectacle and the management of vast numbers of "embrowned" actors and actresses in all manner of processions".[14] (Figure 4.2) The reviews of the play concentrate only on the visual quality of the play. *The New Times* of 28 October 1823 is typical in this regard:

> To the scenery, show and music, the Manager has looked for triumph, and to these we will turn our attention. The opening scene is beautiful: it is by far the handsomest scene in the whole piece, and does the painter, Stanfield, infinite credit. It is a field of battle by moonlight, viewed after a conflict. There are a number of figures in the foreground, and distributed over the stage, which are grouped with admirable effect ...

Yet another review from *The Statesman* of 28 October 1823 justifies the extravagance in representing a scene of Rajput marriage through a procession:

> The first act closes with a grand procession of an immense number of soldiers and females, accompanied with bands of music. They rise from a subterraneous entrance, under the gates of a fortress: and though some architectural objections might be taken, the stage effect is grand in the extreme. After the infantry have arranged themselves, the cavalry appear; and are followed by a triumphal car of great dimensions, and drawn by six horses, richly caparisoned. The skill with which the horses are managed almost exceed belief, and no stage ever presented so imposing an appearance.

Clarkson Stanfield, the stage designer of the play, had been apprenticed to a coach decorator in 1806 and his experience is reflected in the expertise with which he manoeuvred the above-mentioned scene. It is interesting to note that Stanfield had been a marine painter employed in the service of Royal Navy for a while in 1808. His voyage to China and India in 1814, from which he returned with a booty of "on the spot" paintings promptly earned him a career as a decorator and scene painter at the Royalty Theatre at Wellclose Square in London, before he finally joined as resident scene-painter at the prestigious Drury Lane in 1823.

Figure 4.2 Poster by Montague Chatterton and Co. advertising *Cataract of the Ganges,*
or, The Rajah's Daughter at the Theatre Royal, Drury Lane (1873)
Image courtesy: Theatre and Performance Collection, Victoria and Albert Museum

Though the opening picturesque battle scene generated instant applause, the stage scenery was replete with numerous other scenes of Indian architecture and landscapes like the views of Rajah's palace, a Hindu temple, "the Pagoda of Juggernaut", along with representation of private spaces and inner quarters of family life.[15] The tradition of such kinds of stage craft was usually that of closing with the grandest of scenes. The final scene was deemed the most important in narrativising the spectacular development and culmination of the play unto a closure. By the nineteenth century, all kinds of artificial stage mechanisms and new technological innovations were used to create the much-desired *mise-en-scene*. The culminating scene of *The Cataract* was replete with special effects which utterly subordinated the actors and elicitation of emotion to the spectacular effects of the play:

> The change which presented the (final scene) was striking, it was from the "Wood of Sacrifice" to a view of the Ganges, rushing with all its might down a prodigious cataract. The water was real, and it tumbled with headlong fury and in great quantities from the height of the proscenium to the level of the stage. The effect was fresh, dashing and highly interesting. In the midst of the engagement the heroine mounted a charger and ascended the Cataract with wondrous velocity and invincible resolution, to the inexpressible delight of the Galleries, Pit and Boxes.[16]

Stanfield's visit to India helped attach the aura of authenticity to the imagery of colonial fantasy. However, Stanfield was especially known for his vast, moving dioramas, which were highlights of Christmas pantomimes and festivities.

A number of post-colonial scholars have identified in this, what is called a "museological mode" or "exhibitionary complex" when it came to representation of the colonies in theatrical spaces. The same was epitomised in the Great Exhibition of 1851 at the Crystal Palace, arranged by Queen Victoria's consort, Prince Albert, which contained relics or "objects" (such as the Kohinoor) and several daguerreotypes from all parts of the world.[17] The hub that grew up following the Great Exhibition, nicknamed "Albertopolis", which consisted of the museums of South London such as The Natural History Museum, the Victoria and Albert Museum and the Science Museum, provided a space for exhibiting cultures and sciences from across the world. The British Museum had already been established in Bloomsbury in 1759 as a "universal museum" and housed a collection of numerous curiosities and antiquities from around the world, originally belonging to Sir Hans Sloane and bequeathed to King George II. Taking it to another level, the Empire of India Exhibition of 1895, organised by Imre Kiralfi at Earl's Court London, exhibited not only souvenirs but real human beings representing the various tribes of India performing their daily chores.

These spectacles were obviously meant for the gratification of those who had no opportunity of being eye witnesses and so the characteristic distinctions of countenance and dress are represented with utmost faithfulness driven to its last plausible limit, of showing them live. Just as there were visual compendia in circulation about the working classes in Britain in works such as the *The Cryes of the City of London* (1687), *Twelve London Cryes done from Life* (1760), *Costume of the Lower Orders of London* (1820) or the *Etchings of remarkable beggars* (1815) and *Vagabondia* (1817), the colonial others too were dragged into a totalising systematic framework of methodical investigation on the self and its others. They were projected in their everyday-ness, in their mundane settings or their recreated native surroundings. Ironically, these depictions of existing variety were of people who remained as subjects to be studied, observed and gazed at, as opposed to portraits in which there was a mechanism of power in play between the painter and the sitter, who assumed a status of worth. Whereas the sitters in the portrait tradition were at most times identified by name, the staffage in outdoor or landscape scenes remained anonymous and mostly dwarfed.[18]

Transparencies of the empire: the panoramas

The visual culture of the times teemed with pictures, objects and artefacts from all over the world and especially the imperial colonies. The panorama was an exhibitionary devise which struck a responsive chord in the nineteenth century metropolis among its varied art forms.[19] It sought to satisfy an increasing demand for visual information. The vast apocalyptic canvases and the grandiose scenic entertainments, along with providing amusement, also served as a means of education of the general public who were already aware of a world outside their own island through the burgeoning print media. That the painter himself had visited the depicted spot and made the preliminary drawings himself was a necessary feature of panorama advertisements. The traveller or military or naval officer with a pencil and proficiency could, on his return to England, make money or gain public recognition by selling or loaning the sketches to a panorama painter. If he had a reserve of talent he could produce a panorama himself. However, in this medium, though artistic taste was advocated, artistic licence would not be tolerated. Topographical accuracy and authenticity could not be dispensed with at any cost, even for art's sake, and this accuracy was independent of principles of art. A panorama of New Zealand, shown in London in 1849, was "not the work of a mere artist, but of a surveyor whose business it was to explore and set down with topographical accuracy the natural features of the colony".[20] The panorama, in the eighteenth and the nineteenth centuries, provided, as Ralph Hyde says "a substitute for travel and a supplement to the newspaper". In fact, many of the panorama shows created an illusion of virtual tour. Years before the *Illustrated London News* came into existence, the London panoramas had been providing pictorial details to the current affairs across the world. A writer in the *Repository of Arts* writes in a similar spirit:

What between steam boats and panoramic exhibitions, we are every-day not only informed of, but actually brought into contact with remote objects.[21]

In fact, the images of the panorama were at times hailed to be even more detailed and perceptive than a visit to the place itself as a *Times* critic wrote:

There are aspects of soil and climate which ... in great panoramas ... are conveyed to the mind with a completeness and truthfulness not always to be gained from a visit to the scene itself.[22]

In short, the whole panoply of the shows of London such as the panorama and the diorama was able to create an inhabitable space, albeit artificial. The colonies and Britannic war settings, usually on the fringes of Britain's expanding empire, were the select favourites, though European cities like Athens, Rome, Marseilles and Venice also found a place in the London panoramas.

India soon became a great favourite among the London audiences. Numerous military scenes such as "The Storming of Seringapatam" (Figure 4.3) showing the defeat of Tipu Sultan in 1799, the "Defeat of Sikh Soldiers" and of course the Sepoy Mutiny of 1857, depicted the news incessantly pouring in from British India and the ever-expanding empire pushing its bounds overseas, instilling a jingoism in the audience to the tune of "Rule Britannia". Among these, "Seringapatam", a massive 270 degree semicircular display by Robert Ker Porter was enormously popular and could pull a profit of £1202.14. Touring Edinburgh, Dublin, Liverpool, Plymouth, Glasgow, Belfast, Chester, and Tavistock before reaching Philadelphia in 1805, the spectacle encouraged diverse regions to celebrate and participate in the glory of the empire. When Dibdin, another maker of panorama, saw this, he exclaimed:

I can never forget its first impression upon my own mind. It was as a thing dropt from the clouds – all fire, energy, intelligence, and anima-tion. You looked a second time, the figures moved, and were com-mingled in hot and bloody fight ... You longed to be leaping from crag to crag with Sir David Baird, who is hallooing his men onto victory! Then, again, you seemed to be listening to the groans of the wounded and the dying – and more than one female was carried out swooning. The accessories ... rock, earth and water ... half choked up with the bodies of the dead, making you look on with a shuddering awe, and retreat as you shuddered. The public poured in by the hundreds and in thousands for even a transient gaze – for such a sight was altogether as marvellous as it was novel. You carried it home, and did nothing but think of it, talk of it, or dream of it.[23]

Figure 4.3 Panorama key of Storming of Seringapatam by Robert Ker Porter, 1800
Image courtesy: Victoria and Albert Museum

The outstanding success of the panorama images of the military triumph at Seringapatam also gave rise to a market for prints which could be purchased and mounted and hung on the wall. Contemporary theatre practices played upon the popularity of accounts of warfare in exotic lands catering to public imaginations. In 1825, the editor and publisher of *Blackwood's Edinburgh Magazine* begged Captain J.R. McNeill, a surgeon serving as British emissary to Persia, to send him accounts of the battles then being fought along the Persian frontier, for "these pictures of Oriental manners interest everyone and stamp quite a new feature on Maga".[24] Often regarded as "those Noble Exemplars of the True Military Tradition", Eastern warfare was able to captivate and charm the audience. J.W. Kaye, the artillery officer, later a high ranking official of the East India Company, who later collaborated with John Forbes Watson in their monumental photographic project *The People of India* (1868–75), proclaimed that

> Whilst it was somewhat decayed in the West, the poetry of war seems to have its freshness in the East ... the nature of the country, the character of the people, their mode of warfare, their dress – are all surrounded with poetical associations.[25]

This is not to be seen at a remove from homegrown productions like James Walker's *A Picturesque Representation of the Naval, Military and Miscellaneous Costumes of Great Britain* (1807) subjecting various social types to scrutiny and observation.

The pomp and grandeur of the Mughal court never failed to fascinate the British, so much so that they would soon follow the symbolism of the spectacular tradition on assumption of the royal office. The "Durbar" scenes in a Mughal court and processions possessed a picturesque quality and could be exhibited as a panorama. A panorama of a Durbar procession of Akbar II c. 1815 is striking in this respect because it was painted by an anonymous Delhi artist in the manner of native scrolls and wall hangings that were popular at the time. It represented the Mughal Sultan's weekly procession from the Red Fort to the Mosque (Jama Masjid) followed by his sons and the British Resident (probably Charles Metcalf) along with covered bullock carts which carried the ladies, as well as horses, elephants, camels, even a leopard, and a huge trail of courtiers. Mughal India's own version of spectacle is appropriated within a quintessentially British medium of spectacle. Another panorama of the Delhi Durbar showed the British assumption of the same ritual much later in 1911.[26]

As discussed in the preceding chapter, an antiquarianism was noticeable on the part of the British in its delight for ruins and decaying architecture, which fitted into the aesthetic frame of the "picturesque" and whose images were ever growing in popularity. Indian temples, mosques, caves and forts found their place among a host of images from across the world, from pyramids in Egypt, the temple of Karnak, Balbek or Thebes or the city of Jerusalem.[27] "Hindoo Excavations in the Mountains of Ellora" as well as T.C. Dibdin's famous moving diorama based on James Fergusson's architectural trail across India, fall within this paradigm. Tapati Guhathakurta identifies in this kind of pictorial documentation of ancient art and architecture of exotic cultures, an urge to record a graphic history, of architecture in particular and of the civilisation in general. According to her:

> The diorama became the natural technological successor of the "picturesque" landscape, where the atmospheric and illusionist techniques of a painting was transformed into a performative spectacle before an audience in a darkened room, directly drawing the viewers into the object of their view.[28]

Dibdin's diorama had two parts. The first part was a virtual tour following Fergusson's footprints, as it were, beginning with a panoramic bird's eye view of the city of Calcutta, celebrating British architectural genius and then taking the audience away from the centre, through the jungle, into where one could see and marvel at the "Black Pagoda" (the Sun Temple at Konarak, called "black" due to its moss-covered exterior), followed by the "Temple and Town of Juggernaut" (Puri). The second part was called "The Diorama of the Ganges" which began with prospect views of the "Sacred City of Benares", following the river upstream to the Fort of Chunar, reaching city of Allahabad and finally ending at Agra with a view of the Agra Fort and the climactic spectacle of the architectural wonder of the Taj Mahal.[29] The spectacular journey could evoke a foreign land in material terms, the make-believe show sustaining an illusion in the audience that they had actually been on the picturesque trail themselves. The diorama was able to militate against one of the disadvantages the panorama had – that of movement, and therefore it could invoke the thrill and experience of the journey before a London audience. According to Mark Roskill, in the case of British-occupied India, the aura of mystery and the fabulous had fizzled out by this time. Therefore also, the nature of subjects depicted changed substantially. It catered to a need of constantly supplementing knowledge through graphic records that could be the basis for prints and book illustrations:

> Alongside the growing influx of visitors set on travel, the activities of the East India Company in surveying the provinces of Bengal that had come under its rule, and military and judicial expansion more generally helped for such art, as extensions of the picturesque into new or unexplored territory, included temples and ruins, great rivers and waterfalls, mountains and rock faces, as well as roadways and local architecture more generally. The special sanctity attaching to such places for the natives was recognised and considered descriptively

important; but their actual role in everyday life and their religious use was only hinted at within the images by the showing of a few token figures about their business – or else the human presence was left out altogether, in favour of expanse and grandeur.[30]

"Route of the Overland Mail to India" April 1850 to February 1852 (Figure 4.4) was another extremely interesting moving panorama employing forty or so tableaux and a singular travel motif. Though there was a lecturer who was an expert on India to elaborately explain the images, *The Times* spoke of it as elevating the level of panoramas "from a mere source of instruction to a work of art".[31] This one employed the most noted scene painters of the time like Thomas Grieve, William Telbein and John Absolon. The voyage depicted in the panorama began at Southampton Docks. The route was via Osborne, the Needles, Cintra, Gibraltar, Algiers, Malta and Alexandria:

Crossing the desert in vans resembling omnibuses, the travellers were able to observe artificial egg hatching, nomadic encampments, Bedouin tribesmen, Joseph's Well and a dead camel. At Suez they took ship for Jeddah, Mocha and Aden. A voyage across the Indian Ocean brought them to a Point de Galle, Ceylon where the steamer refueled before proceeding to Madras and Calcutta.

And as the journey draws to an end, a notice in *Punch* reads:

Finally, we have reached Calcutta, and by the noise and shuffling are reminded that we have never left London. It is most curious on coming out into Regent Street to find the porters and cabmen are not black, and that persons are riding around on horses instead of camels.[32]

Similarly, "London to Hong Kong in Two Hours" c. 1860, produced at the time of the second Opium War, carried images of India. Pictures and prints used in the panoramas often depicted unique land formations in landscape vistas or a remarkable cultural artefact. Such images are in line with a cult which were also reiterated in travel accounts of the time, that of employing a descriptive category characterised by the term "singularity" which was applied to objects in nature that were unusual and remarkable, and served to bridge scientific interest and aesthetic attractiveness.[33]

"A Sectional View of Mr. Wyld's Great Globe" and "View of the Exterior of James Wyld's Globe" (1857) included a tour all across the globe (Figure 4.5). A mapmaker by occupation, Wyld put up a monster globe with a series of wrought-iron stairs which carried the audience to regions depicted all over the circumference of the globe. The moving panoramas included models and maps of several regions which had recently seen military sieges titled as "A Dioramic Tour from Blackwall to Balaclava" or "The Diorama of the Campaign in

PUBLISHED BY ATCHLEY &.CO. GREAT RUSSELL ST. BEDFORD SQUARE, LONDON.

Figure 4.4 Frontispiece from the *Route of the Overland Mail to India*, (1850–2)
Note: This book consists of scaled-down images from the hugely successful diorama and
engravings from drawings by Grieve, Absalom and Telbein, illustrating the journey from
Southampton to Madras and Calcutta, inspired by scenes of ports and places encoun-
tered by the Peninsular and Oriental Steam Navigation Company
Image courtesy: The Bill Douglas Cinema Museum, University of Exeter

India" or "New Diorama of the War in China" and "Diorama of Russia".
However, performances of the "Diorama of Russia" were alternated with per-
formances of a panorama of Upper India, topical at the time, due to the Sepoy
Mutiny of 1857. Towards the end of the nineteenth century A.H. Hamilton's
"Excursions" similarly portrayed his journeys all across the world. Hamilton's
"Delightful Excursion to the Continent and Back within Two Hours" included
the "Grand Moving Panorama of Hindoostan", originally exhibited at the
Asiatic Gallery in the Baker Street Bazaar.

The high-altitude perspective coupled with the 360 degree view makes it
akin to the aerial magisterial view advocated by the cartographic gaze.
Panoramas, therefore, can be seen as being complicit with the colonising
process, in that they subscribe to a manner of "overlordship". The viewer in
the panorama is coerced into identifying with the point of view of the ruler.
As Comment claims:

> The invention of the panorama was a response to a particularly strong
> nineteenth century need – for absolute dominance. It gave individuals
> the happy feeling that the world was organised around and by them,
> yet this was a world from which they were also separated and

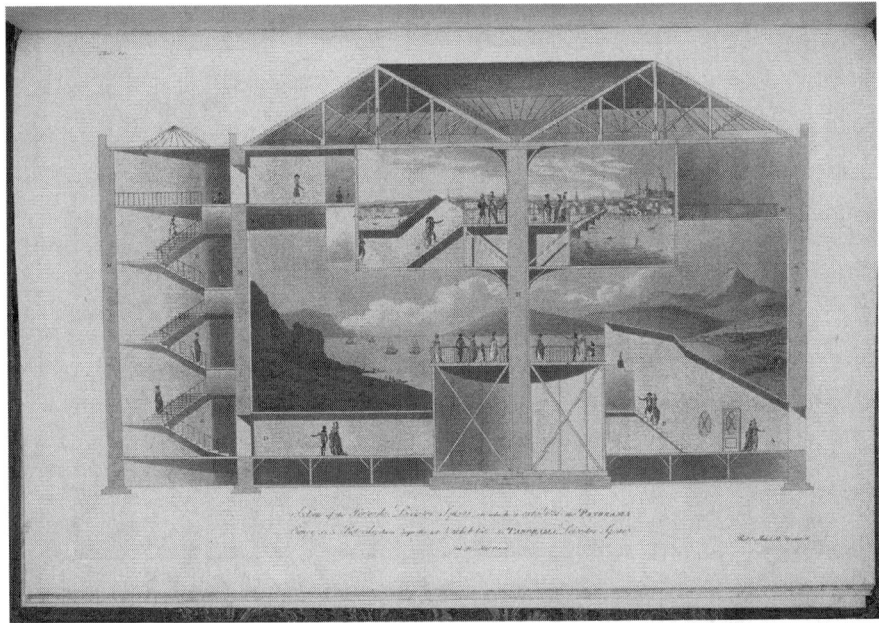

Figure 4.5 Section of James Wyld's Rotunda at Leicester Square, designed by Robert Mitchell, showing the viewing platforms from which viewers could take in two panoramas: large and small. Here the large panorama was a view from the hills around Edinburgh by Robert Barker and the small panorama shows London from the Thames

Reproduced with the permission of the British Library

protected, for they were seeing it from a distance. A double dream came true – one of totality and possession.[34]

On the one hand, they encouraged a sense of identity between the viewer and those in power, and on the other, they excluded by distancing through an assumption of superior station the possibility of identifying with the subjugated. Moreover, unlike a gallery painting, the panorama did not encourage a single point perspective, but its very form, by subscribing to a circular format, invited a collective participation from its spectators.[35]

British painters of the Company Raj

So far we have discussed the appropriation of paintings and prints into a distinctly European visual order. This should logically lead us to the complex issue of colonial picture making which constitutes an interesting study of the existing economy in the colonial society closely embroiled with both the administration and symbolism of the colonial state.[36] Art in colonial India was strongly tied up with colonial institutions in a manner which was unlike that which existed in contemporary Britain or the Indian predecessors in the Mughal court. The real world which existed outside the exhibition or the picture consisted of a complex intermix of political and aesthetic debate and the grappling of native subject within an imperial aesthetic frame. As already mentioned, the growing awareness about India in the imperial metropolis extended to diverse spaces, from polite enclaves like exhibitions at the Royal Academy to the print shops in Covent Garden, influenced by either elite theory of decorum in art or popular Grub Street sensationalism. In fact, be it the exhibition hall or an album series where the works of a painter are arranged according to the location they depict or the stage or panorama sceneries, they display the same metonymic impulse talked of earlier. Therefore, the trend that evolved by the eighteenth century balanced a form of ethnography and academic aesthetic in dealing with the exotic subject.

The return from India of William Hodges and a few years later of the Daniell duo stirred interest in paintings from India in the late eighteenth century and soon travel to India was considered for aesthetic possibilities. The meticulous plates known as *Oriental Scenery,* resultant of seven years' travelling through-out India by the Daniells, was a huge financial as well as critical success. J.M. W. Turner himself observed that the series was "a feast of intellectual and unusual entertainment ... bringing scenes to our fireside, too distant to visit, and too singular to be imagined".[37] Most of the publications in the eighteenth century related to India, such as those of Alexander Dow and Robert Orme, contained set pieces of its history usually translated from Persian sources such as *Firishta* or Abul Fazl's *Ain-i-Akbari.* This was the age, of course, when detailed information was being collected in the fields of linguistics, law, economics and archaeology. More and more information was still required and that too with graphic clarity. The accession to *diwan* of the East India Company and the formation of the British capital at Calcutta now required art to serve the empire, to invoke India as a fabulous Oriental asset. Now also began the phenomenon of a steady influx of painters on commission for the East India Company. The sequence of military and diplomatic events that took place as the East India Company rapidly established itself as ruler of India through conquest and alliance, were appropriated into the much-esteemed genre of the "history painting". Thomas Hickey planned an ambitious sequence of the *Third Mysore War* beginning with *Storming of the Breach at Seringapatam* and culminating with *The placing of the Rajah on the Musnud of Mysore.* Officially, the artist Robert Home had accompanied Lord Cornwallis and his army in the campaigns of the Third Mysore War. Benjamin West, in 1795, submitted a painting in which the First Lord Clive was seen receiving the grant of the *Diwani,* or in other words, appropriating the revenues of Bengal from the Emperor Shah Alam. Thomas Daniell exhibited at the Royal Academy in 1805 an elaborate Durbar scene based on sketches by James Wales, in which the British Resident at Poona was shown concluding a treaty with the Peshwa of the Marathas.[38] The paintings not only roused interest but promised financial gains for their makers.

India seemed to be a fecund field for artistic experiments and success. Back home, artists faced stiff competition to be a Royal Academician. It was virtually imperative for an artist to be a member of the Royal Academy (the body which defined and regulated national art at the time) for it not only provided the artist security but also ensured a respectable clientele. According to statistical data of the time, as many as 800 artists were seeking membership of the Royal Academy which registered only 40 artists. On the other hand, once a Royal Academician, an artist was debarred from exhibiting elsewhere: this caused many of its talented artists including those who travelled to India like Tilly Kettle or William Hodges to give up membership.[39] India was a choice as a destination of profession for many as it represented a land of luxury and plenty. There circulated stories of Company officials returning home rich from India, commonly nicknamed "nabobs" and painters like Johann Zoffany and Willison who too were said to have accumulated enormous wealth during their stay and work in India. On the other hand, the myth of prosperity in India, was countered by existing stories of its being a quintessentially tropical land of hostile climate, diseases and death. It was a great risk, but artists were often undaunted by the risk in favour of seeking a fortune. Those who chose to go to India, therefore, often searched for an alternative idiom away from the institutional regimentation of the Royal Academy which dictated art at home, even to the extent of borrowing from native techniques of art. Though the British would vouchsafe for a standardised pure and unadulterated national art form, trafficking of ideas, styles and forms influenced both British and native art. Francis Swain Ward's landscapes and Tilly Kettle's *Awadhi* portraits were influenced by native art. Kettle probably borrowed from Awadhi artist Mir Chand. Zoffany used local colours and strokes in his paintings. James Wales began using Indian pigments. Sir Charles D'Oyly employed a Patna artist Jairam Das, trained in the Mughal tradition, who was entrusted with his lithographic press. The native techniques and the local touch helped endow the British paintings with a readily identifiable Indian character. These instances, according to Eaton, rule out unilinear structure of British ideas of art and representation in India. The unique colonial pictorial vocabulary which evolved from this transculturation was distinct from metropolitan compositions. According to her:

> (This) further mediates notions of "orientalist empiricism" and limited likeness ... Such a device, even if not directly derived from specific Mughal miniatures, bestowed an aura of expected "character"; Indian rulers' likenesses being primarily known in Europe through the medium of engravings after Mughal portraits.[40]

Prior to the visits from British painters, India primarily existed in print through line engravings inserted in travelogues either from the author's own temperamental sketches or based on Indian miniatures. Many European artists including the famous Dutch painter Rembrandt and the British painter John Flaxman were indebted to Mughal images which were regularly published in scholarly British journals. Eaton points out that, such images continued to exercise an aura of authenticity for a time being, even over the colonial representations which began seeping in later on. Even William Hodges, in his *Travels* places himself within this tradition saying:

> I cannot look back at the various scenes through which I have passed these excursions without almost involuntarily identifying a train of reflections to the state of the arts under this as well as under the Hindoo government.

Colonial art therefore had to straddle popular forms of a newly emerging public art with notions of existing aesthetic idioms of national high art.

How the distinctly European genre of the landscape and country house portraits or its subgenre, the garden conversation piece, underwent an Indian adaptation, is demonstrated in Tobin's assessment of Johann Zoffany's portrait of Warren Hastings, then governor general of Bengal, and his wife. (Figure 4.6) The emergence of the genre, as discussed in the preceding chapter, revolved around ideological (feudal reactionary) issues of land and labour. Famous portraits, like *Mr and Mrs Andrews* by Thomas Gainsborough, epitomise this genre which usually depicted landed families outdoors on terrace, lawns or estate parklands, reflecting the period-specific topographical tradition in depicting the setting and background. Zoffany's *Mr and Mrs Warren Hastings*, painted in the mid-1780s, captures the spirit of the social ambition of the British gentry and the outlook towards ownership of land and property, though the representation of the topography was altered with respect to Indian landscape.[41] In this picture, one sees Hastings posing with his wife and her maid as in the traditional portrait of the landed gentry, amidst a vast green park with a Palladian mansion in the distant background. In this case, though, the land did not belong to him and neither was he part of the British gentry. Ironically, in fact, he was criticised later for being a quintessentially corrupt, unscrupulous "nabob" for his extortionist policies of revenue collection and violating the gentlemanly code of conduct towards land and property. Zoffany's treatment of vegetation is differently handled here. Following iconographic traditions, vegetation played an important role in conversation pieces which usually contained ideologically resonant trees symbolising the sitter's roots in an ancient organic community. For example, Mr and Mrs Andrews pose beneath a massive oak tree signifying the ancient lineage and fortitude of the family. In comparison, the Hastings family poses before neither the culturally significant oak nor the elm but a typically tropical tree, the jackfruit tree. As Tobin mentions, its only function is to add exoticism to the piece thereby undercutting any association with native roots in the foreign environment. Academic attempts to illustrate studies in horticulture resulted in grafting exotic-hued variety onto the parent stock of the established quintessentially British topos. The tree is perhaps emblematic of incorporation of indigenous knowledge within a European aesthetic and scientific canon. Also, Hastings' love for Indian trees was well known.[42] His interest and promotion of natural history, antiquities, habits and customs led to the establishment of many formal academic institutions of imperial scholarship including the early nineteenth century Botanical Garden in Calcutta.[43] As is well known, the East India Company's enthusiasm for oriental phenomena, flora and fauna resulted from the concerted efforts of William Jones and Warren Hastings, concisely summarised in the motto of the Asiatic Society of Bengal: "Man and Nature; whatever is performed by the one and produced by the other".[44] In Zoffany's painting, unlike Mr Andrews or any other sitter in other British pieces, Hastings does not emanate a spirit of mastery and authority over the space or the rest of the people and objects depicted in the piece. In most such pieces the wife is seated while the husband's erect towering figure stands in control at the centre of the scene. In Zoffany's portrait however, Mrs Hastings' upright figure and stature is placed at the centre, between her husband and an Indian maid who is shown slightly bowing, while Hastings points at the mansion behind with his hat in his hand instead of on his head, as a sign of reverence for his wife. Zoffany's reason for designing the picture thus could have been a product of his being privy to Calcutta grapevine gossips on Hastings' affection for his wife and stories of his lavishing wealth upon her. As Tobin points out, where traditionally, British conversation pieces rarely or never contain figures of servants, most such portraits of East India Company officials depict black male or female servants or native soldiers. In certain British exceptions, however, black servants from Africa or Caribs in exotic garments sometimes feature in such pieces as one among many other luxurious commodities available to the lord and lady of the manor and therefore represent them as owners of overseas plantations. This assumption of a new tradition in colonial portraiture where black bodies of native servants abound, only make them objects owned by the imperial official, in marked contrast to the absence of labourers in British conversation pieces who were not seen as part of the property of the manorial lord.[45] In art, the censuring expert gaze back in the metropolis could easily make or break an artist's career and fortune and the artist had to compulsively make claims to authenticity and a standard of empirical precision. Natural history, the Indian climate, exotic flora and fauna together with the distinctive native figures, become authenticating props in the unfolding drama of the British expansion in the colonies.

Figure 4.6 Warren Hastings and his (second) wife, oil on canvas, by Johann Zoffany
painted between 1784–7
By kind permission of the Trustees of Victoria Memorial Hall, Kolkata

Gone were the days when an earlier generation of artists relied on heresay,
fantasies and written documents to reproduce a place. For example, as early
as 1731–2, the landscape and scenery painter George Lambert (1700–65), in
collaboration with Samuel Scott (c. 1702–72), was commissioned to paint a
series of six landscapes of Indian ports and Company settlements to be displayed at the Directors' Court Room at East India House. Since neither of
them had ever visited India it can be assumed that this work was based on
sketches by Company servants and partly from imagination. Many of these
fell in the by now established tradition of country house painting. Their per-
emptory treatment of background detail suggests that they may not have had
access to extensive detail.[46] It was only after the 1750s and Lord Clive's victory in Bengal that any serious enterprise was undertaken to represent the
colony in art. Francis Swain Ward was the first painter who sent pictures
from India, though William Hodges (1744–97) is generally presumed the first
painter of India. Other painters like Tilly Kettle (1735–86) and George Willison (1741–97) soon followed suit. From then on begun a steady influx of
travelling artists from Britain as *in situ* drawings mobilised support from
artists and educators alike.

Most pictures by the first generation of British painters depicted views of southern India as Madras was the first settlement and trading centre. The peninsula tip was soon to be forsaken in order to privilege Calcutta. Only after the Battle of Plassey in 1757 and the subsequent removal of the administrative capital of Bengal from Murshidabad in 1772, did Bengal and particularly Calcutta receive any attention in art. With the loss of the American colonies in 1783, Calcutta soon assumed the status of "second city of empire", second only to London. The architectural construction of the city was minutely recorded by the visiting painters. Views of Calcutta soon became a genre in itself which contributed a great deal to the popularity of India as a subject among British painters. Countless painters in the eighteenth and nineteenth centuries painted views of Calcutta, including William Hodges, Thomas and William Daniell, James Baillie Fraser, George Chinnery, Samuel Davis, Henry Salt, James Moffat, Francois Balthazar Solvyns, William Wood and amateur artists like Sir Charles D'Oyly, Robert Smith, Cornelius Smith, Colonel Jasper Nicholls and the Princeps to name a few. On the architectural front, it was built up as a comparison to London as it boasted of broader roads and palatial buildings which needed to be exhibited to the audience back home. It was primarily based on a metropolitan imagination of a port city on the bank of an arterial river. Calcutta soon acquired a sophisticated manifestation in art as the city space was mythologised and narrativised at the crossroads of three kinds of imperial or civilisational discourses, that of the Roman, Mughal and modern British. Its tradition was invented through reiteration in pictures and prints. Strikingly, as James Rennell's surveys and cartographic activities started from Calcutta, gradually progressing outwards to northern India and beyond, often travel writers and painters too followed the same route straddling an antiquarian interest starting from Calcutta on the trail of temples, shrines and mosques which served as reference points. William Hodges was the first among the British painters to undertake this "picturesque" journey starting from Calcutta and penetrating into the heart of the country, stopping by the Rajmahal Hills and Caravanserai, which had been strategic locations for the Mughals in the past, as they progressed to northern India. Hodges accompanied many significant military expeditions and often painted battle scenes as well as landscapes of regions which were steadily falling under the British. In multiple ways, Hodges was a trend setter and torch bearer for another generation of colonial artists. He is a classic example of how a distinctive viewing of an alien land could be codified in future. His search for vantage points and privileging of sight perception would determine the routine of artistic viewing and reconnaissance of the land in the later times. The track he took was followed about seven years later in 1788 by another set of successful artists, Thomas and William Daniell, and then by numerous other professional or amateur artists like Henry Creighton, Samuel Davis, Charles D'Oyly, James Fergusson, Henry Salt, James Prinsep and many more. While these travelling artists depicted stock scenes from their journeys, the individual bits of contributions fed into a conceptual whole just as pieces in a puzzle fit together, forming the totalising structure of the map. In a digital age, this would correspond to focussing on singular areas within an overarching map in order to gauge the regional distinctions within a given area.

Indian representations by contemporary artists extended the stylised gaze creating an idiom which collated the picturesque with an ideologically motivated and disciplinary aesthetic of "Orientalism". Painters like William Hodges and the Daniells had already popularised this scopic idiom and created a market and audience for "on the spot" paintings which adhered to an aesthetics of travel and "views". These painters established and popularised a tradition of representing India which was followed by later artists of the Company School, where spatial mobility became an integral and significant trope as they joined hands with the cartographic project which was underway at this time. The riverine landscape views constructed visual travelogues: the travel inwards into India beginning at the colonial urban centre of Calcutta.[47] However, where views of roads were concerned, the newly constructed white town of Calcutta dominates these paintings. Travel into the interior was made generally by way of the waterway of the Ganges, capturing numerous topographical and architectural views politically significant as these spaces gradually fell under colonial governance.

Colonial prospects: the town and the countryside

The construction of Calcutta as a city space not only bore architectural imprints of the colonial power but required its continual visibility through charts, pictures, maps and later photographs.[48] Though large pockets of the Black Town remained excluded from mapping, in most paintings, native figures at their daily chores abound, towered over and framed by the monumental European construction in the background. Street vendors, beggars and others involved in menial occupations are frequent in these cityscapes. Simultaneous researches in clothing and physiognomy of the native inhabitants helped artists to populate their images appropriately with native staffage with empirical precision. However, the most dominant feature in these formats of urbanscapes was the architecture. At the heart of colonial architecture in a colonial setting during this time were two buildings strategically placed at the city centre representing the emphasis on defence: a massive fort and a government house. From the very onset, the fort marked the presence of an occupying power. Initially meant for shelter and the protection of a small mercantile enclave, forts underwent renovation and expansion with the strengthening of colonial position. Fort William of Calcutta received its establishment and confirmation after the Battle of Plassey by routing Siraj-ud-daulla. In other colonial cities established by the East India Company too – Bombay, Madras and Cape Town, the fort remains the central structure emphasising a need to be protected from the indigenous people. In all these cases, the fort became the seat of the Presidency

Government, and the design of urban development drew upon this structure as absolutely central around which the city grew and bloomed. Surrounding the very ramparts of the fort grew the white town, distinctly bordered with a high boundary wall, protected from the black town. Meanwhile, the white sector got inscribed with various other European structures, such as mercantile offices, churches, clubs and spacious bungalows. The soaring spires of the churches were not only mere religious symbolism but indicated the growing political power of the English and set out to mark their presence and the superiority of their race and faith on the geography of India. The fort engendered ghettoisation against the subject people who were potentially rebellious and hostile as the incident of the "Black Hole" in Calcutta had shown and later memorialised.[49] The Government House was another structure which provided the clear statement of assertion of power.[50] Metcalf shows how the design of the Calcutta Government House was debated and ultimately a reproduction of Kedleston was decided upon because it was thought to be more suitable to Bengal's climatic requirements. Isolated at the head of the Calcutta Maidan, in its own extensive compound marked by imposing gates crowned with lions, without even trees to obscure the view, the Government House loomed over the city so that all might see and appreciate the powers of the Raj.

Other adjacent structures along the Esplanade, and private houses along Chowringhee Road, complemented the Government House giving a picturesque quality to the emerging city which by the first decade of the nineteenth century, was transformed into the "City of Palaces" with an appearance that was stately, imperial and elegant. The genre of Calcutta paintings celebrated the project and emphasised the triumphal British colonial endeavour which, through sheer pluck and determination, had transmogrified the otherwise mundane wilderness into a majestic spectacle. William Hodges describes the entry by the river route into the great colonial capital:

> The appearance of the country on the entrance of the Ganges, or Houghly River ... is rather unpromising; a few bushes at the water's edge, forming a dark line, just marking the distinction between sky and water, are the only objects to be seen. As the ship approaches Calcutta the river narrows; that which is called Garden Reach, presents a view of handsome buildings, on a flat surrounded by gardens ... these are villas belonging to the opulent inhabitants of Calcutta. The vessel has no sooner gained one other reach of the river than the whole city of Calcutta bursts upon the eye.[51]

This depiction of Calcutta is quite in contrast to the images of the city in British art and literature which perceived the city as a place of gloom, toil, disenfranchisement and exploitation, in marked contrast to the countryside which is romanticised for its innocence and purity.[52] Travellers were characteristically charmed by their first view of Calcutta. This scene of arrival was replayed in numerous accounts as a sudden appearance of a sparkling and glittering golden fairyland emerging out of a homogeneous wilderness. This is also the point which captures the painter in transit as they invariably drew sketches while in passage, for they travelled to India in search of unique subjects for painting and went back home with their booty. In the same breadth, the figure of the travelling artist became ritualised. The recurring presence of diagonal lines in these landscapes captures the fleeting movement of the artists in transit.

In the same way, it became a performative ritual that Company painters located at nodal presidencies across India drew and painted these cityscapes and adjacent sites. Gradually, the entire subcontinent and its physical features came alive as parts of the map were painted red with regions being steadily annexed to the Company empire. In the southern part, painters like Alexander Allan, Thomas Anburey, Richard Barron, Robert Hyde Colebrooke, Thomas and William Daniell, James Fergusson, James Hunter, Henry Salt and Francis Swain Ward painted ancient Hindu temples, churches which had been built, Fort St. George, the Government House and the beaches and harbour. The presidency of Bombay and adjacent western India found shape through the paintings and sketches of Thomas and William Daniell, James Fergusson, James Forbes, Robert Melville Grindlay, Henry Salt, James Wales. Similarly, the Himalayas, northern mountains and desert in the north-west were depicted by painters like James Atkinson, the Daniells, William Edwards, James Fergusson, James Baillie Fraser, Charles Stewart Hardinge, William Simpson, Alicia Eliza Scott and Anne Eliza Scott. Among these, only a few were professional artists from Britain on commission for the East India Company, most of them being amateurs, or otherwise holding various administrative offices. An implicit relationship can be traced in their works between administration, the scientific enquiry involved and the graphic medium under discussion. These sometimes served as route maps, topographical or anthropological surveys providing much required information about the empire to the foreigners, resulting in its material control. However, etchings, drawings, paintings and prints formed a vibrant economy with the mechanical reproduction of images both in Britain and in India.

The large scale architectural commissions and imperial public works in the four presidencies in colonial India were major highlights in colonial paintings. This prominent urban centrism is also evident in such laudatory paintings of the eighteenth and nineteenth centuries is a brazen celebration of imperial power and progress. As artists moved away from urban scenes to capture views from the outposts and countryside the visual dimension of the practices of colonial expansion are brought to light. What characterises these other images is the invocation of the physical encounter with the foreign space. The ideas of "spatio-temporal compression" and erosion of distances, which pervaded the Enlightenment metropolis with the rapid building of roads and railways, travelled to the colonies with added impetus to expedite and facilitate the mobility of trade capital. One of the chief preoccupations of the time was with "dissolving distance" between places as seen in topographic art: "a way in which the world came to be conceived as integrated, as interconnected, and finally, as constituted a global village".[53] Nature, then, was conceived of not as a set of fetters or barriers, but as "prospects" in art or as possibilities to be harnessed, tapped, utilised and consumed materially. Spaces that existed as an abstraction, a vast emptiness or void between places was thus bridged.

The trajectory of this movement of vision from the urban to the rural, which went into making England's first empire and its practices of internal colonisation in the British Isles, is a fruitful path to track when talking about the British Empire in general. The colonies were brought under a cartographic regime in a fashion which mirrored the map making culture in Great Britain.[54] As maps stabilised and naturalised a London-centric expansive vision of Britain as a geo-political reality, so did other cultural forms such as travel and topographical art. The same assemblage was deployed to bring colonial spaces under surveillance. As travel practices shifted from the routine social visits in the European continent to uncharted territories of the East, the "prospects" too moved from being merely English ones to more exotic. Travellers flocked to places like India in search of a wild picturesque which could be brought under visual harness. What John Urry says about contemporary tourism is also true of the period under discussion, wherein the romantic gaze is part of the mechanism of extension of the "pleasure periphery", drawing ever more new spaces into its ambit, constantly in search of new objects of romantic gaze.[55]

George Chinnery and Charles D'Oyly: the search for prospects

I shall at this point discuss the life and works of two British artists: one, a professional and the other, an amateur whose works represent the very ethos that is being described here. Both of them influenced the other to a lesser or greater extent and both painted topographical watercolours around the same time. Seen together, both these painters' works, like their counterparts in Britain, fit into the tradition of landscape as a cultural activity determined by a quintessential Western Enlightenment aesthetics. Their works show that landscape is not merely an imitation but a creation, which is a product of a response. Topographic art in this respect is not merely an objective transcript of nature but a product of a systematically elicited reaction to the outside world and an environment. This response of course is mediated through received knowledge and cultural tradition. Unmediated experience is thus subordinated to a particular form of cognition which processes the natural world in order to make it significant, based on claims of utility and their truthfulness and fidelity to an existent reality.[56]

George Chinnery (1774–1852), a Royal Academician, became known as the artist of India and the China coast. He set sail for India after a more or less failed career in Britain. On the other hand, Sir Charles D'Oyly (1781–1845), the son of a senior merchant of the Bengal establishment, was himself a Company official and was employed at various places in India under various capacities.

Most importantly, of course, he set up his own lithographic press in Calcutta, one of the first of its kind in India. He had an influential role to play in the development of what became known as the "Company School". The two of them met in 1818 in Calcutta. This was when Chinnery, by now a noted portraitist, moved from Madras on a commission by Henry Russell, the then Chief Justice of the Supreme Court. The two briefly became involved together in an informal society of artists called the "United Patna and Gyah Society" or "Behar School of Athens". The choice of title reflects a particular "antiquarian" interest akin to that evoked by the Grecian Elgin marbles in England when bought by the British Museum in 1816.[57] The collaboration was meant to strike a balance between epistemological enquiry and aesthetic subtlety.

Chinnery had been a contemporary of the renowned British landscape painter J.M.W. Turner then a student at the Royal Academy School, and is said to have been a pupil of Sir Joshua Reynolds. He regularly submitted paintings to Royal Academy exhibitions before he decided to move to Ireland. Ireland, before his arrival, offered little prospect for artists. He went there specialising as a miniaturist. Chinnery's visit to Dublin set forth a string of activities like public exhibitions until in 1800 the Society of Artists of Ireland, of which Chinnery was the secretary, reintroduced exhibitions as a regular feature of social life. Chinnery was soon elected a director to the Drawing School of the Royal Dublin Society which was equivalent to the Royal Academy in London. Though Chinnery was seemingly doing very well in Ireland, the nationalist uprising there probably forced him to flee from there. According to others, his unhappy marriage might have been another reason. It was in May 1802 that Chinnery, after many efforts, finally received permission to travel to India, then seen as the land of promise.[58] Those who travelled out as an East Indiaman were generally employed either in the military or in the civil service of the East India Company; if the latter, they were often obliged to pay extravagant sums for their passage in expectation of far greater rewards when they arrived back. To go out in any other capacity such as "free merchant" or "painter" required the specific permission of the directors of the Company for they were given to believe that British India could allow only a specific number of artists.[59]

John Chinnery, George Chinnery's elder brother, was already posted in Madras and was in a strong position and capable of introducing his brother to potential clients. Chinnery first started with portrait paintings on commission from several of his brother's colleagues and their wives. However, the foreign views and people soon caught his attention and he was intrigued by the native people in their daily chores and tasks which he had never seen before. This, as he soon figured out, could earn him a distinguished audience back in England, which by now craved for more and more variety. He would have to go off the beaten track in order to make a mark in a fiercely competitive market. The toned supple dark bodies of the boatmen who dexterously manoeuvred the catamarans, the ripples of their every muscle and sinews caught aglow in the sun, groups of water carriers, bearers, labourers engaged in physical activities, the cattle grazing, all of these required intense study in their varied postures and formations, before perfecting them to be used later as motifs in landscapes (Figure 4.7). His sketchbooks contain many such studies in the manner of John Constable's famous studies in England. In this, he moved away from the Romantic depiction of toil, for such paintings almost always presented bodies in restful repose. However, he had not yet given a serious attempt to topographical drawing. He was still a portrait artist though his exposure to the sensational paintings of Hodges and the Daniells in England possibly had made him cherish a wish for trying his hand at landscape paintings. In Madras, Chinnery did not receive any major commissions and among the better ones he received was that of officially recording a *durbar* in February 1805, when the new commander-in-chief, Gen. Sir John Cradock, formally presented a letter to the Nawab-ud-Daula, congratulating the Nawab on his accession to the *musnud*. Another important, and his most famous, painting during his stay in Madras was the portrait of Kirkpatrick's children born out of his marriage to a local princess.[60] The portrait had captured the attention of many because of the nawabi outfits of the children, their glowing oriental hues clearly marking their hybridity and his own innovative style. This is remembered as his most remarkable work. In 1806, he undertook to produce a series of etchings with explanatory text which he titled "The Indian Magazine and European Miscellany". On 26 November in the same year it was announced that:

> Mr. George Chinnery as Jt. Proprietor of the work, will furnish an etching monthly. The first number will exhibit a view of Madras, from the beach; and every succeeding publication will contain either a landscape from nature or figures illustrative of the character, and occupation of the natives.[61]

The series clearly outlined his interest in depicting not only the natural scenery but the native bodies inhabiting it, becoming part of the exotic landscape, while the Banqueting Hall or the Government House hovered in the distant background. The series ran to nine issues from February to October 1807 before finally being dissolved with Chinnery leaving Madras for Calcutta on 20 June 1807. Calcutta, as the colonial headquarters, held promise of lucrative commissions, and this is where the focus of Anglo-Indian society had now shifted. The shift to Calcutta proved to be lucky for Chinnery who soon rose to the position of being acclaimed as one of the eminent British artists in India. His move most probably was in response to a commission he received to paint the portrait of Sir Henry Russell, the recently appointed Chief Justice of the Supreme Court of Judicature for Bengal in 1807.[62] Though Russell's portrait brought about a reversal of fortune, Chinnery moved away from Calcutta soon afterwards to Dacca in July 1808. The reason for this move, though not quite clear, might have been urged by Sir Charles D'Oyly whom he had met in Calcutta and who was now a collector in the city. Portrait painting, though considered a higher form of art, by this time had reached a crisis point and its sole

Figure 4.7 Study of Native Figures and Oxen Standing by the Wall, pen and ink, George
    Chinnery
Image courtesy: RISD Museum, Providence, Rhode Island, USA

pursuit was generally not considered wise for artists in India especially after the
relative financial failure of painters like Johann Zoffany, Charles Smith and
Ozias Humphrey. According to Natasha Eaton, the portrait market was no
longer the primary recourse for aspiring professional painters in India, having
suffered a set back in the recent past. Talking of the late eighteenth-century
scenario, she points out:

> Although Hastings manoeuvred British artists across India, his succes-
> sors Sir John Macpherson and Lord Cornwallis maintained a far more
> ambivalent attitude towards these portraits. Given this lack of official
> interest coupled with the reforms of the Parliamentary India Act of
> 1784 and economic recession. Calcutta's colonial portrait market had
> collapsed by 1786. Desperate painters left in search of either exotic
> landscapes as the inspiration for print schemes or for portrait com-
> missions at Indian courts.[63]

This might have been one of the reasons behind Chinnery's choice of leaving
Calcutta for Dacca, which was as yet unexplored and unrepresented by any
British painter. Charles D'Oyly convinced him of immense possibilities of the

"picturesque" kind available in Dacca. D'Oyly urged Chinnery just as he later proposed in the prospectus for his folios, *Antiquities of Dacca*:

> The ancient Metropolis of Dacca, on the banks of the Ganges, (is) an interesting part of India not visited by the Messrs. Daniell, nor, it is believed, by any European Artist.[64]

"Stir a little from the banks of Ganges": pastoral prospects

D'Oyly had recently been appointed collector of the city of Dacca and possibly this invitation was motivated towards throwing open to the gaze the beauty of a "virgin" territory. Every action of the British coloniser needed to be recorded visually for brethren to witness, every place he visited needed to be scrutinised and examined for its future potential, every view needed appraisal. D'Oyly himself was at this time an amateur painter of some worth. He was able to identify the scenic quality of places which would draw attention from the British public, for the action of reproducing a space in painting was solely driven towards creating interest in the region and applause for those kinsmen who performed the important task of officiating there. Part of it was the sentimental agony of those who remained away from their country and people, screaming for attention, to be remembered, recognised, memorialised and rewarded. Its resultant effect was to give shape and meaning to the British conquest. It is worth remembering, of course, that imperialism and its legacies are as much cultural products as they are political, social and economic processes, and in their cultural production they came to be influenced by the aesthetic standards of the day.

In July 1808, Chinnery moved to Dacca to work with D'Oyly, already the foremost amateur painter at that time. Chinnery's close association with Charles D'Oyly might have sprung from knowing his father, Sir John Hadley D'Oyly, in Ireland, and his having painted Charles's cousin and sister-in-law in Calcutta. According to D'Oyly, the painter Francesca Renaldi had been to Dacca in 1789, and Robert Home had paid a visit there in the summer of 1799, but neither had tried their hand at depicting the architecture of the place which was native-Mughal in style. It is only through portraying the architecture or, in other words, the "land-marks", that a space could be identified: for it is of course architecture that could create a place out of a space with no distinguishable physiological trait. These could also appropriated in the framework of antiquarian studies which D'Oyly was interested in and which was also the dominant leitmotif in art at the time. The decadent ruins of the Mughal architecture seemingly provided opportunity for Chinnery similar in subject to those of Brueghel or Delacroix in Europe, or Hodges in India before him. Many of the buildings there dated back to the seventeenth century which had seen Dacca's peak of glory when it was the capital of Bengal under the regime of Jahangir. Under the British, it was once a trade centre for cotton textiles but had fallen into disuse in the face of increasing rivalry among European trading nations. Bishop Reginald Heber in his accounts mentions the "stateliness of the ruins … huge dark masses of castle and tower … now overgrown with ivy and peepul-trees" as he travelled to the city by river.[65]

Again, like in Madras, Chinnery returned to depicting native village scenes, huts and cattle, native men and women in manifold chores. (Figure 4.8 & 4.9) His paintings of ancient Mughal mausoleums, mosques and forts would not be a raging success as Hodges' or the Daniells' depictions of Indian marvels were. The pictures did not get published. Even D'Oyly did not seem to be entirely satisfied with Chinnery's paintings, for he ultimately included only three vignettes from Chinnery's *ouevre* at Dacca in his later set of engravings called *Antiquities of Dacca* (1814–27) published from London. On being back in Calcutta by May 1812, where he stayed for 12 years, he was to earn his living as a portraitist but his passion lay in depicting rural Bengal as village scenes were clearly his favourite. The fact that he was an acclaimed artist and spent so many active years in Calcutta must have won him commissions for painting the architectural feats, thoroughfares, public buildings and private manors of Calcutta, but he plausibly militated against all of these for very few paintings of this kind are extant. His fascination for native life reflected itself in his personal preferences: in keeping a native mistress after the contemporary fashion and indulging in local habits like smoking a *hookah*.

Chinnery was desperately seeking a new visual language, which he found in rural Bengal. His village scenes were strikingly original as very few European artists tried representing this facet of India. As seen earlier, more and more visual data of different kinds was being amassed and appropriated with artists being encouraged to contribute to the systematic production and enhancement of the colonial information archive. On the other hand, early nineteenth-century urban consumers were being asked to read and digest different sorts of images from diverse sources of information or art. In this milieu, the tropical views of the lush, thick overgrowth and vegetation of date palms and broad-leafed banana plants held for Chinnery a wild enchantment unlike the carefully constructed British countryside, with its maintained bushes, thickets and trees. This was definitely a different idea and execution of the "picturesque". Also, going with the spirit of natural sciences required different brush strokes: the native vegetation needed to be clearly defined as silhouettes of the palm towering over thatched roofs or a large banana leaf partly hiding a mud dwelling, in contrast to a homogeneous mass of greenery in British scenes. Uncouth scantily clad native figures adorned the scenes, very often granted the foreground with accompanying cattle, gracefully becoming part and parcel of the actual landscape, their labour and toil making them of romantic pastoral significance like those in Brueghel's dramatic pastorals. It is, as it were, Brueghel translated into an antiquarian native tropical landscape. These went on to build up an image of India being quintessentially agrarian. His conscious relegation of both British architecture and the Indian marvels of the

Figure 4.8 A Hut Beside a Tomb by George Chinnery, part of Charles D'Oyly's *Antiquities of Dacca*
Image courtesy: The British Library

great temples and monuments in favour of rural life and scenery foreshadows a larger preoccupation with Asian village community at various points, into its becoming a subject of theorisation over the years for Munro and Maine, or later on, Marx.[66] In their works, the village was the basis of South Asian agrarian history which was thought to be the essential traditional social order truly representative of South Asia. Chinnery's images are a far cry from the popular "panorama" culture in its depiction of space as practised (by the native) rather than as passive, acted upon or "improved' by imperialist annexation. Yet they offered "prospects" for improvement and opened up

Figure 4.9 A Village Scene by George Chinnery, part of Charles D'Oyly's *Antiquities of Dacca*.
Image courtesy: The British Library

prospects for trade and tourism. Also, Chinnery differed from earlier artists like the Daniells whose endeavour was to catalogue the wonders of the Orient in that his range of subjects remained limited to his vision of what he understood as the essential basis of Indian society and life.

It is not to say, that this was a new genre and that Chinnery invented a new tradition through his choice. In England too, rural workers occasioned most interest, peopling scenes of husbandry or harvest. W.H. Payne refers to this peculiarity as the "English pastoral".[67] Andrew Hemingway points out that the market for art in nineteenth-century Britain was a metropolitan one. Therefore, the landscape images circulated within a primarily urban context, and were largely viewed by spectators who were remotely located from the scenes thus represented. The reason that the village scenes were readily and eagerly consumed was because of the geographical and conceptual distance. Interestingly,

Hemingway's research makes him conclude that in the nineteenth century watercolour artists in particular were prone to concentrate on rural landscapes as picturesque motifs.[68] Given the aesthetic priorities of the period, high art was associated with the poetical and ideal, which the countryside and rural scenes were perceived to embody. Rural labour, as in the Romantic tradition, was most often depicted in a state of poised restfulness representing the ideal, whereas urban figures of workers were often incorporated as staffage within a tradition of self-reflexive satire.

Chinnery's more glamorous and colourful paintings belong to his period of stay in China from 1825 to 1852 where he breathed his last. Heavily indebted, Chinnery is said to have fled from India to Macau. He is particularly noted for the paintings he sent from Canton, of ports and harbours, to be exhibited at the Royal Academy. All his life, Chinnery was fascinated by the everyday scenes all around him. His sketches of everything from cows to boats might not have been very commercially successful but they try to provide realistic vignettes of life in the manner of a camera before photography came into vogue. The engravings after Chinnery were published in the "Behar Lithographic Scrapbook" which denotes that D'Oyly had collected examples from his friend before Chinnery left for China in 1825. Chinnery also played an important part in an artistic project undertaken by D'Oyly after he returned to Calcutta in 1818. This was an illustrated book entitled "Tom Raw the Griffin" and the fashionable portraitist, Chinnery, features as the chief subject of the burlesque in Canto V:

Imprimis, o'er the walls are charcoal dashings

Of sudden thoughts or imitative keys,

Hung on a nail – and various coloured splashings –

The shape of frames, houses, horses and trees,

Prismatic circles – five dot effigies;

Notes of short hand – a card for five o'clock,

"Lord M desires the honour of Mr. C's

Company", in conspicuous station stuck,

To shew the deference paid his talent – or his luck![69]

The influence of Chinnery on D'Oyly is hard to define. D'Oyly was never a pupil of Chinnery in a conventional sense but their exploring the *mofussil* together in Dacca in search of rural and picturesque subjects was seminal to D'Oyly's evolution as an artist. Together, they gave birth to an alternative picturesque idiom. In the 1820s, D'Oyly became the most acclaimed amateur artist in British India, of whom Bishop Heber says:

I found great amusement and interest in looking over Sir Charles' drawing books; he is the best gentleman artist I have ever met with. He says India is full of beautiful and picturesque country, if people would stir a little from the banks of Ganges, and his own drawings and paintings certainly make good his assertion.[70]

Charles D'Oyly: the "Gentleman Artist"

From the late eighteenth century onwards, many amateur artists in the East India Company's service explored innumerable opportunities to record the sights of India. As Mildred Archer points out, only a few of the large number of British artists in India were actually professional artists and draughtsmen specially employed by the East India Company. Most available artists were amateurs for whom drawing was a passion or pastime, not always dependent on patronage or financial commissions. According to Archer:

> amateur artists in India could express themselves in two directions. The first was private, when relaxing from their duties as army officers or civilians, they drew landscapes, or painted pictures of life in camp and station or of the people of India with their colourful costume, their many trades and diverse methods of transport. The second was official when, assigned to duties of which drawing was a necessary part, they produced topographical landscapes, records of monuments and studies in ethnography. The skill, technique and attitudes involved were in many cases the same, but the purposes to which their talents were put differed in emphasis and in degree.[71]

Among the most productive of the amateur artists in India was probably Sir Charles D'Oyly (1781–1845). Son of a senior Company official, John Hadley D'Oyly, the Company's Resident to the Nawab Babar 'Ali at Murshidabad, Charles was educated in England from where he returned to India in 1797. He held minor posts in the Company towards the beginning of his career but gradually rose to higher and more responsible positions in the service. His first major appointment was as collector of Dacca from 1808 to 1812. Following this, he returned to Calcutta first as deputy collector and then collector of government customs and town duties, a post he held until 1821 when he was appointed opium agent in Patna. This sinecure he held for ten years. In 1831, he became Commercial Resident at Patna. Thereafter, he took a long leave when he visited Cape Town, South Africa. In 1833, he returned to Calcutta as senior member of the Board of Customs, Salt, Opium and of the Marine Board, until his retirement in 1838 to Italy where he died.

Charles D'Oyly's works are doubly significant as these drawings are not by a professional painter, but by an amateur artist who is also a colonial official posted in various capacities at various locations across eastern India in the early

166

part of the nineteenth century. In D'Oyly's work, one can trace the merger of specific aesthetics and administerial observational schemes of travel, landscape and the map, garnered towards constructing a particularised identity of the space traversed. Following Michel de Certeau's exposition on "spatial stories", one can say that these activities select, link together, draw itineraries and hence organise places. Be it his paintings of Calcutta, Dacca or Gyah or his monumental work titled *Sketches of the New Road in a Journey from Calcutta to Gyah* (1830), a project undertaken on the occasion of the inauguration of a new military road linking the Grand Trunk Road, the plates are meant to celebrate and contextualise the new colonial circulation and public work. One can read his topographies as a kind of spatial practice whereby a fresh idea of space emerges supplanting the space and network of the earlier *ancien* regime.

Charles was a prolific painter and he produced hundreds of paintings of the Indian countryside, many of which are now in the collections of the Victoria Memorial Hall, Calcutta. Some of his paintings are so fine that they had for long been falsely attributed to William Daniell or Chinnery himself. He, of course, was not trained in perspective drawing, and faltered greatly when depicting architectural views. But his association with Chinnery taught him to overlook Calcutta's Palladian architecture, the subject of most British artists at the time, and look for alternatives in rural scenes. D'Oyly's early style shows an amateur crudeness until he came under the influence of George Chinnery. A number of D'Oyly's sketchbooks which survive show studies in the manner of Chinnery, along with finished watercolours and oil paintings. He also produced delicately painted little scenes on embossed cards, many of them meant to be presents for visitors. His second wife, Elizabeth Jane, whom he married after the death of Marion, his first wife, was herself an amateur painter and musician. The D'Oylys often hosted a circle of amateur artists and their enthusiasm and hospitality are recounted in the memoirs of many public figures of the time. In 1824, D'Oyly was the driving force behind setting up a society of dilettanti called the "Behar Society of Athens" with Chinnery as one of its patrons. The proceedings of the society are preserved in a volume including also a huge number of watercolour paintings by D'Oyly. One of these proceedings record the foundation of the society:

> At a meeting of the Sons of Art at Patna the 1st July 1824, it was proposed and after some little discussion unanimously agreed that a society be immediately formed to be entitled the United Patna and Gyah Society, or Behar School of Athens, for the promotion of the Arts and Sciences and for the circulation of fun and merriment of all descriptions.[72]

D'Oyly was elected president. Christopher Webb Smith, another amateur artist whose sketches of birds had particularly made him renowned, was elected vice president and in charge of the Gyah branch and is jocularly mentioned in the proceedings as "Bird Smith". D'Oyly collaborated with Smith to produce a huge number of sketches, drawings and watercolours of varied Indian subjects from nature, wildlife and birds to topography and landscapes, and clothes, customs and rituals of the natives. In their typological ordering, these may be seen as pictographic taxonomic surveys. A sense of purpose and organisational logic is discernibly present in these series. The modus operandi behind establishing a relationship within otherwise disconnected objects was provided by the natural sciences. Publishing a collection as a category or type was already earning acceptance in Britain. The same idea was to be worked on till it reached its culmination in the establishment of museums which would then house diverse objects and club them together as discrete groups based on similarities. Similarly, locations earned their geographical specificity by virtue of identificatory marks, through a documentation of the natural world done by first observing and then carefully recording them according to conventions laid out at that time.

In his style, D'Oyly varied from the picturesque as practised by Hodges or the Daniells. In his correspondence with Hastings, who seems to have admired D'Oyly's drawings, he reveals his passion for Indian scenes completely neglected by other artists. In 1806, when D'Oyly happened to be briefly unemployed following retrenchment after Lord Wellesley's extravagance, he engaged himself in painting views in and around Calcutta. In a letter to Hastings, he speaks of finishing four large drawings which he proposed to engrave before sending them to Hastings: views up and down the river, a nearby mosque and a banyan tree. Of the last, he says:

> These I know you will value particularly the tree. To this wonderful work of nature I devoted four days of the last cold weather and while sitting under the spreading branches I could not help wondering that no painter had been induced to exert his talents in describing this tree as it ought to be – alone. That is that it should not be brought in as a subservient feature of the landscape as Daniell has made it but that it should stand in the picture as it does in nature unrivalled.[73]

Likewise, D'Oyly's particular admiration for Indian nature and native life made him represent these in his drawings with sensitivity and utmost dedication. Having no formal European training, he merely depicted what he saw without much stylistic distraction such as that of the "rococo" which Hodges used to paint vegetation or botanical features, thereby framing Indian or Caribbean scenes in a distinctly European style. Again, "The Great Fig Tree", painted by D'Oyly when he was posted in Patna demonstrates this speciality in him. In this picture he not only gives the tree the centre stage but characterises it as part of the ritual custom of the natives. By now, claims towards fidelity towards the real world was the demand of the times and the visual culture was making a significant contribution to the understanding of the world and shaping it too. Since D'Oyly was not a trained artist, it was not difficult for him to visually record naturalistic data without its aesthetic trappings. Likewise, D'Oyly truly did move away from the banks of the Ganges, as he prescribed in his views expressed to Reginald Heber, and ventured into the black town in search of authentic Indian views: the streets, bazaars, the huts, the rituals, fairs and festivals of the natives were captured by his searching gaze. In his *Views of Calcutta and its Environs*, though published much later after 1848, most of the drawings belong to around this time. In most of his drawings there is a clear demarcation of foreground and background. (Figure 4.10) The foreground is usually and strikingly dedicated to native life. Native bodies abound, usually engaged in their daily rituals. Animals too often share space with humans on the streets, beside ponds or the Hooghly river. The middle space is usually occupied by thatched huts, whereas the background captures towering Indian or European iconic architecture, partly shrouded in thick foliage. Therefore, the trappings of an established picturesque aesthetic convention, such as the prescriptive feature of having a waterbody in the frame, were not entirely absent from his work. Yet, he manages to reinvent it with a quasi-anthropological gaze.

Figure 4.10 Esplanade, from 28 *Views of Calcutta and its Environs* by Charles D'Oyly, and lithographed & published by Robert & Lowes Dickinson & Co (1848) By kind permission of the Trustees of Victoria Memorial Hall, Kolkata

Antiquities of Dacca: aesthetic scrutiny and production of original knowledge

In 1808, when he was appointed collector of Dacca, he wrote to Warren Hastings saying:

> I shall some time hence please God offer you as companions a few of the ruins of the city of Dacca which I assure you are exquisite for their magnificence and elegance and are calculated to tempt the pencil of an artist.[74]

The result was the drawings for *Antiquities of Dacca* (1830), which he had planned as a joint venture with Chinnery. It, however, materialised much later. His sketches and paintings dealing with Dacca were brought out from 1823 onwards in the form of folios from London. Each of these folios had about four to five sketches or paintings in it, together with topical and historical descriptions of them. These brief explanatory notes were submitted by an acclaimed historian, Persian scholar and artist, who was in the military services, military surgeon James Atkinson (1780–1852). The roping together of image and accompanying explanatory letterpress was not new but a relatively established convention at this time. It lent an overarching organisational logic to the collection: that these were varied facets of a given unit. The folios, taken together, later took the shape of *Antiquities of Dacca*. D'Oyly was keen to ascribe greater status to the vernacular humble architecture of an erstwhile important port township. He insists on replacing the hierarchical distinction between places largely established through the preceding artistic ventures of Hodges and the Daniells, with a more fundamental equivalence adjudicating a case for formal aesthetic scrutiny. The *Dacca* paintings too share a similar format with the Calcutta paintings and capture *mofussil* (suburban) life in its plebeian detail. Most depict a Mughal building then in its ruins hovering atop with at times dense, oppressive vegetation taking root in it, so carrying a sense of decay and a passing away of an old order. Also depicted are roads, rivers and *nullahs* (streams), which are scarcely used or have fallen into disuse. This reverse trope in depicting antiquities in the British colony is therefore interesting, as back home the same was used to eke out a far more salubrious response to Britain's own glorious past.

For a substantial portion of the treatise, D'Oyly seems to be filling in information in the gaps left by Tavernier who visited this city in January 1666, or rectifying mistakes appearing to supply specifics and accuracies to generalisations. D'Oyly starts off by recalling the strategic location of Dacca not only because it had once been the capital of Bengal, which rose to pre-eminence during the time of Aurangzebe, but also because of the presence of other European powers prior to the coming of the British and the East India Company:

Long before the English settled at Dacca, the Dutch had established a factory there, and transacted their business through native agents .... The English factory at Dacca, having been preceded by that which Tavernier terms "a tolerably good one", was rebuilt about a century ago by Mr. Stark, with the permission of Iltizam Khan ... previous to which native agents had been employed to purchase cloths, and convey them for sale to Calcutta. It was not till the year 1742 that the French succeeded in getting permission to rebuild a factory here, which is now, as well as that erected by the Dutch, a heap of ruins.[75]

In both his Dacca and Behar paintings, D'Oyly makes his identity as an administrative officer in the colonial service quite prominent in that both make comments on the state of decay and administrative lapse the regions had suffered from in the recent past. The *Antiquities* deal with descriptions and depictions mainly of the architectural past and present, and there is inevitably an attempt to draw comparisons between past Islamic elegance and the poverty stricken present, while drawing copious references to Alexander Dow and his account of Islamic culture and its decadence:

Thus Dacca for more than half a century was the capital of Bengal, and continued to be enriched by the multitudes which crowded to the courts of its governors. The stupendous remains of gateways, roads, bridges, and other public works, which present themselves on every side, sufficiently prove the former grandeur and magnificence of the city.[76]

While D'Oyly points out the deficiency in Tavernier's account, he in actuality takes credit for his own "discovery", the detailed research undertaken and original knowledge produced by British representatives:

It would appear from this account by Tavernier, that almost the whole of Dacca at that time consisted of habitations built of mud, straw, wood, matting and bamboo, such as are constructed by the common people at present. He mentions no public buildings excepting those of the Europeans, although the Great Kuttra, a most magnificent edifice, as well as the Mosque of Syuff Khan, had been erected many years before, and the small Kuttra more recently, though still several years before the celebrated French traveller visited Ducca. These splendid buildings, as well as several others, seem to have eluded the observation or escaped the memory of Tavernier.[77]

The *Antiquities* deals mainly with the past and present architecture in the region, adopting a comparative framework for the work. In describing an old mosque, D'Oyly speaks in the same vein:

Like some of those Venetian buildings which adorn the shores of the Adriatic, and are beheld with so much pleasure in the pictures of Canaletti [sic], this Mosque rises immediately from the margin of the river, with an effect at once stately and picturesque. Its neglected domes and arches are now shattered by accidents, crumbling to decay; yet in the general proportions and character of its architecture, the principles of elegance and simplicity appear to be combined; and the tout ensemble can scarcely fail to impress the beholder with respect for the taste and talent of its architect.

While he takes pride in exhibiting his discoveries through his drawings, he does not miss an opportunity to point out the existing flaws in the survey of a representative of a rival European power. In a move that was crucial to the age, he disentangles what is seen from what is already known, demonstrating a spirit of scientific objectivity and enquiry shared with fellow European artists of the nineteenth century.[78] As with cartographic and other scientific documentation of the time, much of the challenge was to present authentic data and one of the conventions of presenting that entailed the correction or faulting of earlier gleanings. The colonial cartographic age played on producing errata and finer nuances in its ever-evolving claims at truth generation as it consolidated and widened its inclusive vision. It proved British survey and governance as a far more efficient and immaculate system than any other competing governance structures.

However, when he mentions the Daniells and their inability to identify the picturesque significance of the location, he simply gloats in self-glorification in having the artistic vision, and acumen too, to proceed with and complete an unfinished agenda – to reveal to sight an entire terrain which had been so long (and not without surprise) left out. He definitely had a format for the *Antiquities* in mind, for he had to choose from Chinnery's drawings, ultimately incorporating only three of his vignettes from his limited repertoire, which could suit the ruling structure of a comparativist paradigm. He definitely trusted more in himself when it came to representing his own findings – of undertaking a systematic recounting of the stately architecture of a lost world overlooked by all. The first among the three of Chinnery's drawings was that of *An ancient Mosque and Modern Habitations of Dacca* which was supposed to contrast "present poverty with Mohammedan importance, and rusticity with architectural elegance" (see Figure 4.8). The second was a more explicit comment on the decline of Dacca, according to D'Oyly. Though Chinnery might have painted it with utmost passion and admiration, the cottage of "a poor muslin weaver, formed of bamboo, mud, and matting, thatched with straw: his umbrella and a few of his domestic culinary utensils are lying about, and at the right hand corner is part of an old loom"[79] (see Figure 4.9). The third one, *Approach to Tungy*, a hunting scene, has an elephant and a dead tiger slung across its back.[80] The work ends with a high romanticist note contemplating on the passage of time, fall of empires and vicissitudes of life and human endeavours:

To the noise of mariners and shipwrights which once resounded along the nulla – to the bustle and pomp of commerce and princely equipage – has succeeded a degree of loneliness and silence.... passion is lulled; and the imagination, willingly enthralled by feelings of melancholy pleasure, is instinctively led to compare the vicissitudes of human power and opinion, and mutabilities of human art, with the permanencies of Nature herself. ... The bridge before us is fast following its predecessors .... Though now mutilated and mouldering under the effects of time and neglect, and the ruder dilapidations of war, it is still an interesting object to the eye of the landscape painter and poet.[81]

It is in comments such as these that we can detect the connections between the medieval and the modern, nature and industry, to the detriment of the latter. The ruined mill of an earlier colonising power proves purposeful in establishing a continuum, granting a timeless quality to the scene and the place, in a classic Gilpinesque way. The seemingly romantic statement obliquely smacks of the failure and falling apart of the earlier system of circulation, thus submitting itself to the possibility of repair by a stronger force. D'Oyly is able to straddle the timeless and the particularist with marked deftness. The same idea reiterates in a number of his paintings in the same series, in, for example, the depiction of collapsed bridges like *Paugla Pool*. (See Figure 4.11) At another level, the romantic streak that his work elicits, definitely due to the time and social climate he grew up in, is linked to the latent anxiety in his work to be remembered (and rewarded) in future through his paintings, despite the brevity of human life, as he says in a prior passage:

While we sympathise with the warm consciousness of an artist who hopes to live in his works, we are involuntarily apt to indulge ourselves for a moment in the vain wish that we could have preserved his name also from oblivion.[82]

According to Gilpin's guidelines, a picturesque scene could be granted a sublime status if it could nobly inspire the artist with elevated thoughts. Topographic representations framed by philosophical and sentimental utterances such as the above could reaffirm the sublime prospect of the place while going on to reflect that the designs of scientific amassment of naturalistic data was not wholly incongruous with the aesthetic predilections of the day.

Colonial prospects: optics of control

The systematic unfolding of past history through recounting the spatial configuration and architecture of the region gives the space an identity and grants it a coda of registers for ready recognition and eventually produces a place. The British colonial regime with its growth of mercantile capital,

Figure 4.11 Paugla Pool by Charles D'Oyly, part of his *Antiquities of Dacca*
Image courtesy: The British Library

commercial manufacture, and a drive towards monetisation of social relations moved for a "considerable intensification and social re-contextualisa-
tion of circulatory practices – in other words, a transformation of the
circulatory regime." The decline in commerce and attrition of urban centres
in the early phase of British colonisation of India was therefore generally
seen as the breakdown of a circulatory regime, the notion of which can be
used as a conceptual tool of periodisation, through identification of discontinuities and ruptures in the production system. Of course, production is
only complete when the product travels spatio-temporally and reaches the
market place, and the existence of any fissure in this process is seen as a
failure of a regime. Many of the arguments foregrounding the importance of
transport infrastructure, of roadways, navigation and later of the railways in
British India, were based on this foresight of bringing products and capital
into a disciplined and efficient system of circulation.[83] D'Oyly's next major
project, *The Sketches of the New Road* celebrated the initiation of this idea
of geographical improvement as a crucial signature of British triumph, and in
that sense, this work pertains to both the ideas of the "prospect" discussed
earlier. This also meant that the areas surveyed by the artist's eye was also
brought under a cartographic surveillance and military administration, and
subordinated under the missionary zeal of scientism.

The final publication of the *Sketches* contained a series of only 19 to 20
lithographs from the Asiatic Lithographic Press, Calcutta, though there were 80
pen and ink drawings in this collection. Technical processes involving aqua-
tinting, lithography or photolithography were far more advanced in London
and were executed with far more efficiency. It was novel and unusual for official
reports involving illustrations to be printed and published in India till the late

nineteenth century. The *Sketches* were published with the support of the Company government and produced with the help of the relatively new printing process of lithography, which had recently been invented by Alois Senefelder of Munich in 1798.[84] With the acquisition of his new press in Patna, which D'Oyly called Behar Amateur Lithographic Press, during his tenure there as the opium agent in Patna, he published his paintings as lithographs. He published a huge number of paintings by himself or by his circle of artists, in various scrapbooks and volumes devoted to ornithology and hunting. Several volumes called the *Behar Amateur Lithographic Scrap Book, The Feathered Game of Hindustan* (1828), *Oriental Ornithology* (1829), *Costumes of India* (1830), *Indian Dead Game* (1830) came out of this new press, which taken together, can be seen as an encyclopedic archive dealing with multifarious facets of Indian natural history. However, his most serious lithographic work from this phase were the *Sketches of the New Road* published in 1830.[85]

The task in the *Sketches* is apparently that of naming and marking out those landmarks individually which could be seen in the course of a journey by the newly constructed road. With James Rennell's surveying operations surging outwards across India, while being focalised in Calcutta, large parts of India which were so far outside the scope of colonial information order, were fast brought under optical control. Writing about the utility of knowledge of travel routes for the administrator, Rennell observes in his book, *A Description of the Roads in Bengal and Bahar* (1788):

> The utility of such a work in any country must strike everyone: much more in a country where the people employed by government are mere sojourners, and from the want of local knowledge must depend on the information of Guides, who often mislead them either through ignorance or interested motives. At best those guides know only the most frequented roads; so that in crossing the country no information whatever can be derived from them: and as for the peasantry, or ryots, they cannot be supposed to know the roads beyond the circle of the markets which they frequent.[86]

Where a panoptic knowledge of routes was sought from the administrative class, the peasants or the peddlers were not expected to know or travel beyond their limited closed circuit. This itself is fallacious as many historical studies today have shown how local trade took place beyond narrow demarcations imposed by the British and continued to exist despite numerous attempts to curb these in order to bring them under a strict revenue regime.[87] Rennell lists four major centres, Calcutta, Moorshedabad, Patna and Dacca: "the first being the seat of government, and the others either the capital military stations, or factories or both". Following the grant of the Dewani to the English East India Company by the Mughal Emperor Shah Alam II in 1765, the seat of power shifted from Murshidabad to Calcutta, which now needed to be connected with all major stations across the region. The said road was constructed in order to connect Calcutta to the present Grand Trunk Road, one of South Asia's longest and oldest roads, near Sherghati in Gaya district in Bihar.[88]

The *Sketches* simulated an ideal tour: a coherent visual experience for an interested audience. In this series, D'Oyly adopted the European style of the "prospect", by a then widely accepted format, for his panoramic views. For him, this was the ideal form to portray an entire social world. Taking a bird's eye view of an entire terrain (as of the estates in Britain), such views could include with ease a social scenic view of fields, forests, roads and rivers, sometimes towns and cities, travellers on the road, tiny against the scenic landscape. As discussed earlier, "prospect", as the term signifies, charted both time and space, suggesting a view into the future as well as the distance, often, of an improving world.[89] The format of the prospect suited D'Oyly's *Sketches of the New Road* accurately, for fusing a society and territory in this way it clearly defined a form of cultural identity for the entire region. The road, interlaced with semaphore towers, as it were, paved the way to India's modern future.

Frontal views of temples occupy the foreground in the Gaya pictures, at times framed by the banyan or peepul tree, with unidentifiable faceless people occasionally visible. They are primarily pastoral subjects or traders who are seen using the road. Mixing the two forms here is interesting which can take us to J.M.W. Turner's Royal Academy lectures emphasising the need to distinguish different visual languages in the representation of architecture:

> geometrical elevations and perspective views of buildings function very differently, and that their alternative codings of architecture should not be confused. Elevations designed to show accurately the proportional relationships and dimensions of a structure, offer an ideal view and should not be subjected to pictorial effects; perspective views, on the other hand, show relative proportions as an experiential phenomenon, as the building would be seen in life.[90]

On the road through forests and over hills, the prospect could allow a sweeping cartographic gaze. Once in Gaya, which was an important centre of religious practice, the need was to be more grounded, to take stock of the environment and be perceptive to ordinary practices in the area.

In the early colonial era, many new temples went up in the Bihar region, encouraged by the upwardly mobile. In fact, as Anand Yang observes, temples and forts signified the twin hallmarks of power and authority of an existing feudal class of local landlords: "While forts towered over the huts of the subject people, temples proclaimed their founders' religious faith and moral authority".[91] Therefore, these characters, which were signs of hegemonic power and legitimacy became the Indian counterparts to the English estates, specifically its castles and manors in the British landscape paintings. Among other things, Yang points out the complete obliteration of the Muslim ruling elite from the

Bihar region, mainly based in Patna, as a consequence of the changes brought in by the overarching colonial administration. Gaya was one of the most important Hindu pilgrimage sites on the subcontinent, sometimes considered even more important than Varanasi. Being such an important site, the spot was chequered with many existing roads, the most important being the one via Patna which connected with the old Mughal highway. In the colonial era, Gaya became a crucial link between north and eastern India, and thus a hub of economic activity and trade, since with the advent of new roads, middle class pilgrimage increased manifold. Pilgrimage or *Tirthayatra* had its own unique significance in a person's life. It was a spatial gathering determined by a shared religious almanac based on special astrological-astronomical events. Unlike in colonial records and illustrations, which were characterised by faceless and nameless masses always depicted on the road, the Tirtha was a unique, memorable and once in a lifetime experience in the individual's life. Moreover, in the nineteenth century, a body of laudatory literature was in circulation which evolved from the Sanskrit genre of *Mahatmya*, celebrating and advertising particular pilgrimage spots. These jointly shared and experienced spaces when described through a perspective of faith would be markedly different from when seen through the lens of removed objectification.

The tenth sketch in the series is *View from the Summit of the Kutcumsunder Pass* (1828). *Hazareebaugh*, epitomises D'Oyly's landscape structure (Figure 4.12). It is described as "situated on a flat table land of considerable extent which is gained by a gradual ascent to the top, but in itself owns no temptation to the pencil of an artist." D'Oyly accordingly, uses a cartographic draught-like plain form. Its topography is recreated meticulously in the drawing supplemented with this written letterpress:

> The country about Hazareebaugh is a dead flat and not until you travel about 10 miles beyond it, do you again behold picturesque scenery. Arrived however, at the Kutkumsunder Pass, the descent from the Table land; - one of the finest and most extensive prospects; at once gratifies the eye. From the summit is observed a low green beautifully dotted valley, from which to the right gradually rise hills covered with rich folliage, and the horizon is bounded by a range of hills, varying in their tones by recessions, till they are lost in haze

Similarly, the quasi ethnographic details in the fourteenth sketch, *View of the Summun Boorje at Gyah* (Figure 4.13), sets up a resemblance of the shrine to a European structure, only to dissolve it:

> The first appearance of this singular building strikes the beholder by its resemblance to an old European Castle, but it is in fact merely a shell to a Rocky ascent behind, composed of small narrow rooms inhabited by Gyawabs, the priests and descendants of the priests of Gyah,

177

personages of great reputed sanctity who amass considerable wealth from the rich and liberal pilgrims resorting from all parts of Hindoo- stan to worship at the different shrines. These persons have large pos- sessions in Gyah and form a kind of fraternity sending out emissaries to collect pilgrims on their way of becoming their guides in the per- formances of their religious rites.

D'Oyly was perhaps referring to the extensive and efficient system of pilgrim hunting where scouts intercepted pilgrims and led their way to Gaya. As Yang mentions, the development of more efficient communication and transportation not only facilitated travel by pilgrims, but also these pilgrim hunters who gathered pilgrims from across India in exchange for a commission.[92] Again, in this series, reappear the figures of native travellers, visible with their horses, camels, bullock carts and elephants. They appear as tiny, dwarfed figures pitted against the towering scenic landscape in the known Romantic style which D'Oyly embraced, making them look powerless and vulnerable. The region was prone to anti-colonial resistance and therefore saw a number of military cam- paigns. The diminutive scale of the staffage indicates an impression of admin- istrative regulation and control. These, too, followed upon the designs of a map in projecting a vision of a rebel-free universe. An inhabited space is refashioned until an alternate imagined space is fabricated through colonial practices of image-making, travel and mapping, catering for a metropolitan audience. On demystification, D'Oyly's numerous compositions appear to be diverse expres- sions of a single political outlook that creates a new circulatory regime through colonial intervention into the *mofussils*. Connecting the colonial metropolis with the hinterland and rural countryside appears to be a crucial objective behind spatial practices of the colonial system whose interface with its own aesthetic domain is manifestly obvious. The penetrative new circulatory regime not only pushed forward an efficient system of continent-bridging movement of trade capital but also knowledge capital with drafts of maps, travelogues and images efficiently reaching the imperial epicentre in London to be printed and disseminated as movable digests.

D'Oyly's fascination with the Attic movement a la Romanticism in con- temporary England was quite obvious in his naming the informal society of dilettanti that he formed with other amateur painters and poets, the "Behar Society of Athens".[93] Evidently, this quintessential breed of English antiquar- ianism was reconstructing Britain's national past and history, launching it upon a medieval Roman inheritance. The celebration of Britain's Gothic architecture in paintings and poetry was an offshoot of this vision. As dis- cussed in the last chapter, a categorical and systematic documentation of Britain's history through a thorough cataloguing of its architecture was enabled through a use of the visual language of the picturesque. This language becomes incrementally inclusive cutting across regions and demolishing spatio-temporal distance. It is interesting to see how ideas generated in the

Figure 4.12 Kathkamsandi Pass, north-west of Hazaribagh, *Sketches of the New Road*
by Charles D'Oyly
Image courtesy: The British Library

Figure 4.13 Summun Burj, Sketches of the New Road, Charles D'Oyly
Image courtesy: The British Library

English climate travel, translate and appropriate Indian nature, flora and fauna in its own terms and for its own ends. Ideas of primitive atavism, views of pastoral harmony and simplicity as ideal (innocent childhood of mankind), were thereafter made subservient to racially endorsed notions of bringing these far frontiers within the bracket of advanced modern governance via colonisation. The genre of landscape and topographical views were still not regarded with the high esteem associated with fine arts, but their status was certainly raised from mere sensational exotic pictures to that of scholarly enterprise in being connected with the history of various countries and providing valuable views of the manners and habitats of mankind with prized truthfulness. Carefully balancing poetic vision with an accurate, scientific, naturalistic detailing, this genre was able to counter the dogma against it and made greater truth claims toward human history in a colonial climate which nurtured it. They wholly contributed to the rhetoric of empire in appropriating contested terrains and wresting debatable lands through representations while projecting British rule as an effective machinery for controlling and governing the space presented.

Notes

1 Wordsworth. William. *The Prelude: The Four Texts*. (1750). UK: Penguin, 1996. ll. 256–64.

2 Ogborn, Miles and Charles W. J. Withers. "Introduction: Georgian Geographies" in Ogborn, Miles and Charles W.J. Withers ed. *Georgian Geographies: Essays on space, place and landscape in the eighteenth century*. Manchester: Manchester University Press, 2004. p. 2–23.

3 Eaton, Natasha Jane. "Imaging Empire: The Trafficking of Art and Aesthetics in British India c. 1772 to c. 1795". Vol. 1. (Ph.D. Thesis, University of Warwick, 2000). p. 14.

4 Eaton, Natasha Jane. "Imaging Empire: The Trafficking of Art and Aesthetics in British India c. 1772 to c. 1795." Vol. 1. (Ph.D. Thesis, University of Warwick, April, 2000). p. 16.

5 Bratton, J.S. et al. *Acts of Supremacy: The British Empire and the Stage 1790–1930*. Manchester: Manchester University Press, 1991. p. 22.

6 See Breckenridge, Carol A. "The Aesthetics and Politics of Colonial Collecting: India at World Fairs". *Comparative Studies in Society and History*, Vol. 31, no. 2, 1989, pp. 195–216.

7 The scenery was painted by Richards, Phillips, Lupino, Hollogan and Blackmore based on drawings of Indian scenery by Thomas Daniell.

8 Quoted in Hogan, Charles Beecher. *The London Stage 1776–1800: A Critical Introduction*. Carbondale: Southern Illinois University Press, p. 1968. p. lix.

9 The oil on canvas exhibited in the Victoria Memorial Hall (R2856), captioned *View of the Esplanade, Calcutta, Garden Reach*, c. 1785 is, however, attributed to Thomas Daniell. However, comparing this to other *in situ* paintings by Hodges composed at the same place will inevitably reveal obvious similarities, indicating that it is indeed by Hodges.

10 Steube, Isabel Combs. *The Life and Works of William Hodges*. London: Garland Publishing Inc, 1979. p. 220.

11 See Starke, Mariana. *The Widow of Malabar: A Tragedy* as it is performed at the Theatre Royal, Covent Garden. London: J. Barker, 1799. Scenes of Sati were also presented in other plays like *The Burmese War; Or Our Victories in the East* (1826).
12 Hodges, William. *Travels in India During the Years 1780, 1781, 1782 and 1783.* London, 1794. pp. 81–3. Though the manuscript was published in 1794, the paintings were already in circulation and must have reached the London audience much earlier.
13 See O'Quinn. "Torrents, flames and the education of desire: Battling Hindu super- stition on the London stage". In Michael J. Franklin ed. *Romantic Representations of British India*. London: Routledge, 2006.
14 O'Quinn, Daniel. " Torrents, flames and the education of desire: Battling Hindu superstition on the London Stage". In Michael J. Franklin ed. *Romantic Repre- sentations of British India*. London: Routledge, 2006. p. 73. The playbills of the play *Cataract of the Ganges, Or The Rajah's Daughter* has the following settings and scenery: "the shores of Guzerat, in the Gulph of Cambay" by Stanfield, "Hindu Terrace and Gardens" by Roberts, "exterior of the pagoda of Brahma in Jaggernaut" by Marinari, 'sanctuary of the idol Brahma" by Marinari, "encampment of the Mahrattas" by Roberts and "the pile of sacrifice on the borders of the Ganges" by Stanfield and Andrews, culminating in the final climactic scene of "the conflagration of the wood' and ultimately, the "terrific cataract of Gangotri". See https://www. antipodean.com/pages/books/24773/theater-india/cataract-of-the-ganges-theatre- wolverhampton-february-16-1824-playbill/?soldItem=true
15 According to Daniel O'Quinn, such depictions of inner domestic spaces come from the impulse of British policy makers to regulate private spaces.
16 *The New Times*, 28 October 1823.
17 See Mitchell, Timothy. "The World as Exhibition". *Comparative Studies in Society and History*, Vol. 31. No. 2. (April 1989). pp. 217–36; Breckenridge, Carol A. "The Aesthetics and Politics of Colonial Collecting: India at World Fairs." *Comparative Studies in Society and History*, Vol. 31, no. 2, 1989, pp. 195–216. White, Daniel E. "Imperial spectacles, imperial publics: panoramas in and of Calcutta." *The Words- worth Circle* 41.2, 2010: 71–81.
18 Smiles, Sam. *Eye Witness*. Routledge, 2017. pp. 77–100.
19 See Hyde, Ralph. *Panoramania: The Art and Entertainment of the All-Embracing View*. London: Barbican Art Gallery, 1988–9; Comment, Bernard. *The Panorama*. Revised and expanded ed., Reaktion, 1999; Ellis, Markman. "Spectacles within Doors: Panoramas of London in the 1790s". *Romanticism: The Journal of Romantic Culture and Criticism*, Vol. 14, no. 2, 2008, pp. 133–48; White, Daniel E. "Imperial Spectacles, Imperial Publics: Panoramas in and of Calcutta". *The Wordsworth Circle*, Vol. 41, no. 2, 2010, pp. 71–81. Also see Special online collections titled Pic- turing Places of the British Library: https://www.bl.uk/picturing-places/articles/the-sp ectacle-of-the-panorama, sourced on 12.07.2018.
20 Hyde, Ralph. *Panoramania: The Art and Entertainment of the All-Embracing View*. London: Barbican Art Gallery, 1988–9. p. 37.
21 Quoted in Hyde, Ralph. *Panoramania*. p. 37–8.
22 Hyde, Ralph. *Panoramania*. p. 38.
23 Dibdin, Thomas Frognall. *Reminiscences of a Literary Life*. (2 Vols) Vol. 1. London: John Major, 1836. p. 146–8.
24 Blackwood to McNeill, 26 August 1825. Quoted in Douglas M. Peers. "Conquest Narratives: Romanticism, Orientalism and Intertextuality in the Indian writings of Sir Walter Scott and Robert Orme". In Michael J. Franklin. *Romantic Representa- tions of British India*. London: Routledge, 2006. p. 242. (Maga was the nickname of Blackwood's Edinburgh Magazine.)

25 J.W. Kaye. "the Poetry of Recent Indian Warfare". *Calcutta Review*, ii, (1848):222. Quoted in Douglas M. Peers. "Conquest Narratives: Romanticism, Orientalism and Intertextuality in the Indian writings of Sir Walter Scott and Robert Orme". In Michael J. Franklin. *Romantic Representations of British India*. London: Routledge, 2006. p. 243. In their photographic project, *The People of India*, the views and profiles of various tribes and castes of India, they detail typical modes of warfare and expertise in particular weapons.
26 See https://commons.wikimedia.org/wiki/File:Panorama_of_a_durbar_procession_of_Akbar_II_(Retouched).jpg
27 The panorama images of the architecture of the "Near East" were born directly out of the expedition undertaken in 1833 by Frederick Catherwood, himself an architect and engineer as well as an archaeologist.
28 Guha-Thakurta, Tapati. *Monuments, Objects, Histories: Institutions of Art in Colonial and Postcolonial India*. Delhi: Permanent Black, 2004. p. 18.
29 Guha-Thakurta, Tapati. *Monuments, Objects, Histories*. pp. 18–19. See Fergusson, James. *History of Indian and Eastern Architecture*. (1876) Cambridge: Cambridge University Press, 2013.
30 Roskill, Mark. *The Language of Landscape*. Pennsylvania: Pennsylvania State University Press, 1997. p. 104.
31 Quoted in Hyde, Ralph. *Panoramania*. p. 152.
32 Hyde, Ralph. *Panoramania*. p. 140.
33 Roskill, Mark. *The Language of Landscape*. p. 108.
34 Comment, Bernard. *The Panorama*. Revised and expanded ed., Reaktion, 1999. p. 19.
35 Wood, Gillen D'Arcy. *The Shock of the Real: Romanticism and Visual Culture, 1760–1800*. Palgrave, 2001. p. 103.
36 Eaton, Natasha Jane. "Imaging Empire". p. 6.
37 Quoted in Archer, Mildred. *Early Views of India; The Picturesque Journeys of Thomas and William Daniell 1786–94*, London: Thames and Hudson, 1980. p. 224.
38 Conner, Patrick. *George Chinnery. 1774–1852: Artist of India and the China Coast*. Suffolk: Antique Collectors Club, 1993. p. 82.
39 Eaton, Natasha Jane. "Imaging Empire". p. 42.
40 Eaton, Natasha Jane. "Imaging Empire". p. 69.
41 Tobin, Berth Fowkes. *Colonizing Nature: The Tropics in British Arts and Letters, 1760–1820*. Philadelphia: University of Pennsylvania Press, 2005. pp 95–9.
42 After Hastings' retirement following his impeachment, many plants were brought from India to landscape his India-themed estate in Daylesford. See British Library, Warren Hastings Papers, Add MSS 29232ff. 393–410.
43 Axelby, Richard. "Calcutta Botanic Garden and the colonial re-ordering of the Indian environment". *Archives of natural history* 35.1 (2008): 150–163; Thomas, Adrian P. "The establishment of Calcutta Botanic Garden: plant transfer, science and the East India Company, 1786–1806". *Journal of the Royal Asiatic Society* 16.2 (2006): 165–177; Raj, Kapil. "Colonial encounters and the forging of new knowledge and national identities: Great Britain and India, 1760–1850." *Osiris* 15 (2000): 119–34.
44 Keay, J. *India Discovered: The Achievement of the British Raj*. Leicester: Windward, 1989; Axelby, Richard. "Calcutta Botanic Garden and the Colonial Re-ordering of the Indian Environment." *Archives of Natural History* 35.1 (2008): 150–63.
45 Ibid. p. 99; Also see Barrell, John. *The dark side of the landscape: the rural poor in English painting, 1730–1840*. Cambridge: Cambridge University Press, 1980.
46 Allen, Brian. "East India Company's Settlement Pictures: George Lambert and Samuel Scott". Pauline Rohatgi and Pheroza Godrej eds. *Under the Indian Sun: British Landscape Artists*. Delhi: Marg Publications, 1995. pp. 7–11.

47 See for example, Forrest, Lt Col. *A Picturesque Tour Along The Rivers Ganges and Jumna, In India*. New Delhi: Niyogi, 2016 [1824].

48 Teltscher, Kate. "India/Calcutta: city of palaces and dreadful night". 2002. In Peter Hulme and Tim Youngs eds. *The Cambridge Companion to Travel Writing*. Cambridge: Cambridge University Press. 2002. pp. 191–206.

49 The "Black Hole" of Calcutta was a small dungeon in Fort William where troops of the Nawab of Bengal, Siraj-ud-Daula, held British prisoners of war under fearful conditions after the capture of Fort William on 20 June 1756. There is doubt over whether this event happened at all: many believe the fearful event was fabricated by John Zephanniah Holwell, a Company official, to malign Siraj and his soldiers. See Teltscher pp. 196–7.

50 Ibid.

51 Hodges, William. *Travels in India: During the years 1780, 1781, 1782 and 1783*. Delhi: Munshiram Manoharlal, 1999. p. 14.

52 Williams, Raymond. *The Country and the City*. 1973. New York: Oxford University Press, 1975.

53 Duncan S.A. Bell. "Dissoving Distance: Technology, Space, and Empire in British Political Thought, 1770–1900", *Journal of Modern History*, Vol.77, 2005, p. 159.

54 Also see Nilanjana Mukherjee. "A Desideratum More Sublime: Imperialism's Expansive Vision and Lambton's Trigonometrical Survey of India", *Postcolonial Studies*, 14 (2011), 429–47.

55 John Urry, *Consuming Places*. London; New York: Routledge, 1994. p.139.

56 Smiles, Sam. *Eye Witness: Artists and Visual Documentation in Britain 1770–1830*. Ashgate, 2000. p. 151.

57 Eyre, Giles. "Foreword" in Shellim, Maurice. *Oil Paintings by Sir Charles D'Oyly, 7th Baronet 1781–1845*. London: Spink, 1989.

58 On 2 June, the same year, another British painter set sail for India – Henry Salt. However, Henry Salt had already secured a position of secretary cum draughtsman to Lord Valentia. Chinnery had no such position or occupation to boast of when he set out for Madras on 11 June on board the *Gilwell*.

59 See Conner, Patrick. *George Chinnery 1774–1852: Artist of India and the China Coast*. Woodridge, Suffolk: Antique Collectors' Club, 1993; Conner, Patrick. "The Poet's Eye: The Intimate Landscape of George Chinnery" in Pauline Rohatgi and Pheroza Godrej eds. *Under the Indian Sun: British Landscape Artists*. Bombay: Marg Publication, 1995; Rohatgi, Pauline and Pheroza Godrej eds. *India: A Pageant of Prints*. Bombay: Marg Publications, 1989.

60 For a detailed recounting of the affair, see Dalrymple, William. *White Moghuls: Love and Betrayal in Eighteenth Century India*. UK: Penguin, 2002.

61 *Madras Courier*, 26 November 1807. Quoted in Patrick Conner. *George Chinnery. 1774–1852: Artist of India and the China Coast*. p. 79.

62 The portrait would hang in the town hall and was to an extent derived from the tradition started by Sir Elijah Impey, the first chief justice of Bengal, whose portrait was painted by Johann Zoffany.

63 Eaton, Natasha. "Between mimesis and alterity: Art gift and diplomacy in colonial India". In Michael J. Franklin ed. *Romantic Representations of British India*. London: Routledge, 2006. pp. 98–9. In an interesting exposition, Eaton discusses the complete reversal of the portrait-gifting tradition in colonial India from the one existent in England at the time wherein the portraitist received payment from the person who commissioned the portrait and not the sitter who was generally the one to whom the portrait would be gifted. Generally, British painters were presented to the nawabs by the Company Residents who would then paint portraits of the native nawabs as gifts from the Company appropriated in the "nazr-khilat" symbolism for

which they were expected to receive payments, not from the Company, but from the native ruler who was the sitter. Many a time a raja or a nawab's default in payment to the painters resulted in extortion and land annexation in lieu of the unpaid debt. Often, the thus retrieved amount would not be recovered by the actual creditor i.e. the painter himself, who by now had already left India for Britain.

64 Quoted in Conner, Patrick. *George Chinnery 1774–1852: Artist of India and the China Coast*. p. 89.

65 Heber, Reginald. *Narrative of a Journey through the Upper provinces of India*, 2 Vols. London, 1844. Vol. 1, p. 90.

66 Sir Thomas Munro (1761–1827) held various posts in colonial administration in India and had introduced the "Ryotwari" system of land revenue. See Maine, Henry Sumner. *Village Communities in the East and West* (1871). See Inden, Ronald B. *Imagining India*. Oxford: Basil Blackwell, 1990 for an exposition of Marx's translation of cast as forces and relations of production and his conceptualisation of the "Asiatic mode of production".

67 Pyne, W.H. *Etchings of Rustic figures for the Embellishment of Landscape*. London: James Rimmel and Son, 1815. p. 6.

68 Hemingway, Andrew. *Landscape imagery and urban culture in early nineteenth-century Britain*. Cambridge University Press, 1992.

69 D'Oyly, Charles. *Tom Raw the Griffin*. V: XXV, London: R. Ackermann, 1823. p. 121.

70 Heber, Reginald. *Narrative of a Journey through the Upper provinces of India*, 2 Vols. Vol. 1, London, 1828. p. 238. Quoted in Shellim, Maurice. "Introduction". *Oil Paintings by Sir Charles D'oyly, 7th. Baronet 1781–1845*. London: Spink, 1989.

71 Archer, Mildred. *British Drawings in the India Office Library*. Vol.1: Amateur Artists. London: Her Majesty's Stationary Office, 1969. p. 1.

72 Losty, J.P. "Sir Charles D'Oyly's Lithographic Press and His Indian Assistants". In Pauline Rohatgi and Pheroza Gandhi eds. *India: A Pageant of Prints*. Bombay: Marg Publications, 1989. p. 137.

73 Quoted in Losty, J.P. "A Career in Art". p. 83.

74 Qouted in Losty, J.P. "A Career in Art: Sir Charles D'Oyly". Rohatgi, Pauline and Pheroza Godrej eds. *Under the Indian Sun: British Landscape Artists*. Bombay: Marg, 1995. p. 87.

75 D'Oyly, Charles. *Antiquities of Dacca*. Nos. 1–4. (With engravings by J. Landseer from drawings by Sir C. D'Oyly). London: John Landseer, 1830(?) p. 6.

76 D'Oyly, Charles. *Antiquities of Dacca*. p. 14.

77 D'Oyly, Charles. *Antiquities of Dacca*. p. 14.

78 Gombrich, E.H. *Art and Illusion: A Study in the Psychology of Pictorial Representation*. London: Phaidon, 1962. p. 12

79 D'Oyly, Charles. *Antiquities of Dacca*. p. 14.

80 Conner, Patrick. *George Chinnery 1774–1852: Artist of India and the China Coast*. pp. 89–90.

81 D'Oyly, Charles. *Antiquities of Dacca*. n.p.

82 D'Oyly, Charles. *Antiquities of Dacca*. n.p.

83 Ravi Ahuja, *Pathways of Empire: Circulation, Public Works and Social Space in Colonial Orissa, C. 1780–1914*. Hyderabad: Orient Black Swan, 2009. pp.72–4.

84 J.P. Losty, "Sir Charles D'Oyly's Lithographic Press and His Indian Assistants", in Pauline Rohatgi and Pheroza Gandhi eds *India: A Pageant of Prints*. Bombay: Marg, 1989. p.135.

85 Charles D'Oyly, *Sketches of the New Road in a Journey from Calcutta to Gyah*. Calcutta: Asiatic Lithographic Press, 1830.

86 Rennell, James. *A Description of the Roads in Bengal and Bahar*. London: East India Company, 1778.

87 Chatterjee, Kumkum. *Merchants, Politics, and Society in Early Modern India: Bihar, 1733–1820*. BRILL, 1996; Markovits, Claude, ed. *Society and Circulation: Mobile People and Itinerant Cultures in South Asia, 1750–1950*. London: Anthem Press, 2006; Wilson, Jon E, "A Thousand Countries To Go To": Peasants and Rulers in Late Eighteenth-Century Bengal", *Past & Present*, 189 (2005); Markovits, Claude. "The Political Economy of Opium Smuggling in Early Nineteenth Century India: Leakage or Resistance?", *Modern Asian Studies*, 43 (2009), 89–111.

88 For details on the construction of the new Military Road see Mukherjee, Nilanjana. "Drawing Roads, Building Empire: Space and Circulation in Charles D'Oyly's Landscapes of India". South Asia: *Journal of South Asian Studies,* Vol. 37, II, 2014. pp. 339–55.

89 Stephen Daniels, "Thresholds and Prospects", in Nicolas Alfrey, Stephen Daniels and Martin Postle eds. *Art of the Garden*, London: Tate, 2004.

90 Smiles, Sam. *Eye Witness*. p. 150.

91 Anand A. Yang, *Bazaar India: Markets, Society, and the Colonial State in Gangetic Bihar*. Berkeley and Los Angeles: University of California Press, 1998. p. 144.

92 Anand A. Yang. *Bazaar India*. p. 136.

93 Athens was the capital of Greece and a key city dominating the Attica region. It is considered a landmark for cultural heritage and Classical and Hellenic architecture.

Part III

NARRATIVISING TRAVEL

This section talks about travel as a cultural activity and writing about it as a practice garnering support towards constructing a particularised identity of a space. Following an identical scheme to that in the preceding section, the first chapter looks at English travel narratives, which primarily aimed to fashion English society and space, and then progressed towards coagulation of a national identity of the British conglomerate. Though the recording of the experiential facet of the activity is encoded within various generic conventions, that the author-narrator retains an individual voice is not to be elided. However, the construction of space in the way that John Urry calls "place myths" depends on literary representations following aesthetic codes born under specific socio-cultural and economic patterns in Britain. The technology of spatial construction developed through travel literature is then extended further beyond the nation to encompass a wide array of spaces across the globe. The English narratives of travel by the British in India similarly constructed the subcontinent in its terrestrial materiality. The construction of spatial identity through writing thus involved specific aesthetic and observational schemes. The narratives then supplied information into individual grids within the overarching frame of the map. While the map designated the extents and bounds of the geographical unit, landscape paintings and especially travelogues supplied information regarding the appearance of the space.

5

PLACE AND IDENTITY
Travel narratives in the making of Britain

> No place, not even a wild place, is a place until it has had that
> human attention that at its highest reach we call poetry.
>
> *The Sense of Place*, Wallace Stegner[1]

An entry in the *Penny Magazine*, brought out by the Society for the Diffusion of
Useful Knowledge, dated 16 June 1832, states:

> Everyone says that geography is one of the most useful things that can
> be learnt; yet nothing is learnt so ill, because nothing is taught so ill.
> Look into any of the elementary books of geography, and read what is
> said about England. First, we are told that it is divided into forty
> counties, then perhaps follows an account of the several law circuits
> and then after some short notices about religion, government, produce
> and manufactures, there are given lists of the chief towns, mountains,
> rivers and lakes. But all these things are given without any connexion
> with each other, and it is a mere matter of memory to recollect what is
> no more than a string of names. And if a man thus recollects them, still
> he is not much the wiser for them; he has got no clear and instructive
> notions about the country, but has merely learnt his map, and knows
> where to find certain names and lines upon it.[2]

Clearly then, there was another way of doing geography, and that had to be
one which drew connections between names and the physical terrains of the
regions identified, and between the "hills" and the "rivers".[3] Narrative
recounting of travels in those regions could secure the required connections and
give a voice to the map. This chapter will try to find out how literature (especially of travel) is used for cartographic goals: to control, order or limn a place.
In instituting an analogy between the discipline of geography and the art of
writing, literature of travel pinpoints location and ushers untrammelled places
into recognition by orienting them according to a coordinate system that unifies
the globe. Whereas the study of cartographic procedures of mapping reveals
processes of control of land and technologies of governmentality, symbolic

representation of landscapes in literature can deal with land as perceived and experienced adding dimensions to the scaled down replica of regions. One would tend to believe that what is missing in planimetric maps, would likely be captured in such space–time based narrative strategies, such as socio-economic structures, emotions, tastes, smells and noises. "Literary geography" which has recently emerged as a critical method can prove to be quite helpful in under-standing travelogues as rupturing the cartographic synchrony.[4] For a space to come into being, a complex intersection between cartography and literature is required to chart the interior topography and the exterior landscape, sculpting boundaries and forging distinct units. However, this chapter will demonstrate how English travel texts moving horizontally through space are syntactically linear and narratological, seldom becoming anything other than a practice in cartographic reasoning.

In post-colonial scholarship, most discussions about travel writing and its relationship with the construction of "West" and "East" have followed Edward Said and his concept of "Orientalism". Said, of course, looks at a gamut of lit-erary forms over the ages, from the classical age onwards, to show how they function as technologies of colonialism even before any overt act of colonisation had formally been incepted. Later works, like that of Mary Louise Pratt's *Imperial Eyes*, while sharing Said's premise, assert that narratives of travel about "other" places and their people are marked with Western privilege and superiority.[5] Said's works, especially *Culture and Imperialism*, have alerted us to the primacy of geography that underlies Western fictional and literary work and the immanent ideology of control over territory.[6] Moreover, these narra-tives work as discursive forces contributing to the provisional construction and consolidation of that privilege. The study of travel writing has developed in a truly interdisciplinary manner in which specialists benefit from each other's methodologies and insights. Their arguments have revolutionised historical and literary scholarship and usefully complicated understandings of the relations of power and knowledge in the expansion of Europe. Much recent work, like that of Derek Gregory, Felix Driver, Miles Ogborn and John Urry, use methods of cultural geography to counter textuality in order to show how "travel writing is intimately involved in the staging of particular places".[7] While the same is the underlining objective in this section, it is also curious to see how travels within the British Isles shaped, modified and circulated ideas of the British nation. At this point, it is also necessary to draw attention to the fact that although the works of Said and Pratt have made critical examination of colonial travel lit-erature fashionable in post-colonial studies, not much is said about recountings of travel within Great Britain. Therefore, following Simon Gikandi, I would like to maintain that texts, especially of travel, were tools of rule as well as instruments for mediating and mitigating colonial and domestic crises and post-colonial studies should provide tools to question both nativism and colonial epistemology.[8]

The genre and geography

The literature which encompasses the world of travel is substantial and stretches far back in the historical records of many centuries and of many societies. Though there have been travels in spaces and by people other than from the West, these are largely marginalised in scholarship on travel writing: Eurocentrism has dominated this field, undisputed, until recently.[9] My objective here is to see English travel in the eighteenth and nineteenth centuries as experiments in spatial construction of Great Britain, successfully implemented in the homeland and metropolis, and simultaneously extended to spaces outside. Travel literature has a long generic history. More so, in Britain it followed certain generic conventions which arose out of centuries-long practices. In fact, travel literature could not expunge itself of these conventions even in the high age of imperialism. In the age of Enlightenment, Britain witnessed a subtle change in the way travel narratives were constructed. "Travel" in this period severed ties with the idea of "travail" (ordeal) with which it used to be etymologically linked, marking the evolution of what is generally viewed as "romantic travel".[10] Travel as an activity, ceased to be regarded as an uncomfortable and hazardous necessity, or a trial by ordeal or rite of passage, all of which are associated with the word "travail". These were the moral and educational arguments in favour of journeys made right from the early modern period in England to the eighteenth century: a long period over which England saw the burgeoning of the popular phenomenon of the Grand Tour in Europe. In fact, travel was an essential prerequisite for the all-round growth and development of the personality of an English gentleman and its writing was narratologically linked with the *bildungsroman*.[11] Romantic travel literature, therefore gave vigorous new life to age-old figurative meanings of the journey, focussing as much on the spiritual experience as on material exploration. In passionately seeking novelty and intensity in encounters with places, what is often evoked is a strong emotional and aesthetic response to its landscape. Although the vocabulary of such a response is rarely unique to the travel writer, a convincing display of the rhetoric of landscape appreciation is nonetheless a sure means for travellers to perform their romantic credentials.[12] This brings us to the central idea of the present work, i.e. the representational and evocative facet of such writings about land, nature and geographical spaces, much in the same manner as landscape paintings, are processes of spatialisation.

According to Lefebvre, spatial practices range from individual routines to the systematic creation of zones and regions. Spatial practices materially concretise environment and landscape over time. These are representations of space which signify collective experience of space including symbolic differentiations and collective fantasies surrounding a space. As travel is a movement in space, therefore it follows that its literature, which itself is a cultural construct, employs intellectual, metaphorical and cultural skills as vehicle of geographical and topographical knowledge. That the nature of travel writing

191

encountered a sudden change in the period of colonial expansion is today largely known. However, this sudden proliferation could be read as an extension of practices underway in England for a long time in the genre's own evolutionary history. While on the one hand, the author's personal introspection and subjectivity makes way for application of collective fantasies, the cultivation of the form modelled on letters and journals at another level has been the marker of authenticity: a veritable proof of genuine knowledge. Therefore, though authenticity is at times suspect, still, as with "on the spot paintings", truth claims of the genre are quite high precisely because of its first-person narrative format. These forms derive their strength and credibility from their being premised on sight. So they are as much ocular centric as are maps and paintings.

The question to be answered is: does the description of a place recreate that space? According to Michel de Certeau, a description of a place "oscillates between the terms of an alternative: either seeing (the knowledge of an order of places) or going (spatializing actions)".[13] "Seeing" involves cartographical rationality or the knowledge of the order of places, and is the privilege of the map: the projection onto a plane which exposes to full view, a totality of spatial relations. "Going" belongs to the order of the itinerary or tour, which involves exploring space through movement and operative action.[14] Travel writing fluctuates between both seeing and going where there coexists a desire to conceptually and visually control the space as well as capture the lived experience manifested principally through its social dimension depending on the existence and livelihood of its inhabitants. It provides a textual map to navigate the material experience of a particular place. Travel writing cannot be held apart from the material consequences of the act of travel itself, and therefore, ideologies of class, gender or race structure experiences by creating iconic objects associated with particular places. It brings people and places into being through its own discursive mechanisms which are very often permeated by unequal social and environmental relations.[15] There could be a tension arising from irreconcilable contradictions between the narrator-traveller's embodied experience in the land traversed, and its representation in text and image, complicated and aggravated by demands of the aesthetic and literary cultures of the time, such as those conventions of the Romantic era which influenced the depictions of landscapes through paradigms of picturesque and sublime. In order to understand the process of construction of place through travel literature, a brief recounting of the genre's gradual evolution as it developed in post- Renaissance Europe is essential before further deliberations on the constructions of British geographies.

Charting moral geographies: narratives through the ages

At one end of the spectrum of this literature was the guide book, produced to assist potential tourists on their journey (developed on medieval linkages to religious pilgrimages and crusades to the Holy Land). The sixteenth-century growth in travel and trade within and beyond Europe resulted in a massive surge in travel accounts. Reading about travel became a staple diet for the educated elite by the later seventeenth and early eighteenth century. During the latter century, the pervasive influence of these accounts in literary culture can be seen in its role in the development of the novel where plot, structure, character and style all reveal links with travel literature traceable in the works of Swift, Defoe, Fielding, Smollett, Johnson and Sterne.[16] As in the rest of Europe, travel literature of England before 1600 was primarily focussed on travel outside the country. In fact, there was no stable idea of the country at the time and, as seen in the preceding chapter dealing with landscape paintings, travelogues too form part of the discursive process giving a geographical shape to the nation. These early Western travelogues, memoirs and journals collectively participate in the active making of a global geography. They constitute the framework of sinews which British imperialism was to build upon in the later years.

Travel narratives began in England before the advent of writing, with the retelling of the great Germanic epics by bards connected to the courts of regional kings.[17] However, in such epics and romances, the landscapes through which characters travel are invariably imaginary and fictive. The medieval movements of crusade and pilgrimage generated the largest number of non-fictional narratives, including writings about both England and the Middle East. Chaucer's *Canterbury Tales* uses the device of the journey to tie together disparate tales and characterise the tellers who came from various walks of medieval English life.[18] In such writings, spaces are charted as metaphorical moral-scapes. Similar to its representations in landscape paintings, the sites of Christian significance are often idealised, where, on the other hand, non-Christian spaces are demonised. Later Western travels often deploy similar topoi with respect to travels outside Christendom, prominent in the representations of the New World and the European colonies. David Harvey notes while tracing the transition of the European world from disjointed pagan worldviews to a unified Christian ethos:

Since space beyond the bounds of Christendom "lost its positive qualities", the extension of Christian space became part of an often virulent struggle to wrest space from the forces of evil.[19] *The Travels of Sir John Mandeville* (1357?) was one of the first accounts to weld the structure and information of the pilgrimage narrative with the geographical accounts of the medieval "encyclopedias" which were compendia of exotic wonders of the Eastern world complete with elaborate description of its flora, fauna, climate and human life. Though, now largely believed to be inauthentic and fantastic, the work was deemed to be non-fictional at the time and till the fifteenth and sixteenth centuries, explorers including Christopher Columbus used *Travels* as a source of reference for their voyages. The emergence of a naturalistic and ethnographic paradigm begins, with the necessity to respond to the credibility of belief. This could prompt lay writers to use the traditional form of the allegorical journey of the pilgrimage to explain theological dogma or to assert religious vision. The response was to embrace empiricism in order to establish a universal truth about people and places. Most importantly, the offshoot of this kind of moral literature was narrative truth and growth of ethnography. In fact, later missionary travel accounts with primarily religious aims stimulated the creation of an empirical, rather than spiritual, travel literature:

> the issue about him was no longer salvation, not even moral wisdom, but rather the reliability of knowledge, a kind of science of nature and mankind. ... the authority of the traveller replaced that of the book; the book was only authoritative if the traveller whose report it contained was authoritative too.[20]

Where pilgrimage is the dominant kind of travel to be associated with the medieval times, exploration characterises travel writing during the Renaissance. Undoubtedly, Richard Hakluyt's *Principall Navigations, Voyages and Discoveries of the English Nation* (1589) is the most important English collection of travel narratives in the sixteenth century. Functioning as geographical compendia, the collection contains not only travellers' narratives, but also maps and other related documents. As is identifiable from the title itself, the purpose of this collection was to celebrate the success of English exploratory and trading enterprises and to stimulate further endeavours in colonisation. Sir Walter Raleigh's *The Discovery of the Large, Rich, and Beautiful Empire of Guiana* (1596) is another celebratory narrative which highlights the spirit of optimism and opportunity of the Renaissance. Early English travel writing, therefore, adopted a ritual format that straddled possibilities of trade and scientific information about distant posts. Navigation was central to the mission and writing was central to navigation:

> The course a ship took and where it anchored was a matter of "reading" economic possibilities, natural aids and hindrances, and political forces. In finding suitable anchorages, keeping a ship out of range of the guns of a castle or fort was as much a consideration as the potential trade available or the depth of water and solidity of the sea bed. Company navigation was a promiscuous mixture of forms of knowledge and practice. It included astronomical and solar observation. It involved consulting written accounts and maps, either from Dutch and Portuguese sources or produced by previous East India Company pilots, captains and generals.[21]

Travel, exploration in the age of Enlightenment, not only incorporated scientific consciousness, but also religious faith, becoming a curious mix. In what looked like "scriptural narratives" the author's foregrounding of his persona was a notable transformation:

the production of landscape narratives was not an unproblematic description of geography, but was a calculated and collective act of representation. Narratives functioned not only to bring landscape to the readers' attention, but, more importantly, to service the claims – geographical, scientific and reputational – which their authors wished to make.[22]

Travel writings grew to be an eclectic genre, aspiring to draw into its fold many classes of readership. In this respect, travel narratives bear the imprint of the intersection between science and the Christian religion in that the physical world and natural phenomena were to be studied to reveal God's designs and biblical truths. Whether narratives of travels to the Holy Land or in the overseas colonies, travelogues aimed to place science in the service of religion by trying to unearth a truth which lay in concordance with holy scriptures.[23] Together, the genre whetted an appetite for "distant reading", where, as Moretti says, "distance is not an obstacle, but a specific form of knowledge".[24]

The Grand Tour: fashioning self and the nation

The Early Modern period was not only the age of the explorer. It also saw the humanist leaning towards old and new learning and the resulting educational forms. This was the era also, when travel became part and parcel of formal education and learning, which was uncommon earlier.[25] The phenomenon called the Grand Tour emerged as a particular type of journey which would gravitate towards the traveller's personal development and initiation into the world. It soon became a popular and almost compulsory social and diplomatic institution which took the English elite to certain stipulated destinations which would invoke in the observer awe and admiration towards classical culture and aesthetics as part of the classical revivalist movement of the times. In what was meant to be professional training, travellers were generally taken to France and Italy and also to Germany and Switzerland. Italy usually claimed top priority among places visited, as the cradle of European culture and civilisation. The tour served to build an encyclopedic collection of all kinds of knowledge thought to be beneficial for England in its relationship with other nations. This included observations on climate, trade, agriculture, fortification, drainage systems and many other details of everyday life. The impact of these tours was great on England, largely shaping its social and public spaces. In his *Instructions for Forreine Travell* (1642), James Howell underlines the traveller's obligation to his nation while on such tour, commissioned or otherwise, and reminds them of the service such travellers had given their nation in yesteryears:

The most materiall use therefore of Forreine Travell is to find out something that may bee applyable to the publique utility of one's own

countrey, as a Noble Personage of late yeares did, observing the uni-
forme and regular way of stone structure up and down Italy, hath
introduced that forme of Building to London and Westminster, and
elsewhere.

Another seeing their Dikes, and draynings in the Netherlands, had
been a cause that much hath beene added, to lengthen the skirts of this
Island.

Another in imitation of their aqueducts and fluces, and conveyance
of waters abroad, brought Ware-water through London streets[26].

Therefore, right from its inception, such travel was a mode of self-fashioning
not for the traveller alone, but also for the entire community. Because of the
service done to the nation, such travellers were also known as patriotic travellers. Chapter 3 of this book referred to visits to continental locations to see
how such exposure influenced English art and architecture and a quintessentially landed class's aesthetic sense in general. The Grand Tour was not considered to be of national import only during the humanist era but continued to
be so till the nineteenth century when the universalist, pan-European spirit of
the Enlightenment fostered travel as a means of moral and intellectual
improvement. Several political factors, such as the Peace of Utrecht (1713–14),
granted impetus to the Grand Tour in shaping it as a popular institution.
Touring, which was once an aristocratic male preserve, extended after the
Napoleonic Wars, following the opening up of the erstwhile blockade of the
continental system. Once the floodgates were opened for all classes and genders,
travel soon materialised into mass tourism.

With these journeys, writings of travel diversified. Bacon's essay recommended the guidance of an experienced tutor and the keeping of a detailed
diary. The need for an older and experienced tutor emerged out of the notion
that there were chances for a young man to be led entirely astray once he found
himself outside domestic paternal surveillance. Such a view of the Grand Tour
had already emerged in the late sixteenth century and can be seen in Thomas
Nashe's short work of prose fiction *The Unfortunate Traveller* (1594). Many
journals kept by travellers focused on the lures of prostitutes and other such
immoral temptations prevalent in those societies. Where travellers were not
accompanied by tutors, there was extensive apodemic literature available as a
substitute, from the 1670s, in order to safeguard the youngster from such dan-
gers. These outlined a "profitable" and methodical way of travelling and of
recording the travel experience. Among the most influential works regarding the
Grand Tour was James Howell's *Instructions for Forreine Travell* (1642). Some
of the works, like Leopold Graf Berchtold's *An Essay to Direct and Extend the
Inquiries of Patriotic Travellers* (1789), were extremely detailed and mechanical
which was designed for a plethora of questions concerning all conceivable
aspects of the country traversed. These and the huge amount of apodemic literature produced during the seventeenth century regarding the proper way of

recording, also meant that the narrative style was highly impersonal and objective. However, even though the writing subject never appeared in the forefront, they inversely defined what was British and what was outside "British" culture. Simultaneously outlined was what was to be considered home and what outside.

Britain as a unified territory and Britishness as a characteristic identity were concepts which first emerged as staple ideas only by the mid-seventeenth century. In the earlier half of the century a clear identity of the British polity had not been forged despite attempts by James I based on several kinds of visual symbolism.[27] Travel had a very important role to play in the construction of the British nation as a geographical entity. According to Doris Feldmann:

> It is a critical commonplace that travel writing is grounded on the (both fictional and non fictional) analysis of one group by another, an analysis which tends to produce a collective notion identifying "self" against "otherness". Most critical assessments of this cultural phenomenon focus on foreign travel and the literature that records it. The discursive strategies of travel accounts, which encourage the notion of a cultural hierarchy while permitting – indeed reinforcing – the illusion of national unity, are, however, also at work within domestic travel accounts.[28]

Chorographies of home

According to most scholars, home tours did not come into prominent vogue in England till the eighteenth century, or rather, few scholars have looked at texts written prior to that time to examine their imaginative potential in configuring the national territory. As has been pointed out, "Elizabethan geography encompassed both an expanding globe and an enclosing nation".[29] Texts like Thomas Blundeville's *Exercises* (1595) and Robert Devereux's sonnet draw up the striking image of an Atlantic Britain, celebrated in its insular separation exposing a desire to identify its special geographical location with its unique status:

Seated between the Old world and the New,

A land there is no other land may touch.[30]

In a period of an emerging national consciousness and pride, the English developed an interest in their own country as well as for lands abroad and the former became an object of meticulous study.[31] As in the Grand Tour journals, domestic travel in the sixteenth and seventeenth centuries are recorded with precision and exactitude in an encyclopedic manner. Typical of this period was a generic by-product generally termed "chorography" or "earth writing" which provided topographically organised accounts of individual regions. One of the noteworthy texts in this corpus is John Leland's (1502–52) notes made on his journey through England between 1535 and 1543. He was a royal librarian under Henry VIII. He was gathering material for his extensive work "History and Antiquities of this Nation", a multivolume chronological description of Britain, which never materialised. In a promotional tract presented to the king, which was later edited by John Bale, he described himself as "totallye enflamed wyth a love, to see throughlye all those parts of thys your opulent and ample realme".[32] The extent of his travel seemed to have been exhaustive:

> there is almost neyther cape nor baye, haven, creke or pere, ryver or confluence of ryvers, beeches, washes, lakes, meres, fenny waters, mountaynes, valleys, mores, hethes, forests, woodes, cyties, burges, castels, pryncypall manor places, monasteryes, and colleges, but I have seane them, and noted in so doynge a whole worlde of thynges verye memorable.[33]

Leland's project clearly centred on the land itself as a source for a description capable of capturing the essence of Britain. He was incapable, however, of collating all the data that he had accumulated and is said to have turned insane over his vast material. His notes were circulated in manuscript form and published as *Leland's Itinerary* between 1710 and 1712. Despite his failure, Leland inspired subsequent chorographical writing. Concerned with space rather than time, with place rather than person, late Tudor and early Stuart chorography and travel writing took after the antiquarianism of Leland. He was acknowledged as the central source in two major national projects of the time: William Harrison's *Historicall Description of the Islande of Britayne* (1577) and William Camden's *Britannia* (1586). Harrison, an armchair geographer, through his "poeticall voiage" offers a broadly imagined geographical, social, and institutional account of the country. It largely follows a medieval model of prefacing historical narratives with topographical introduction. It starts with the spatial setting but instead of being a topographical account of Britain it addresses several social, economic and political issues. *Description* looks at topics as diverse as England's architecture, its inns, markets and fairs, forests and parks, law courts and palaces, thus offering a multi-faceted spatial reality. Alongside, it makes elaborate observations about the resident population's dietary habits and local customs and also bitterly criticises vagaries in fashion and lifestyle. There is also a resounding tone of covert criticism disguised under a dominant patriotic stance. The travelling metaphor is able to recreate a dynamic space with a neural network of roads and rivers which is able to replicate the idea of movement through variegated space. According to Bernard Klein, the text provides scope for alternative narratives to the dominant royal perspective that the massive chronicle tries to endorse.[34]

The peregrination of the scholar William Camden actually yielded one of the most important early surveys of Britain. His *Britannia* (1586), which first appeared in Latin and was later translated by Philemon Holland, is one of the greatest achievements in English chorographic tradition. According to Bernard Klein, *Britannia* is quite a conceptual alternative to Harrison's *Historicall Descriptions*, for Camden's conception of national space corresponds to the synthesis of a unified cartographic order. In this work Camden collated, county by county, historical and topographical information incorporating every detail about "the Flourishing Kingdoms of England, Scotland, and Ireland, and the Islands adjacent". It is for the first time that the three regions come under a common integrated identificatory system. By the 1580s, Christopher Saxton (c. 1543 to c. 1610) had already completed the monumental project of mapping England and Wales. Saxton's atlas embarked on the county as the basic unit of survey. In the same manner, *Britannia* seizes on the county as the central unit of chorographical description, devoting full chapters to each. In fact, Saxton's maps were included in *Britannia* from the 1607 edition onwards. Camden acknowledged the visual supplement provided by the maps which successfully subjected visual space to Camden's political perspective, portrayed through his textual descriptions. In the preface to this later edition, he accepted that his work was weak for want of visual substitutes and therefore had suffered criticism:

> Maps were not adjoined, which doe allure the eies [sic] by pleasant portraiture, [they] are the best direction in Geographicall studies, especially when the light of learning is adjoined to the speechlesse delineations.[35]

According to Camden therefore, maps were a necessity only because they satisfied an aesthetic side, their "pleasant portraiture" served the purpose of embellishment. It is the written word which provided reason and explanation and shed the "light of learning" to the otherwise meaningless and puerile "speechlesse delineation". To quote Bernard Klein:

> If the nation is defined in and through space, then its cartographic representation needed to concur with the precepts of the dominant political narratives surrounding its induction.[36]

John Speed, too, realised this necessity and accordingly followed up his atlas with five volumes of English history in his *The Theatre of the Empire of Great Britain* (1611). Moreover, recognising the importance of the written word, a great deal of antiquarian information was crammed into the margins of the maps of the counties, each of the plates for which were invariably followed by a two-page annotative and descriptive text. According to Klein, with the antiquarian information that Speed spreads over the surface of his maps in the form of portraits, heraldic symbols, battle scenes, historical medallions, ancient coins or Roman inscriptions, "an autonomous land becomes the protagonist of a story of historical continuity".[37] Helgerson notes that the proliferation of spatial representation of this period is far from ideologically neutral. He argues that the chorographical and cartographical works produced during this time allowed the English, for the first time, to take "effective visual and conceptual possession of the physical kingdom in which they lived".[38]

Fynes Moryson (1566–1630), an English traveller and writer, spent most of the decade of the 1590s travelling on the European continent and the eastern Mediterranean lands. He was also a keen traveller of his native England and of Scotland. In1617, he published the first three volumes of *An Itinerary: Containing His Ten Years Travel Through the Twelve Dominions of Germany, Bohemia, Switzerland, Netherland, Denmark, Poland, Italy, Turkey, France, England, Scotland and Ireland*. The *Itinerary* was originally intended to consist of five volumes, out of which only three were published. The unpublished fourth volume happened to be preserved in the Corpus Christi College, Oxford and was published much later in the twentieth century. Volumes III and IV have short chapters on customs and institutions in England. Volume II, on the other hand, is devoted to affairs in Ireland from 1599 to 1603.[39]

Occasionally, in the sixteenth and seventeenth centuries, there also appeared short, entertaining pieces of travel writing which were directed at a wide audience and published in the inexpensive pamphlet form. Among the popular pamphleteers was John Taylor (1578–1653), who was generally known by his pseudonym "the Water Poet", which arose from his main occupation as a London waterman, a member of the guild of boatmen who ferried passengers across the River Thames. He was a prolific, if rough-hewn, wit who had more than 150 publications in his life time. He was a multi-talented person and wrote both in verse and prose form about multifarious issues. Among these, there were accounts of his travels which he made to places on the continent like Prague and Hamburg. However, what is of interest here is his travel through England, Scotland and Wales. Many of Taylor's works were published by subscription. He would propose a book, ask for contributors and write it when he had enough subscribers to fund the printing and publishing cost. Records show that he had more than 1600 subscribers for his *The Pennylesse Pilgrimage; or, the Moneylesse Perambulation of John Taylor, alias the Kings Majesties Water-Poet; How He Travailed on Foot from London to Edenborough in Scotland, Not Carrying any Money To or Fro, Neither Begging, Borrowing, or Asking Meate, Drinke, or Lodging* published in 1618.[40] Taylor wove spectacular difficulties into some of his journeys, which he advertised in his prospectus in order to attract sponsors for his enterprise. Taylor proposed that he would be able to overcome his self-set obstacles, and the event of the difficulty being mastered and the sponsors agreed to fund Taylor for the trouble of the journey by buying his account of it. The work is more of a socio-cultural documentation in verse than a geographical or topographical delineation of the region, as he makes clear in his dedication right at the onset:

and now Reader, if you expect

That I should write of cities' situations,

Or that of countries i should make relations:

Of brooks, crooks and nooks; of rivers, bournes and rills

Of mountains, fountains, castles and hills

Of shires, piers and memorable things,

…

Lay down my book, and but vouchsafe to read,

The learned Camden, or laborious Speed.[41]

His travel within the British Isles was as much of a cultural encounter as was any overseas voyage. He explicitly suggests in his writing that travel within the British Isles was at times every bit as strange as a visit to foreign lands:

But you shall hear of travels, and relations,

Descriptions of strange (yet English) fashions.[42]

The same feeling was reciprocated towards him by the residents of these distant places:

As if some monster sent from the Mogul,

Some elephant from Africa, I had been,

Or some strange beast from the Amazonian Queen.[43]

Taylor had planned and marketed his work in accordance with a pattern tried and tested by an earlier traveller: in 1599, the actor William Kemp, said to have morris-danced from London to Norwich, had published a day to day account of his journey in a pamphlet titled *Kemp's Nine Daies Wonder* (1600). Taylor follows the same vein of humour in his entertaining account which was loved by the readers and was able to capitalise on popular interest. Even though the *Pennylesse Pilgrimage* does not fall within the bracket of learned chorographic writing prevalent at the time, it served a particular patriotic intent of a different order. It is a discovery of Britain for Taylor – a discovery of his native land which was as yet culturally divided within. It reflects an idea that there was need for better means of communication in order to unify the territory. Thus his journey by river in a wherry from London to Salisbury, as depicted in his tract, "A Discovery by Sea, From London to Salisbury", was intended to show that, for the national interest, measures had to be taken to make the English rivers navigable. Being a waterman by profession, this sort of keen understanding was only expected of Taylor. Taylor's account of his journey to Scotland informed his audience about a region which was still largely unfamiliar to the English, even if the two dominions had been merged under a common monarchy since 1603. Therefore, Scotland, which was seldom toured during this time, was described in all its resplendent grandeur with its castles and teeming social life and economy. Laid bare are a contested terrain and rifts between ethnic groups hidden beneath an apparent shroud of unity created by a political abstraction. Scotland was still a place where one would not go for leisure. Writers, who wrote about their visits to the Scottish Highlands at around this time, had made their trips as traders or soldiers, and had all mentioned it as a terrain fraught with difficulties. Scotland was still a place waiting to be discovered, waiting to be emplotted into the British cartographic schema. The picture was to alter markedly in the eighteenth century when Scotland attracted numerous tourists after the pacification of Culloden. The advent of the cult of the Ossian and the travels of Samuel Johnson and Boswell would popularise it as a picturesque and safe place to travel.[44]

These chorographies laid down a structural network of routes, mainly constituted through interconnected water channels, across the dynamic topography of the newly unfolding nation space. The spatial configuration emerges in traversing through the seemingly fluid regions and navigating through diversities, making an unchallenged monolithic conception of the space nearly impossible. Yet, the metaphor of travel (even when some chorographies were entirely stay-at-home productions) was able to create a spatial continuity by adopting a single point perspective of the central ruling power. In hindsight, when turning back on one's own travel, a cartographic totality could be superimposed on the landscape traversed, making it a legible place of seeing and reading, making possible its incorporation into the steadily expanding ocular regime.

Nation and narration: imagining the country and its countryside

As discussed earlier in the book, the early seventeenth century was the age when the great scheme of nation making unfolded under the aegis of James I. The trajectory would of course reach its culmination with the formal unification of the Scottish territory with England in 1707. The process involved the active participation of texts with both written and pictorial content through which the construction of a unique geographical identity of Great Britain could be forged. The place thus imagined, which existed at a level of abstraction, could now be concretised and granted material palpability. The works of Leland, Harrison, Camden and Taylor participate in this project of nation making. Knowingly or unknowingly, they are appropriated into the national trajectory of territorialising the body politic with cartographers like Saxton, Speed and Nowell imposing geometrical rationale on the visionary fabrication. According to Klein, till the seventeenth century, the geography of Britain served more as a discursive than a physical terrain:

If the emergent nation appears as an unstable signifier in contemporary maps and texts, if neither its outward form and inward structure find their definite shape, these uncertainties only serve to demonstrate that there was more than one Britain, more than one conception of national space, to be imagined or discovered in the productive arena of early modern geographic discourse.[45]

Moreover, the selection of the county as the basic unit of spatial conglomeration, validates the above argument. Such a device only deflected any singular history of a nation and served as a basic premise for writing narratives of individual ownership of spaces and constructed an order of private property. This sort of spatial representation opened up a way to look upon space as open to appropriation for private use.[46] According to Richard Helgerson, the focus on the county as the basic unit of Renaissance mapping and chorographic accounts, strengthened the sense of individual and local powers within a framework of national loyalties, while uprooting earlier identities based on dynastic loyalty.[47] Gradually, these very units developed into sites for aesthetic meditation and contemplation, implicated in a national debate of how best to imagine these privately owned areas of countryside known as "The Country", i.e. Britain. In the years to come, the debate reached its culmination in the very identification of Britain with its land, and Britishness with its landscape. According to Stephen Kohl, imagining the countryside as "The Country" was a class-defined spatio-temporal exercise which assumed a timeless version of Britain's past while sketching a polemical map onto a yet evolving geographical territory.[48]

The self-discovery, which had been in progress in the sixteenth century further gained momentum in the late seventeenth century and eighteenth century. Following the end of the Civil War, the Restoration and the revolution of 1688, there was a period of reawakening of national feeling, seeing a great boom in multifarious geographical activities and surveys. Antiquarians constructed the English past, statisticians and demographers estimated population, while topographers and road surveyors consolidated the geographical details. Camden's *Britannia* resurfaced during this time when publication of route and tour guides was extremely profitable and writing travelogues was considered highly fashionable.[49] At this time, the scope of domestic tourism and its literature increased manifold. It is not to say that tours to Europe stopped altogether. On the contrary, both coexisted harmoniously and each was able to demarcate concepts of "interior" and "exterior". Thus in the late seventeenth and eighteenth century, British patriotism was expressed not only through reviewing and celebrating Britain's natural wealth and prosperity but also by subjecting cultures external to Britain to fierce criticism. For example, in her accounts, Celia Fiennes (1662–1741) clearly considers the primacy of home tours over foreign ones as well as their utility towards developing a true perception of what was home and what was foreign:

It (home tour) would also form such an Idea of England, add much to its Glory and Esteem in our minds and cure the evil itch of over valueing foreign parts; at least furnish them with an equivalent to entertain strangers when amongst us, or inform them when abroad of their native Country, which has been often a reproach to the English, ignorance and being strangers to themselves ... But much more requisite is it for Gentlemen in general service of their country at home or abroad, town or country, especially those that serve in parliament, to know and inform themselves the nature of Land, the Genius of the inhabitants, so as to promote and improve Manufacture and Trade suitable to each and encourage all projects tending thereto.[50]

Travel writing, in this respect, became an important site for construction of the idea of Britishness and for fashioning the nation "as an imagined homogeneous community".[51] According to Barbara Korte, there was a steady shift in the perspective in travel narratives at this time. Domestic travellers of this period did not merely emphasise topography, history and antiquities, which were the favourite subjects dealt with in the previous century, but also made commentary on contemporary socio-economic aspects. Accounts of domestic travel, even when private or semi-private, focussed on intellectual, political, historical, technical, scientific and aesthetic developments in Britain as national achievements.[52] Before the language of map making became subsumed by one of technical surveying and mathematical calculations, the job of a map was to provide a gamut of information about the socio-economic and cultural aspects of a place. The demands were gradually removed from map making, which progressively chased corrective and reformist goals in accuracy. It fell upon chorographic travel writing to include all such detailed information that a map only indicated through symbols. These were clear signs of the approaching Enlightenment order whereby England was gradually waking up to its famous maxim of science, rationality and progress. By the eighteenth century, the significant expansion of territory brought about by the Act of Union with Scotland meant that a home tour did not merely restrict itself to travelling within England or Britain, but what increasingly came to be known as Great Britain. The "heterogeneous patchwork" of ethnicities like Scottishness, Welshness and Englishness which existed at the beginning of the eighteenth century gradually gave way to an overarching unifying structure of the nation called the United Kingdom or Great Britain. The differences which existed during the time of Taylor's journey were gradually ironed out with the progress of communication, connectivity and internal trade. Daniel Defoe's (1659–1731) *Tour through the Whole Island of Great Britain* (1724–6) demonstrates these transitions and the unifying cultural forces at work in what is generally seen as his economic travelogue. Modelled on the structure of popular guide books of the times, it contains a detailed survey of the roads in England in the light of the new Turn Pike Acts.[53] It is meant to operate at a level similar to a virtual tour, a "facsimile representation", where the armchair traveller is able to traverse the entire nation and participate in public discourse based on his fictive travel:

> We give the reader ... a view of our country, such as may ... qualify him to discourse of it, though he stayed at home.[54]

Even though Defoe's *Tour* tries to keep up the semblance of an authentic first-person travel narrative, for the most part, Defoe derives his knowledge from other written documents. As the journey moves further and further away from London, the less authentic it becomes, and has to rely heavily on secondary guide books for the information it tries to disseminate: the reason being the dominance of London from the Restoration onward as the centre of Englishness. This created a distinctive type of domestic travel literature which involved the touring of Britain's length and breadth but was consciously or unconsciously underscored by an imaginative compass point, of a clear demarcation of circumference and centre. This centrist bias is very obvious in Defoe's work, for whom London remains his benchmark of worth.[55] In fact, one way of understanding the Romantic movement of the late eighteenth century, is to see it as a reaction against this metropolitan sophistication, to escape the dominance of urban space in the cultural imagination of the nation. That, however, had its own trappings which contributed to the nation's linear narrative of space. Moretti has significantly drawn attention to the prevalence of the genre of "village stories" and the dominant trope of country walks in the latter part of the eighteenth and the early nineteenth century. Drawing on John Barrell's idea of landscape, Moretti shows how two divergent geographical perspectives developed side by side: the circular and the linear.

> a different geography according to who was looking at it: thus, for those of its inhabitants who rarely went beyond the parish boundary, the parish itself was so to speak at the centre of the landscape .... For those inhabitants accustomed to moving outside it, however, and for those travellers who passed through it, the parish was defined ... not by some circular system of geography but by a linear one.[56]

It is of course the linear cartographic rationale which dominates, ultimately framing the spatial reality of modernity: "a map of ideology emerging from a map of *mentalité*, emerging from the material substratum of the physical territory."[57]

Nature travel and aesthetic travellers: aesthetic construction of Britain

After a brief spell of anti-pastoral invectives on British urbanities in literature, there was a drastic return back to nature in the eighteenth century. With Romanticism in the offing, nature travel or scenic tourism became an important part of travel itself. As discussed earlier in the chapter, it is in the writing about scenic tourism that travel and mainly domestic travel, tend to become more and more subjective. Following the literary and artistic tradition of the times, the home tour became entrenched in the aesthetic paradigms of the age and descriptions of landscape became an essential part of such literature. The works of Edmund Spenser and Philip Sidney had earlier provided a model for conceptualising the English nation as an Arcadian, Edenic space. In the eighteenth century, such travels continuing the long-established tradition of the English pastoral, brought to the fore for the first time the so-long ignored spaces and helped to consolidate them within the meta structure of the nation. The same is exemplified in Samuel Johnson's *A Journey to the Western Islands of Scotland* (1775). This was the time when the Celtic fringe of the British Isles was brought to the attention of the wider public through a plethora of travelogues. Thomas Pennant, a Welsh naturalist, antiquarian and traveller had already made several tours to Scotland and the Hebrides before Johnson, where, in a way, he chalked out the itinerary for future travels, making his servant, Moses Griffith, document the entire journey through sketches and drawings.[58] Johnson was reading Pennant's *Tour of Scotland and Voyage to the Hebrides* (1772) while writing his own account and was at all times cautious of comparisons.[59] Johnson's *Journey*, however, became the most popular among these, participating in this movement of nation making. An antiquarian spirit is dominant in his writing and the records of other travellers which tried to capture the final phase of the fast-vanishing antique charm of the Scottish Highlands. Scotland was thought to be rapidly modernising, therefore urging travellers to catch a last glimpse of the swiftly transforming society before it ultimately disappeared altogether. Ironically, similar anxiety is to be noticed in the perceptions of "pre-modern" societies outside Europe and in its colonies, which Bernard Cohn articulates as the "museological modality" of travel.[60] Interestingly enough, there was a conspicuous "othering" perceivable even while charting out the home territory. Popular and canonical travel literature such as:

> Johnson's tour was very much a part of the exciting geographical exploration taking place in 1773. Johnson was surveying Scotland when Cook crossed the Antarctic Circle for the first time, Constantine Phipps sailed for the North Pole, and James Bruce returned from Abyssinia ... the Highlanders are treated as if they were Eskimos, Siberian nomads, American Indians, and Pacific Savages.[61]

Johnson's detailed ethnographic descriptions of the Highlanders are identical to accounts of the exotic places and people overseas. His travelogue hovers on the margins of anthropological treatises on the "savage" Pacific Islands of the kind Cook wrote. The eighteenth century was a time when British identity emerged albeit fraught with uncertainties: there occurred a conscious or unconscious alienisation and othering based on regional differences, ending up in the dominance of London, the English Southeast and the East Midlands, with all other locations shoved to the periphery.[62] On the other hand, with the rise of regionalism, travels to remote areas was not to affect a cultural separatism of the kind that Pittock mentions, but was part and parcel of the act of imagining the nation in its variety.[63]

Travel thus became a performed art in itself and joined hands with other aesthetic practices of the age like landscape painting, landscape gardening, nature poetry etc.[64] Landscapes were considered of particular aesthetic value: if a view appeared to be fitting into the frame of a landscape painting, the spot became much prized and sought after. There grew up literary landscapes made famous through poems and travel journals that served to develop place-myths around specific locations. The Lake District is one such literary landscape popularised in literature through the writings of Ann Radcliffe, Thomas Gray, William Wordsworth and the other Lake Poets. The literary transition is noticeable in the perception towards the Lakes which, for example, invoked "infinite delight in the grandeur of its landscapes" in Ann Radcliffe, whereas, a century earlier, Celia Fiennes did not find anything here striking enough to comment on, other than the numerous species of fish to be found in the Lake Windermere, or a purely geographical inquisitiveness about the lake's having a natural outlet.[65] In the *Short Survey* of England in the 1630s, the place was described as "nothing but hideous, hanging Hills". Also, Daniel Defoe described the district as "the wildest, most barren and frightful of any that I have passed over".[66] Urry notes that it was in some sense natural in the early eighteenth century, that the hills and mountains represented "unhospitable terror": following which, the area had to be first discovered and then interpreted as appropriately aesthetic and finally managed into a scenery suitable for millions of visitors.[67] The painter, William Gilpin's theory of the "picturesque" promoted nature travel and inspired a keen perception of landscape.[68] Gilpin's essays, especially "On Picturesque Beauty", "On Picturesque Travel" and "On Sketching Landscape" clearly define what constitutes the "picturesque". His works do not merely record his own particular journeys, but are intended to give the readers advice for their own practice of the art of travel. His writings inspired many to carry sketching aids like the Claude Glass.[69] The poet, Thomas Gray is known to have carried a Claude Glass with him wherever he went. The theory of the picturesque drew attention to the ruins of castles and abbeys, the remains of English Reformation, vegetation such as the English oak and typical English atmospheric conditions such as the mist and fog, giving birth to a distinctive physiognomy of the English countryside landscape. As a result, a discursive structure was established which fixed precepts of aesthetic judgment on nature and tried to define its experience and interpretation elsewhere as well. Significantly enough, nature travel of this sort saw the birth of many iconic places in Britain like the Lake District, which were, as Urry notes:

> not part of England until it was both visited in significant numbers and some of those visitors began to write first in a somewhat mannered picturesque style and then in what is known as English Romanticism.[70]

In all these cases, the little place-myths fed into the larger myth of the homogeneous coherent nation. Moreover, Gilpin had proclaimed in his foundational writing, *The Three Essays on Picturesque Beauty* (1794), that:

> the province of the picturesque eye is to *survey nature*: not to *anatomize matter*. It throws its glances around in the broad cast stile. It comprehends an extensive tract at each sweep. It examines *parts*, but never descends to *particles*.[71]

The same understanding. of the "broad cast stile" was effective in erasures of dissidence and therefore in projecting a homogeneous terrain. Significantly, Nigel Leask comments:

> With a preference for the glowering mountainscapes of the Celtic fringes, the metropolitan picturesque also played its part in composing the irregular variety of Scots, Welsh, Irish, and English identities into the *concordia discors* of the imperial British state. Envisioned in this way as the visual setting for "imagined community", it struck a compromise between the smiling beauty of English tillage and pastorage and the sublime "terror" of Scottish or Welsh mountain landscapes.[72]

The travelogues of the early nineteenth century, by describing the land and landscape in particular, try to define the country in general. The multifarious aesthetic and literary models of the picturesque, the sublime and romanticism try to imagine the nation in their own respective ways. In so doing, there is an unavoidable overlap of the two ideas: the rural terrain and the nation.

On the other hand, Ireland is posed as a wasteland, a fallow space which would only reach its natural apotheosis on merging with the nation space of Britain. As Klein shows, throughout the sixteenth and seventeenth centuries, the Irish space existed as a conceptual opposite of the national territory in a carefully constructed literary geography which posited it as empty and chaotic, encouraging the narrative of military and political conquest.[73] Later writings of travel only strengthened the trope. Wordsworth repeatedly recounts the landscape as pleasant but "uninteresting" and finds the region entirely devoid of romantic prospects or of sublime possibilities to trigger his imagination or creativity. While touring Ireland in 1829, Wordsworth wrote of its landscape:

> I have yet seen nothing in Ireland comparable to what we have in Wales, Scotland and among our Lakes. The celebrated vale of Avoca and the Glen of the Dargle are both rich in beauty, the latter in character something between Wharfdale and Fascally in the Highlands, where the Garry and the Tummel meet below the pass of Killiecrankie; superior to Wharfdale, but yet in a greater degree inferior to the Scotch scenes.[74]

In another letter to his wife, Mary, in the same year, he writes: "upon the whole ... I was rather disappointed with the scenery of Ireland." It was so plain and uninspiring that Wordsworth is said to not have composed a single poem out of his travels in Ireland.[75]

A homogenising gaze: Britain sees itself

However, as an obvious resultant of increased travel activities, came the further development and consolidation of the conception of identity, place and nation. During the Tudor and Stuart periods the centres of culture lay within the preserve of aristocracy. This meant that travel for cultural enjoyment lay outside the bounds of a large cross section of society. In the eighteenth century, travel as a cultural activity was vastly democratised. Country houses, ruins and natural phenomena received appreciation as tourist attractions like never before under the new vogue of the picturesque gaze. Domestic travel increased during the Seven Years' War and the French Revolution when travel on the Continent became more difficult for the British: this directed energies to home touring which led to a further consolidation of national consciousness. While at the beginning of the seventeenth century, Elizabeth I called herself "mere English", by the latter half of the eighteenth century, George III, expressing similar sentiments, claimed to "glory in the name of Briton". Knowledge of Britain increased and the new-found sense of knowing the nation was propelled by the literature of travel. Writing at the close of the century, William Mavor claimed in *The British Tourists* (six volumes, 1798–1800) that tourism had made Britain British.[76]

In the nineteenth century, domestic travel became a metaphor for conceptualising a homogeneous nation. The improvement of roads, the construction of bridges and canal waterways, and the introduction of new modes of transport such as the railways made travel not only quicker, cheaper and more comfortable, but acted as an assimilative force. The easy mobility encouraged not only travel for leisure but also migration to different places in the country for work under rampant industrialisation. The advent of a plethora of new technological developments in the nineteenth century, such as the railway, resulted in new conceptions of time and space or what Harvey calls "time-space compression".[77] Such compressions of geography, space and time differentials, subdued and unified space resulting in an imagined "small" place. Distances reduced instilling a sense of the smallness of Britain and the brevity of time consumed in reaching one spot from another, collapsed regional differences and created the possibility of understanding Britain as a whole.[78] Once the centuries-old endeavour reached its desired culmination, excitement with home travel gradually decreased. Literature of the period, as can be found in the popular fiction of Dickens, Hardy and George Eliot, is characteristically marked with nostalgia for a Britain comprising of cultural and regional distinctiveness fast reaching a point of dissolution. George Eliot reflects this sentiment of nostalgia for the old-world charm of travelling the countryside:

209

but the slow old-fashioned way of getting from one end of the country to the other is the better thing to have in the memory. The tube-journey can never lend much to picture and narrative; it is as barren as an exclamatory O! Whereas the happy outside passenger seated on the box from the dawn to the gloaming gathered enough stories of English life, enough of English labours in town and country, enough aspects of earth and sky, to make episodes for a modern Odyssey.[79]

Fast-engulfing industrialisation not only drastically changed the British countryside with its visible signs of modernisation of village and farmyards but also homogenised the landscape to a large extent. The dwindling availability of unadulterated nature turned the gaze elsewhere to find and forge new "romantic" or "literary" landscapes. What survived was a culture of mass tourism fast developing into an industry with agencies like Thomas Cook organising what can be called "packaged tours" and the mushrooming of thousands of small boarding houses. Even at the beginning of the nineteenth century, the fashion of picturesque travel and the literature about it thus attracted scorn from a number of writers. Thomas Rowlandson's caricatures, which function as illustrations to the satirical poem, "The Tour of Dr. Syntax in Search of the Picturesque" (1809) by William Combe, are among the most famous of examples. The disgust and discomfort with visible changes to the agrarian economy is obvious in early nineteenth century travelogues like in William Cobbett's overtly pessimistic tract, *Rural Rides* (1830).

In search of the exotic

By this time, the Grand Tour, too, had degenerated into a simple tour of sights or a routine of social visits. In this context, travellers from the middle of the century tried to escape the predefined route of the tour. They ventured into uncharted territories driven by a greed to "discover" new places in Europe. James Boswell, for one, who was one of the most proficient travel writers of the century, not only went to France and Italy in the traditional format of the Grand Tour, but ventured into the interior of Corsica, and this part of his journey was indeed the only one he chose to write about, in his *Account of Corsica* (1768). In 1773, Patrick Brydone published his *Tour through Sicily and Malta*. The Iberian peninsula too became a hot favourite as an exceptional destination exemplified in Robert Southey's *Journals of a Residence in Portugal* (1800–1). Mary Wollstonecraft published her *Letters Written during a Short Residence in Sweden, Norway, and Denmark* (1796) from another unusual destination. A conscious defiance of the conventions of the tour is also perceivable in Tobias Smollett's *Travels through France and Italy*, which he took as an aged man to recover his health and not as a young man as part of his education. In 1792, Arthur Young, in his *Travels in France and Italy*, which is primarily an agronomic account, also discusses the disappointment he encountered as against his expectations.

Dissatisfaction with the canonised Grand Tour is particularly apparent in the account of William Beckford who set out on a tour in 1782 with a visit to the Netherlands followed by visits to Germany and Austria, finally finishing off with Venice. Beckford, whose Gothic novel *Vathek: An Arabian Tale* (1786) was popular and reflected the mood of the times; it revealed his dissatisfaction with pre-set itineraries and is driven by his wish to experience unknown and faraway lands. His actual experience of countries in Europe only makes him visualise and long for more remote and exotic lands. In Antwerp, he dreams of an imaginary Orient, of "Arabian happiness" which disappears when he wakes up. This desire for the Orient is even more prominent in his outlook towards Venice, which was thought to be the closest European city to Eastern culture: "Asiatics find Venice very much to their liking and all those I conversed with allowed its customs and style of living had a good deal of conformity to their own." He loses himself in his imaginary escapades in search of authentic Arabian pleasures as he braves "the vapours of the canals", and ventures into

> the most curious and musky quarters of the city, in search of Turks and Infidels, that I may ask as many questions as I please about Damascus, and Suristan, those happy countries, which nature has covered with roses.[80]

It is quite clear that there was a rising discontent and surfeit with places oft-visited, spoken or written about. Mass travel had rendered the Grand Tour mundane and familiar, and the readers yearned to hear and learn about places they would not have the opportunity to visit. Also, the romantic notion that the self is found not in society but in solitudinous contemplation of nature, collided with the ever-rising British phenomenon of mass tourism. There also emerged a class angle of travelling as practised at the time. Romantic solitude and romantic travel were largely middle-class notions as opposed to working-class enjoyment of sociability and group travel. The kind of travel one indulged in was a class marker with the latter form of travel being a subject of elite contempt and derision. This probably was the reason behind the birth of the "traveller-explorer". What John Urry says about contemporary tourism is also true of the period under discussion, wherein the romantic gaze is part of the mechanism of extension of the "pleasure periphery", ever drawing new spaces into its ambit, constantly in search of new objects of romantic gaze.[81]

In fact, chronologically, the Romantic period coincided in broad terms with the rapid expansion and consolidation of European colonialism in India, Africa, the South Seas and the newly independent American states. Once a saturation point was reached with continental touring, romantic travel stretched beyond known territories to embrace newer horizons. Under such circumstances, the representation of such places, which is mediated by the self, very often collapsed into the "imperial self" in the Romantic period. The individual gaze unknowingly became a part of a collective gaze and also contributed to the making of that collective gaze. Inevitably, travel writing was one discursive practice among others that, in this context, played a part in the process whereby dominant groups represented other cultures to themselves and convinced themselves of their right to intervene in countries they thought they had "discovered". On the other hand, there is a purposeful courting of what Heideggar calls "active discomfort", which distinguishes the traveller-explorer from a mere tourist. Such travels put the romantic traveller through tests, which produced discomfiture with self and thus became an enabling metaphor of selfhood, thereby empowering the traveller over all those who are not travelling.[82] Travel in this period and in this manner was not just a source of enjoyment but was balanced by a desire for information and education too. Motivation for such travel lay in its danger and the exoticism of the place. With this, travel comes a full circle from where it began as patriotic travel where a booty of foreign knowledge was brought back to the homeland to redefine its vista of cultural systems. There came about a synthesis of views and visions within a new conception of the natural world. This was more "observational" than "experiential", starting the new wave of naturalism which completely took control of travel literature in the nineteenth century. However, even though scientific observation and naturalism were emergent in the eighteenth century, the experiential visual code was never eclipsed. Significantly enough, the experiential code of "sentimentalism" was the dominant recognisable format for travel writing in the eighteenth and nineteenth centuries. Even naturalist's observations in exotic locales came clothed in this distinctly English package. One should not forget that the heroes of sentimental novels like those in Smollett, Laurence or Fielding's works also adopt the frame of the travelogue to narrate the adventures and experiences of the "sentimental" picturesque hero in hostile environments. Also, sentimentalism became the overarching scheme through which to view and ponder about natural scenery: sentimentalism was the English prism for viewing nature. By the end of the eighteenth century, as the precarious Enlightenment balance between science and sentiment began to make way for the subjectivity of the Romantic period, travel writing veered towards the self and individuation. Travel literature of the period is as much about the exploration of places as it is about self-discovery.

By the early 1800s, travel writing about exotic places had become a cultural trend so widespread that it seemed a sign of the times. The expanding market for books, reviews, newspapers and magazines had made explorers famous across Europe with a new rapidity, and brought exotic realms within the grasp of the reading public:

> Sitting at home, readers encountered the Indians of North and South America, learnt of the ancient culture of Persia, absorbed customs of the Hindus, scrutinised the conditions on the Caribbean slave plantations and searched for the Northwest Passage.[83]

Regions east of Europe soon became hot sites for travellers to make their careers in.

At another level, Nigel Leask points out:

> As the century progressed ... the extra European antique lands became subject to the quantifying and typifying scrutiny of colonial surveillance, as well as aesthetic judgment.[84]

The period between 1770 and 1830 saw a huge amount of diversity in travel narratives, during which period there were a number of political upheavals in Britain itself, beginning with the end of Seven Years' War, loss of Britain's American colonies, enactment of the Reform Act and the accession of Queen Victoria. The travel narratives produced during this time, in voicing the colonial self, were both influenced and driven by areas of cultural consumption. The poetry and prose of Wordsworth, Coleridge, Austen, Byron, Radcliffe and Scott were clearly influenced by the exploration narratives belonging to this period. Secondly, travel narratives fed the professionalism of sciences, especially in the specialities of botany, anthropology, palaeontology, and theories of origins of races. Thirdly, commissioned travels and their literature were often acts of reconnaissance before actual military expeditions. Fourthly, all those regions already brought under the scope of colonisation needed to be charted and made visually amenable through a host of ethnographic and cultural information in order to govern them. Another vein of travel literature comprising of evangelical and missionary journals worked to superimpose moralistic frameworks on newly explored landscapes which could then be brought under Christian proselytising operations. Even those who journeyed independently, opened up a path for the forces of colonialism and contributed to the assumption of power. Travel writing in such cases not only provided the factual data but also crystallised an imaginative topoi which furthered conceptual control over spaces and people. Where on the one hand, facts, myths, fictions and topoi fed poets, on the other, the same also influenced policies. Travel soon became a patriotic service where careers could be carved out and medals could be won. For example, Captain Cook was deemed to be a national hero and a martyr as he had been instrumental in claiming land for the British Crown. His personality inspired many young men to undertake journeys to faraway lands, therefore making travel a widespread and attractive activity of the times.

Construction of a knowledge empire

The British Empire was a varied one and accordingly incorporated very different ways of imagining itself. According to Miles Ogborn:

imperial identities were based on particular conceptions of geography that involved various imaginings of Britain and its relationship to the world.[85]

There had been many forces which shaped British identity and, thereafter, imperial geography. The Act of Union between Scotland and England passed in 1707, discussed earlier, was only one of these. Britain had also to carve its identity against a continental other which was often based on centuries-old religious sectarian strife between post-Reformation Anglicanism and continental Catholicism. France and, to a lesser extent Spain, being fierce competitors of Britain, were again forces against which Britain needed to construct its identity, based on ideological differences. Such conceptions of Britishness were also affected by the struggle between European powers for land grab and global hegemony. These questions of imperial identity metamorphosed into representations of empire through words and images invariably related to space, place and landscape, fundamentally linked with understanding and governance of self as well as that of the other.[86]

In order to help gauge the geographical variety and extent of the British Empire, a brief list of occupied regions is provided here. In the early eighteenth century, with the Treaty of Utrecht of 1713, Britain gained access to Gibraltor and Minorca in the Mediterranean, St. Kitts in the Caribbean, and Hudson Bay, Nova Scotia and New Foundland in North America. In the mid-eighteenth century, it gained control over Florida, Canada and Cape Breton in North America, the Caribbean islands of Tobago, Grenada, St Vincent and Dominica, and parts of West Africa and in India, primarily as to gather taxes from. By the early nineteenth century, the empire extended to the Pacific, with control over many of its islands, settlements in New South Wales and Van Damiens Land (Tasmania). The conceptual and material control over these varied places spreading over the Pacific, Americas, Asia and Africa met with serious challenges when it came to customising the conventional modes of spatial representations as practised within Britain itself. Textual and visual depictions such as maps, paintings and travel writing had to cope with different sets of extra-European natural and environmental surroundings as they were encountered and experienced. Much of this depended on empirical knowledge arising from the eighteenth-century conception of the body as a recording instrument where the experiential angle dominated representation. In order to handle such varieties, there originated the tools of imaginative and conceptual topographies in an effort to synthesise views and visions within a new conception of the natural world. The most enduring of these was the notion of "tropicality", formulated as a foil for temperate nature, which stood for European landscapes. According to Driver, the appeal of tropicality provided a powerful imaginative foundation and a stimulus for a variety of scientific, aesthetic and political projects.[87] Driver, in his work, *Tropical Visions*, attributes the rise of scientific instruments to the need to record naturalistic data during travel and exploration: it catered to the need for verifiability of the information gathered in the distant lands. Taking into consideration the mediated nature of these "imaginative geographies", there would often be doubts over the plausibility and trust-worthiness of the archive: more so, because the navigators and explorers often lacked the credentials of the gentleman-philosopher. The inventory of knowl-edge of foreign spaces accumulated "on the spot", had to be transmitted in a formulaic manner acceptable at home. The application of new scientific instru-ments partly resolved this problem: it also meant that observation on the ground had to conform to fixed "protocols of scientific enquiry" before it could be received in the European knowledge economy.[88] This also meant that special skills, techniques and materials had to be acquired by those seeking to represent foreign spaces. There emerged elaborate literature on what to observe and record and how to communicate results.[89] "Science", therefore, was nothing more than a pattern, format or a proforma through which new, extra-European realities could be interpreted, comprehended, made sense of, and finally appro-priated into a knowledge bank located at the global centre, London.

Beginning with the need to pin down the space according to accurate latitu-dinal and longitudinal determination, the scientific requirements and meth-odologies were endless. The science of optics and invention of the telescope led to both the navigator's double-reflecting sextant and the natural scientist's microscope. New tools and empirical techniques of analysis enabled the mod-ernisation of the older sciences, such as astronomy and botany, and the devel-opment of new ones, such as geology, hydrography, oceanography and human ethnography. The new science of geography which is often called the "science of imperialism" sought better understanding of the world through rational quantification and explanation. The mushrooming of botanical gardens, like Kew Gardens in London, and the establishment of scholarly societies like the Royal Geographical Society, played a crucial role in the promotion and spon-sorship of scientific exploration.[90] Geographers, exploiting these fruits of the scientific revolution, worked in concert with merchants and governments and, incidentally, provided an aura of intellectual respectability to such pursuits as mapping all coastlines, seeking new trade routes, and penetrating unknown territories in search of resources, markets and potential colonies. Maritime expeditions to chart the world's coastlines and seaways were increasingly par-alleled by overland surveys to explore continental interiors. Highly accurate observation and measurement entailed an obligation to record and publish findings. Hence an enormous number of official, semi-official and unofficial reports of scientific expeditions joined the already widely popular genre of travel narratives eagerly consumed by an interested public.

The general reading public avidly consumed these travel narratives as enter-tainment, mostly relishing the tales of hardship, heroism, survival and exoti-cism; while the government often commissioned the engraving of plates and maps to give them a serious documentary feel. Some or all of these elements appeared in the most successful and popular books, which went through multiple editions such as Captain Cook's travel diaries, or underwent translations like Mungo Park's (1771–1806) account of his Niger expedition, *Travels in the Interior of Africa* (1799). The ethnographic content of this genre brought other cultures and peoples to light. Elaborate details about the sexual lives of these people and the writer's own sexual escapades added the element of eroticism and titillation.[91] The interracial liaisons described by John Gabriel Stedman (1744–97) on his adventure in Surinam, and French naturalist, Francois Le Vaillant's (1753–1824) romantic account of birding in Africa very likely increased their sales.[92] The passages of Mungo Park's private experiences, according to Casey Blanton:

> were hugely popular not only [because they] secured his reputation as a widely popular travel writer at the end of the eighteenth century but also served to shift the emphasis in travel writing from descriptions of people and places to accounts of the effects of people and places on the narrator. By the early nineteenth century, travel writing had clearly become a matter of self-discovery as well as a record of the discovery of others.[93]

Distant reading

On the publication and reception side, as the demand for literature on travel and exploration was increasing, there came about two very different streams of travel writing: one, which followed the narrative style of a sentimental novel and secondly, the official survey reports. With an eye firmly on their market, publishers like Murray or Bentley had often freely edited travel accounts to supply or delete any element they wished.[94] On comparing actual field notes with the magisterial folios or printed volumes by the same author one finds several changes brought in by publishers. Through such "redaction", the factual field surveys produced as reports of officially commissioned scientific expeditions, on the other hand, were devoid of any such element of romance or adventure.[95] As the British public gorged on travel literature which flooded the market, market forces were bound to influence commissioned survey literature brought out by scholarly research-oriented houses as well, like the Royal Geographical Society. Even Roderick Murchison (1792–1871), himself a pioneering geologist and an avid promoter of scientific exploration, could not remain aloof from the demands of the public for best sellers of this nature. From his commanding position as president of the Royal Geographical Society and later head of Geological Survey, he encouraged and published accounts of heroism of David Livingstone (1813–73) and Henry Morton Stanley (1841–1904), and sponsored expeditions by the flamboyant and controversial explorer and ethnographer Richard Francis Burton (1821–90).[96] His policy however, was criticised by a section of the rational science-thirsty elite who lamented the eclipsing of science beneath non-serious cavalier activities. Murchison was opposed by some fellows of the Society like J.D. Hooker (1817–1911), a noted botanist and explorer of the century who, in a letter, wrote:

> I hate the claptrap and flattery and flummery of the Royal Geo-graphical, with its utter want of science and craving for popularity and excitement, and making London Lions of the season of bold Elephant hunters and Lion slayers, whilst the steady, slow, and scientific sur-veyors and travellers have no honour at all.[97]

Therefore, the visceral, the aesthetic, the intellectual and the scientific, together, brought about a genre which was able to recreate and represent spaces to the reader located at the metropolitan centre. The composite result of this litera-ture, was of course, as Charles Darwin (1809–82) summarised in the final chapter of his own travelogue, which was the outcome of a five-year-long sci-entific mission, *Voyage of the Beagle* (1839):

> The map of the world ceases to be a blank: it becomes a picture full of the most varied and animated figures. Each part assumes its proper dimensions: continents are not looked at in the light of islands, or islands considered mere specks, which are, in truth, larger than many kingdoms of Europe.[98]

The study of travel literature, therefore, is crucial in determining the visual character of colonialism, governance and thereby, of power. By visual char-acter, I mean the collective gaze of an entire community towards spaces con-sidered for occupation. John Urry distinguishes between two types of gazes in a modern society: the romantic and the collective, which, according to him, impact on the consumption of space in different ways. However, in the time period under study here, the romantic or the individualistic gaze contributes to the making of the collective gaze. That is not to say that this romantic gaze is itself isolated and free of collective opinions and perspectives. The frame for the gaze is pre-schematised and designed. There is an underlying grammar beneath all the modes of perceptions of space, i.e. the scientific, the aesthetic and even the sensory or visceral. These followed a formulaic paradigm estab-lished at the centre of circulation in order to convey the space's "otherness" through conceptual tools wielded by that centre. The need to portray the otherness becomes crucial in view of hierarchising and subordinating spaces to a central control. Travel writing, then, as it evolved from the Early Modern period onwards to the post-Enlightenment age in Europe, in general and England in particular, combined these aspects of representation, i.e. visceral, aesthetic and scientific, all of which would, however, generate a calculable, studied response. An ensemble of these formed a conglomerate and operated in tandem with one another to communicate a palpable space to a public who had never seen it. Moreover, these spaces were sought in order to attach them

217

to the colonial metropolis as global extensions of it, and incorporated within its ever-growing empire. Travel literature was adopted as an effective spatial mechanism to produce, circulate, familiarise and consume unknown and new spaces, within Britain as well as outside of it (as already seen in the course of the present chapter) through an active dissemination of knowledge of those "other" spaces, and finally subjugating them to an Anglocentric or much rather, London-centric control. All the aforementioned methods manufactured an imagined territory waiting to be accessed, harnessed and governed by the imperial centre. This act of manufacturing is itself an exercise in power as it is a potent means of both epistemological and material conquest of a land. In the following chapter, I shall study the case of India as a space constructed and manufactured in the collective British imperialist consciousness, to be studied and controlled in the public sphere of London.

Notes

1 Stegner, Wallace. "The Sense of Place". In *Where the Bluebird Sings to the Lemonade Springs: Living and Writing in the West*. New York: Random House, 1992. pp. 201–6.
2 "How to Understand Geography", The Penny Magazine, 16 June 1832. p. 112.
3 "How to Understand Geography", The Penny Magazine, 16 June 1832. p. 112.
4 See Moretti, Franco. Graphs, Maps and Trees: Abstract Models for Literary History. London: Verso, 2007. Huggan, Graham. *Territorial disputes: maps and mapping strategies in contemporary Canadian and Australian fiction*. Diss. University of British Columbia, 1989; Van Noy, Rick. *Surveying the interior: Literary cartographers and the sense of place*. University of Nevada Press, 2003. pp. 1–5.
5 Pratt, Mary Louise. *Imperial Eyes: Travel Writing and Transculturation*. London: Routledge, 1992.
6 Said, Edward W. *Culture and Imperialism*. London: Vintage, 1994.
7 Duncan, James and Derek Gregory eds. *Writes of Passage: Reading Travel Writing*. London: Routledge, 1999. p. 117; Driver, Felix. *Geography Militant: Cultures of Exploration in the Age of Empire*. Oxford: Blackwell, 1999; Ogborn, Miles and Charles W.J. Withers. *Georgian Geographies: Essays on Space, Place and Landscape in the Eighteenth Century*. Manchester: Manchester University Press, 2004; Urry, John. *Consuming Places*. London: Routledge, 1995.
8 Gikandi, Simon. *Maps of Englishness: Writing Identity in the Culture of Colonialism*. New York: Columbia University Press, 1996.
9 For an archival study of central and south Asian travels and their records, see Muzaffar, Alam and Sanjay Subrahmanyam. *Indo-Persian Travels in the Age of Discoveries 1400–1800*. New Delhi: Cambridge University Press, 2008.
10 Jarvis, Robin. "Romanticism". *Literature of Travel and Exploration: An Encyclopedia. ed.* Jennifer Speake ed. 3 Vols. Vol 3. New York: Fitzroy Dearborn, 2003. pp 1022–3.
11 Francis Bacon, in his essay, "Of Travel", elaborates the influence and benefits of travel in the life of an English Renaissance courtier: "Travel, in the younger sort, is a part of education, in the elder, a part of experience. He that travelleth into a country, before he hath some entrance into the language, goeth to school, and not to travel. Bacon, Francis. *The Essays and Counsels, Civil and Moral, of Francis Ld. Verulam Viscount St. Albans*. 1601. n.p.

12 Jarvis, Robin. "Romanticism". *Literature of Travel and Exploration: An Encyclopedia*. ed. Jennifer Speake ed. 3 Vols. Vol. 3. p. 1022.

13 de Certeau, Michel, *The Practice of Everyday Life*. California: University of California Press, 1984. p. 119.

14 Klein, Bernard. "Constructing the Space of the Nation: Geography, maps and the discovery of Britain in the Early Modern Period". *Journal for the Study of British Cultures*. Vol. 4/1–2, 1997. p. 24.

15 MacDonald, Kenneth Iain. "Issues of Ethics". In Jennifer Speake ed. *Literature of Travel and Exploration: An Encyclopedia*. 3 Vols. Vol. 1. London: Fitzroy Dearborn, 2003. p. 404.

16 For example, Jonathan Swift's *Gulliver's Travels* (1726), Daniel Defoe's *Robinson Crusoe* (1719), Henry Fielding's *The History of Tom Jones, a Foundling* (1749), Tobias Smollett's *The Adventures of Roderick Random* (1748) and *The Adventures of Peregrine Pickle* (1751), Samuel Johnson's *The History of Rasselas, the Prince of Abyssinia* (1759), Laurence Sterne's *The Life and Opinions of Tristram Shandy, Gentleman* (1759–69) and *A Sentimental Journey through France and Italy* (1768).

17 The greatest extant Germanic epic, *Beowulf*, with its account of the hero's journey to the hall of Hrothgar, can be cited as an example of the metaphorical and allegorical use of the motif of travel in epics.

18 Speake, Jennifer ed. *Literature of Travel and Exploration: An Encyclopedia*. 3 Vols. Vol 2, London: Fitzroy Dearborn, 2003. p. 394.

19 Harvey, David. *Justice, Nature and the Geography of Difference*. Cambridge: Blackwell, 1996. p. 214.

20 Elsner, Jas and Joan-Pau Rubies. "Introduction". Elsner and Rubies eds. *Voyages and Visions: Towards a Cultural History of Travel*. UK: Reaktion, 1999. p. 37.

21 Ogborn, Miles. "Writing travels: power, knowledge and ritual on the English East India Company's early voyages". p. 162.

22 Keighren, Innes M. and Charles W.J. Withers. "The spectacular and the sacred: narrating landscape in works of travel". *Cultural Geographies*. 19: 1. 2012. p. 25.

23 Ibid. pp. 11–30.

24 Moretti, Franco. *Graphs, Maps and Trees: Abstract Models for Literary History*. London: Verso, 2007, p. 1.

25 Korte, Barbara. *English Travel Writing: From Pilgrimages to Postcolonial Explorations*. Basingstoke: Macmillan, 2000. p. 41.

26 Quoted in Korte Barbara. *English travels: From Pilgrimages to Postcolonial Explorations*. p. 43

27 See the chapter on British landscape paintings. The Jacobean masques and Inigo Jones's theatre scenery during the time of James I tried to forge a vision of Britain as a singular geographical unit in symbolic terms embodied by the monarch.

28 Feldmann, Doris. "Economic and/as Aesthetic Constructions of Britishness in Eighteenth Century Domestic Travel Writing". *Journal for the Study of British Cultures*. Vol. 4/1–2, 1997. p. 31.

29 Klein, Bernard. "Constructing the Space of the Nation: Geography, Maps and the Discovery of Britain in the Early Modern Period". *Journal for the Study of British Cultures*. Vol. 4/1–2, 1997. p. 12.

30 Devereux, Walter Bourchier. *Lives and Letters of the Devereux, Earls of Essex: in the Reigns of Elizabeth, James I, and Charles I. 1540–1646*. John Murray: London, 1853. p. 502.

31 The British Isles were being travelled with a specific historical and geographical interest as early as the sixteenth century as a Tudor phenomenon. Better roads and improved cartography made travel easier and safer. However, the driving force behind this kind of travel was a pride in one's native country, that is, Tudor

England, and a curiosity both in the historic roots of that greatness and its contemporary manifestations. Refer to Korte, Barbara. *English Travel Writing: From Pilgrimages to Postcolonial Explorations.* p. 68.

32 Leland, John. 1549. Quoted in Klein, Bernard. "Constructing the Space of the Nation: Geography, Maps and the Discovery of Britain in the Early Modern Period". *Journal for the Study of British Cultures.* Vol. 4/1–2, 1997. p. 22.

33 Leland, John. "The Laboriouse Journey and Sei'che of Johan Leylande, for Englandes Antiquities, geven of hym as a newe yeares gyfte to Kynge Henry the VIII in the XXXVII yeare of his raigne". 1549. Quoted in Klein, Bernard. "Constructing the Space of the Nation: Geography, Maps and the Discovery of Britain in the Early Modern Period". p. 22.

34 Klein, Bernard. "Constructing the Space of the Nation". p. 24.

35 Quoted in Klein, Bernard. "Constructing the Space of the Nation". p. 25.

36 Ibid. p. 25.

37 Ibid. p. 26.

38 Helgerson, R. "The land speaks: Cartography, chorography and subversion in Renaissance England". *Representations,* Vol. 16, 1986. pp. 51–85.

39 Korte, Barabara. *English Travel Writing.* p. 67.

40 Taylor, John. *The Pennylesse Pilgrimage; or, the Moneylesse Perambulation of John Taylor, alias the Kings Majesties Water-Poet; How He Travailed on Foot from London to Edenborough in Scotland, Not Carrying any Money To or Fro, Neither Begging, Borrowing, or Asking Meate, Drinke, or Lodging.* London: Edward Griffin, 1617.

41 Ibid. p. iv.

42 Ibid. p. 1.

43 Ibid. p. 8.

44 Battle of Culloden (16 April 1746) was the last battle fought on British soil between French-supported Jacobites comprising of Scottish Highlanders seeking to restore the House of Stuart to the British throne, and the Hanoverian British Government. Charles Edward Stuart, the pretender to the throne, was eventually defeated and fled to Rome. The cult of the Ossian was initiated by James MacPherson who claimed to have found and translated Scottish poems by Ossian which he believed were the last residues of an extinct Gaelic culture of Scotland. This caused a literary revival of Scottish bardic verse. Samuel Johnson's *A Journey to the Western Islands of Scotland,* London (1775) and the diaries of James Boswell's (who himself was born in Scotland) tour of Scotland with Samuel Johnson in 1763 plausibly helped to uproot English biases against the Scots. Johnson, Samuel. *A Journey to the Western Islands of Scotland.* London: David Price, 1775; Boswell, James. *The Journal of a Tour to the Hebrides with Samuel Johnson LL.D.* 1775. http://www.gutenberg.org/cache/ep ub/6018/pg6018-images.html. Date of access 6.6.2018.

45 Ibid. p. 26.

46 Harvey, David. *The Condition of Postmodernity: An Enquiry into the Origins of Cultural Change.* Oxford: Basil Blackwell, 1989. p. 228.

47 Helgerson, Richard. "The land speaks: Cartography, chorography and subversion in Renaissance England", *Representations.* Vol. 16, 1986. pp. 51–85.

48 Kohl, Stephen. "Imagining the country as 'The Country' in the 1830s: William Cobett, William Howitt, William Turner". *Journal for the Study of British Cultures.* Vol. 4/1–2, 1997. pp. 113–127.

49 Feldmann, Doris. "Economic and/as Aesthetic Constructions of Britishness in Eighteenth-Century Domestic Travel Writing". *Journal for the Study of British Cultures.* Vol. 4/1–2, 1997. pp. 31–45.

50 *Journeys of Celia Fiennes* (1703?). p. 32. Quoted in Korte, Barbara. *English Travel Writing: From Pilgrimages to Postcolonial Explorations.* p. 70.

51 Ibid. p. 31.
52 Korte, Barbara. *English Travel Writing: From Pilgrimages to Postcolonial Explorations*. p. 70.
53 From the first in 1663 and with an expansion 1750–70, there were thousands of trusts and companies established by acts of Parliament with rights to collect tolls in return for providing and maintaining roads. A General Turnpikes Act 1773 was passed to speed up the process of expediting these arrangements.
54 Defoe, Daniel. *Tour through the Whole Island of Great Britain*. p. 239–40. Quoted in *Feldmann, Doris*. "Economic and/as Aesthetic Constructions of Britishness in Eighteenth-Century Domestic Travel Writing". *Journal for the Study of British Cultures*. Vol. 4/1–2, 1997. p. 32.
55 Baylis, Gail. "England, Seventeenth and Eighteenth Centuries". *Literature of Travel and Exploration: An Encyclopedia*. ed. Jennifer Speake ed. 3 Vols. Vol. 1. London: Fitzroy Dearborn, 2003. p. 395.
56 Moretti. Franco. *Graphs, Maps and Trees*. p. 38–9; Barrell, John. *The Idea of Landscape and the Sense of Place 1730–1840: An Approach to the Poetry of John Clare*. Cambridge: Cambridge University Press, 1972. p. 95.
57 Moretti. *Graphs, Maps and Trees*. p. 42.
58 Pennant, Thomas. *A Tour in Scotland*. (1769) London: John Monk, 1771; Pennant, Thomas. *A Tour in Scotland and Voyage to the Hebrides*. (1772) London: John Monk, 1774. Pennant entered Scotland via Berwick-on-Tweed and proceeded via Edinburgh and up the east coast, continuing through Perth, Aberdeen and Inverness, returning through Fort William, Glen Awe, Inverary and Glasgow.
59 See Jenkins, Ralph E. "And I travelled after him: Johnson and Pennant in Scotland". *Texas Studies in Literature and Language*. Vol. 14. No. 3, 1972. pp. 445–62.
60 Cohn, Bernard. *Colonialism and its Forms of Knowledge*. Delhi: Oxford University Press, 1997. p. 9.
61 Curley, Thomas M. *Samuel Johnson and the Age of Travel*. Athens: University of Georgia Press, 1976. pp. 184–5. Quoted in Barbara Korte. p. 65.
62 See Pittock, Murray G. H. *Inventing and Resisting Britain: Cultural Identities in Britain and Ireland 1685–1789*. Basingstoke: Macmillan, 1997.
63 See Brewer, John. *The Pleasures of the Imagination: English Culture in the Eighteenth Century*. London: HarperCollins, 1997.
64 See Barrell, John. *The Idea of Landscape and the Sense of Place 1730–1840*. Cambridge: Cambridge University Press, 1972.
65 See Radcliffe, Ann. *A Journey made in the Summer of 1794, through Holland and the West Frontier of Germany, with a return down the Rhine: to which are added observations during a tour to the Lakes of Lancashire, Westmoreland and Cumberland*. P. Wogan: Dublin, 1795. p. 449. See Fiennes, Celia. *Journeys of Celia Fiennes*. Cresset Press: London, 1947. pp. 166–7.
66 Cited in Urry, John. *Consuming Places*. London: Routledge, 1995. p. 193.
67 Ibid. p. 193.
68 Cross reference Chapter 3.
69 See preceding chapter on landscape paintings for a detailed discussion on techniques and devices widely used in the eighteenth century.
70 Urry, John. *Consuming Places*. p. 194.
71 Gilpin, William. " Essay 1 on Picturesque Beauty", *Three Essays on the Picturesque*.
72 Leask, Nigel. Curiosities and the Aesthetics of Travel Writing, 1770–1840. Oxford: Oxford University Press, 2002. p. 176.
73 Klein, Bernard. *Maps and the Writing of Space in Early Modern England and Ireland*. Palgrave Macmillan, 2001. pp. 171–5.

74 Wordsworth, William. Letter to Christopher Wordsworth, September, 1829. William Knight ed. *Letters of the Wordsworth Family From 1787–1855*. 3 Vols. Vol. 2. New York: Haskell House, 1969 [1907]. p. 387.

75 Hewitt, Rachel. Wordsworth and the Ordnance Survey in Ireland: "Dreaming O'er the Map of Things". The Wordsworth Circle. Vol. 37. No. 2. 2006. pp. 80–5. p. 81. Also see the online resource mapping Wordsworth's journey in Ireland: https://uploa ds.knightlab.com/storymapjs/05489351d2507e423715fc1443163ad3/wordsworths-ir ish-tour/index.html, sourced on 3.8.2018.

76 Baylis, Gail. "England, Seventeenth and Eighteenth Centuries". *Literature of Travel and Exploration: An Encyclopedia*. ed. Jennifer Speake ed. 3 Vols. Vol. 1. London: Fitzroy Dearborn, 2003. p. 395.

77 Harvey, David. *Justice, Nature and the Geography of Difference*. Cambridge: Blackwell, 1989. pp. 242–7.

78 William Turner's famous painting of the fuming steam engine epitomises the transition that Britain was going through.

79 Eliot, George. *Felix Holt, The Radical*. London: William Blackwood and Sons, 1866. p. 7.

80 Beckford, William. *The Travel Diaries of William Beckford of Fonthill*. p. 100. Quoted in Korte, Barbara. p. 59–60.

81 Urry, John. *Consuming Places*. London: Routledge, 1995. p. 139.

82 Majeed, Javed. "Nationalism, Travel and, Modernity". University of Delhi Lecture Series. Arts Faculty Building. Department of English, Delhi. 18 December 2008.

83 Fulford, Tim and Carol Bolton eds. "Introduction". *Travels, Explorations and Empires: Writing from the Era of Imperial Expansion1770–1835*. 4 Vols. Vol. 1. London: Pickering and Chatto, 2001. p. xiii.

84 Leask, Nigel. "Introduction: Practices and Narratives". *Curiosity and the Aesthetics of Travel Writing, 1770–1840: from an Antique Land*. Oxford: Oxford University Press, 2002. p. 3.

85 Ogborn, Miles and Charles W.J. Withers eds. *Georgian Geographies: Essays on Space, Place and Landscape in the Eighteenth Century*. Manchester: Manchester University Press, 2004. p. 8.

86 Ibid. p. 9

87 See Driver, Felix and Luciana Martins. *Tropical Visions in an Age of Empire*. Chicago: Chicago University Press, 2005.

88 Ibid. pp. 9–10.

89 As early as 1665, England's Royal Society published a set of "Directions for Seamen Bound on Far Voyages", and Robert Boyle, the great scientist and East India Company director, produced an entire volume to teach travellers and navigators how to record economic, ethnographic, and scientific data. The German traveller, Leopold von Berchtold composed *The Patriotic Traveller* (1789) with 2443 prepared questions in 37 different categories. See Distad, Merrill. "Scientific Travelling". *Literature of Travel and Exploration: An Encyclopedia*. pp. 1022–3. Also see Driver, Felix. *Geography Militant: Cultures of Exploration in the Age of Empire*. Oxford: Blackwell, 1999.

90 The Royal Botanic Gardens at Kew, established in 1759, by 1899 stood at the centre of an informal network of 57 other botanical gardens throughout Britain and its empire. Other European nations also established botanical gardens like the Jardin du Roi in Paris. There was a series of geographical societies set up all across Europe in the nineteenth century: France's Societé de Géographie in 1821, Germany's Gesellschaft für Erdkunde zu Berlin in 1828, and Britain's Royal Geographical Society in 1830, through a merger of older bodies that promoted explorations in Africa and the Near East, and numerous others, including in Rio de Janeiro in 1838, St Petersburg

in 1845, New York in 1851 and Washington in 1888. By 1890, more than 100 geo-graphical societies had been established in 21 countries and the International Geo-graphical Union was founded in 1922 to coordinate their activities.

91 The eighteenth-century travel journals related to the Grand Tour also carried similar traits, the most famous of these being Boswell's diaries, maintained throughout his tour and known for their frank personal details. Also the Orient was frequently the site for deviant erotic pleasures. Numerous examples of eighteenth and nineteenth century pornographic literature had the Islamic East with its harem, Hindustan or the Far East as their settings. The stage too deploys narrative strands of bravery, heroism, love affairs and sexual liaisons often found in travelogues about exotic locales and reinvent these exotic sites through stage scenery. See chapter III.

92 In the first edition of John Gabriel Stedman's *Narrative of a Five Years' Expedition against the Revolted Negroes of Surinam* (1790), his editor made changes to the ori-ginal manuscript in order to "insulate his readers from shock" at accounts of inter-racial relationships, torture and ill treatment of slaves in Africa. John Barrow in his own travelogue, *An Account of Travels into the Interior of Southern Africa, in the Years 1797 and 1798* (1801–4), wrote that Le Vaillant did not describe regions of the interior of Southern Africa as inhabited by colonialists, as that "would have dimin-ished the interest he intended to excite". See Tolbert, Jane T. "Censorship and Travel Writing". *Literature of Travel and Exploration: An Encyclopedia.* Vol. 1. p. 207.

93 Blanton, Casey. *Travel Writing: the Self and the World.* London: Routledge, 2002. p.15.

94 John Murray was a famous British publishing house established in 1768 by its pro-prietor of that name, renowned for the authors it has published of the likes of Jane Austen, Byron, Arthur Conan Doyle, Johann Wolfgang von Goethe and Charles Darwin.
   Bentley Publishing House was a Victorian publishing concern set up by Richard Bentley.

95 Withers, Charles and Innes M. Keighren. "Travels into print: authoring, editing and narratives of travel and exploration, c. 1815–c. 1857". *Transactions of the Institute of British Geographers.* Vol. 36: 4. 2011. pp. 560–3.

96 David Livingstone was one of the most popular national heroes of Victorian Britain. He had a mythic status as Protestant missionary-martyr, that of working class "rags-to-riches", that of a scientific investigator and explorer, and advocate of commercial empire. His fame derived from his status as the first European to see the Mosi-oa-Tunya which he named the Victoria Falls after the ruling British monarch. He is said to have initiated the European obsession for the search for the source of River Nile. Henry Morton Stanley was a Welsh journalist and explorer who acquired acclaim for his exploration of Africa and search for David Livingstone. Sir Richard Francis Burton was an English explorer, translator, soldier, linguist, poet and writer. Bur-ton's best known achievement was his travelling to Mecca disguised as an Arab. Among his other adventures was his travelling with John Hanning Speke as the first Europeans guided by Sidi Mubarak Bombay, said to be Africa's greatest explorer-guide, to the Great Lakes of Africa in search of the source of Nile. He was engaged by the Royal Geographical Society to explore the east coast of Africa and led an expedition to Lake Tanganyika. He was a prolific writer as well and wrote on topics like human behaviour, sexual practice and ethnography and translated the *Kamasu-tra* and *The Book of One Thousand and One Nights.*

97 Cited in Distad, Merrill. "Scientific Travelling". *Literature of Travel and Explora-tion: An Encyclopedia.* p. 1065.

98 Cited in Casey, Blanton. *Travel Writing; the self and the World.* London: Routledge, 2002. p. 16.

6

NARRATING INDIA

Hodges, Heber, Fraser and Hooker

> Now would I have a book where I might see all characters and
> planets of the heavens, that I might know their motions and dis-
> positions. ... Nay let me have one book more, – and then I have
> done, – wherein I might see all plants, herbs, and trees, that grow
> upon the earth.
>
> Doctor Faustus, II.i[1]

In this chapter, I shall consider some of the literature of travel produced by
colonial personnel who had been to India on various missions during the period
1750–1850, in the light of the arguments outlined in the preceding chapter.
Imperial travel literature dealing with the colonies, recast their own homegrown spatial notions in seeking to articulate from this basis, its territorial
bounds. While the culturally divided regions of England, Scotland, Wales and
Ireland (to a certain extent) could be subordinated to the singular spatial organisation of a nation, so too, various other locations spread around the globe
could be manoeuvred and assimilated into the great spatial trajectory of the
empire. In the present chapter, I shall study the writings of a few selected
representative colonial travellers in India from among a vast array of such
writings, in order to validate the foundational arguments put forward in the
earlier chapters in which, following Lefebvre, I see British landscape paintings,
travel writings and maps as spatial architecture and their practice as "spatial
architectonics". Textual methods rendered space determinable, describable and
finally, legible, thereby culturally creating a normative affective region emplotted onto a vacuous map. I have, for this purpose, selected a painter, a missionary, an administrative retinue and a botanist, who describe and inscribe
physical and moral character onto the space traversed by dramatising the story
of travel.

Travel literature, together with being an effective means of acquiring and
transmitting information about India for able governance, was also a calculated
mechanism towards fulfilling another crucial political aim. From the eighteenth
century, travels in various parts of an amorphous conceptual terrain came to be
located within a single spatial frame called India, interchangeable with the name

Hindostan. Surveyors might have struggled with uneven terrains and mathematical dilemmas, but the end result, the map, held claims to apparent neatness and accuracy. Travel literature, being a discursive attempt to shape space, contributed to that coherence, by either eliminating differential landmarks or behaviour or highlighting similarities, as it criss-crossed the land, inscribing a set of moral and spatial presumptions onto it. In other words, travel literature, by offering topographical and ethnographical details, was able to "subjugate space[s] by its transformation into place".[2] The boundaries of this amorphous region derived clearer definition through the travels of Europeans undertaken in the manner of surveys, transcribed in the form of written dossiers documenting every possible cultural and geographical minutiae. It is interesting to see how the British colonial design, following an expansionist logic, was to continually shift and push the peripheries of its territory till it received stable frontiers which the British thought of as safe and maintainable. Together with this, travels into the interiors of the core domain meant consolidation of knowledge about them: the rationale of the endeavour being the translation of this knowledge for imperial administration and thus into power. While describing the flora, fauna, vegetation, physiology, land rights, social behaviour and religion of the people, routes of communication etc. of a specific location with increasing accuracy, such literature sought to detect certain overarching similarities pervading throughout and across South Asia. The Enlightenment rationality, with its totalising drive, sought to create a certain unified homogeneous space while fixing its bounds and its identity at the same time.

Talking about the travel routes of the British travellers in India, Bernard Cohn points out that:

> The questions that arise in examining this (travel) modality are related to the creation of a repertoire of images and typifications that determined what was significant to the European eye. It was a matter of finding themselves in a place that could be made to seem familiar by following predetermined itineraries and seeing the sights in predictable ways.[3]

Therefore, Cohn specifies two or three predominant travel itineraries patronised by these travellers depending upon the routes and circumstances that brought them to India. The travel accounts of the seventeenth-century British travellers emerging from their status as primarily merchants, followed the route which first brought them to Gujarat by sea and then proceeded along the west coast to Ceylon and finally up the Bay of Bengal. Later on, in the first half of the eighteenth century, the British itinerary changed: British ships generally anchored either in Madras or Calcutta and travel inland usually began from these two cities. From the latter half of the eighteenth century, Calcutta was the primary port used; so first-time British travellers first encountered the magnificence of the white town of "City of Palaces" before they headed deeper into the terrain. The itinerary curiously replicated the pattern of the conquest's movement inland, away from the original coastal trading areas. Similarly, natural history led Enlightenment travellers of the nineteenth century, in search of scientific knowledge, away from the coast and into the interior where science could exercise control over nature in the remotest areas inland. Routine tours involved the journey by boat up the River Ganges, covering the essential and iconic sites of Banaras, Oudh, a visit to the palace of the Nawab of Oudh and then going on to Delhi and finally reaching its epitome with the visit to the Taj Mahal at Agra. They then either proceeded south-west through Rajasthan and Gujarat to Bombay, or moved north-west to Punjab and Sind. Viewing and experiencing these sites was thought to be appropriate for reading and experiencing many other sites and generally the whole of what was to be demarcated as India. Moreover, recounting the experience of being to these specific sites in the narratives of various European authors not only constructed a gaze but also cultivated and circulated place-myths around these sites which remained unaltered through centuries even when mediated by the transient socio-historic intellectual and aesthetic traditions discussed in the preceding chapter. Conventions of viewing India had already been established by the nineteenth century through a long tradition of iconography and narratives.

The exploration narratives since the East India Company's establishment in 1600, which subsequently started off trade missions, often narrated journeys to the Mughal court. Numerous Portuguese and Italian travellers had been visiting parts of India during the fifteenth and sixteenth centuries. With the triumph over the Spanish Armada, the English spirit of competition was kindled and the prospect of lucrative trade relations with the East came about as a legitimate temptation. The stories about the wealth and riches in the court of Akbar, whom the west romanticised as the "Great Mogul", whose religious tolerance was phenomenal, and the stories of his warm welcome to Christians, was too attractive for the British to resist. However, long before the foundation of the East India Company, there were Englishmen who visited parts of India. The first Englishman who is known to have set his feet on Indian soil was Thomas Stephens, who went to Goa in 1579, which was already a Portuguese stronghold, and became rector of the Jesuits' College in Salsette. His correspondences back home stirred considerable enthusiasm in England and Purchas preserved one of his letters.

The first Englishman to have credited the appellation of a traveller, however, was "Master Ralph Fitch, Merchant of London" as Purchas spoke of him. An outline of his itinerary serves to demonstrate the framing of the travel route for the British travellers who approached by sea for centuries to come. He started his voyage in 1583 and returned to England in 1591. With two other Englishmen, called Newbery and Leedes, he travelled to Ormuz and from there crossed the Indian Ocean to Goa. On their way to Goa the three travellers visited Diu, from where no vessels could pass without a Portuguese permit. From thereon, the trio travelled down south to places like Bijapur, Golconda and, after traversing a number of villages including Burhanpur, finally reached "the country of Zelabdin Echebar" or Jalal-ud-din Akbar. Fitch noted the habits of the natives of the region and talked of practices like cow worship, child marriage and prevalence of Sati as well as cremation of dead bodies. On reaching Agra, which Fitch describes as "a very great citie and populous, built with stone, having fair and large streets with a faire river running by it", he moved to Fatepur, the residence of "the Great Mogor". Here the three parted ways: Newbury started for Lahore and Leedes entered Akbar's service as a jeweller, whereas Fitch, himself, set out on further peregrination. On his way to lower Bengal via Benaras and Patna, he makes several observations about the rites and customs of the people residing in the Gangetic plains. On completing a long itinerary through Satgaon, Hugli and several villages in Bengal, he took a 25-day journey to what he calls "Country of Conche," "not far from Cochin China" and then on to Pegu, Macao, Malacca and Ceylon, finally returning to Cochin in March, 1590.[4] The authenticity of the perspective through which cultural remarks are made may not be vouchsafed for. Many travelogues were in circulation and highly popular in Europe, so had already constructed the gaze and informed the travellers about interesting things to look for in the region. He probably only emulated the experiences and adventures of former Portuguese and Italian travellers whose narratives created uproar and stirred curiosities in England.

There existed another alternative route from Europe to India: that of overland journey through Eastern Europe via central Asia, Persia and Afghanistan. The itinerary of the second English traveller demonstrates this alternative. In 1599, John Mildenhall, who served Richard Stapers, one of the first members of the board of directors of the East India Company, made an overland journey to India "in some fiduciary capacity". His objective for his journey was to negotiate with the "Great Mogul" in order to make inroads into Hindustan to establish diplomatic and commercial relations between the dominion of Akbar and that of Queen Elizabeth. He went by sea to Aleppo, and travelled overland through Armenia, Kurdistan, Persia and Afghanistan and finally, after passing through Candahar and Lahore, reached Agra in 1603.[5]

The seventeenth century saw a rise not only in trade missions to India but also a number of English written accounts by sailors, merchants, diplomats or accompanying chaplains. Among these, the narratives of figures like William Hawkins, William Finch, Sir Thomas Roe and Edward Terry, were incorporated and published as part of *Purchas his Pilgrimes* (1625). The narratives with a mercantile orientation often represented India as a land of enormous riches and as flowing with milk and honey and therefore bearing favourable prospects for trade. There is little reflection on the region's geography. Edward Terry's narrative, for one, draws a paradisal image of the region of being full of fruit-laden orchards and gardens bearing scented blossoms. However, his position as pastor forces him to implant numerous "venimous and pernicious Creatures", reptiles and insects, or talk about fierce monsoons in this idyllic space in order to refute the quasi-Arcadian status that he himself attributes it with. Kate Teltscher sees in Terry's attempt to "moralize geography", allegorical and mythic dimensions popular in the Elizabethan and metaphysical literature, like those of Spenser and Sydney, not bereft of an implicit intention of drawing comparison with his own homeland.[6] He can thereby be seen as crafting a technique of spatial othering on lines of religious principles and inherited dogma. The narratives of Hawkins and Sir Thomas Roe delineate Hindustan on similar lines as a land of "stupid idolatries" and irrationalities of both "Mahometans and Gentiles".[7]

According to Teltscher again, till this time English opinions about India were framed by earlier and contemporary European accounts both of civilians and missionaries, which were often translated or generally transmitted by hearsay or word of mouth. Therefore, according to her, an authentic and original British gaze towards India had not been formulated till the middle of the eighteenth century.[8] Some of the iconic European texts to have shaped the idea of Hindustan in contemporary seventeenth-century English society, were Francois Bernier's Mughal history translated 1671–2 and Jean-Baptiste Tavernier's *Six Voyages* translated in English as *History of the Late Revolution of the Empire of the Great Mogol* (1677). Even in the early eighteenth century, missionary narratives, usually of Jesuits, formed the main source of information about India. Together, these were able to consolidate as well as validate a perspective of viewing Indian society as fundamentally different from the western civilisation and as a spatial manifestation of heathen abstractions. These oft-repeated traits would then in the years to come demarcate what constituted India.

From the mid-eighteenth century onwards, there came about a more identifiable British rhetoric in the travel narratives produced by visitors to India. They were marked by clear intellectual and aesthetic principles, followed for home tours as well as for European travels outside Britain. To be more precise, the change was effected with the British accession to Diwani in 1765 and henceforth the East India Company's altered status as not merely a commercial player, but a controller of the land.[9] This was also the year when Robert Clive assigned James Rennell, the naval officer turned surveyor, the task of making a general survey of the newly acquired Bengal territories. Rennell's *Map of Hindostan,* in framing India's new boundaries, reflected and reinforced the Company's ambitions to extend its power beyond coastal entrepots to operate throughout India. When Warren Hastings became the governor general of Bengal from 1772–1785, he modelled the civil administration on governmental practices in Great Britain. According to him, successful administration required drawing up a kind of Domesday Book of the Company's territories. From then on, distinct features of "survey modality" as well as "surveillance modality" became pre-eminent in travel documents which began to involve an aspect of keen examination and observation in the process of travel and its representation in literature.[10] Even when following the same routes as their predecessors, as discussed earlier in the chapter, the narratives of travellers from the latter half of the eighteenth century, reflected a discernible British territorial ambition which persisted right through the nineteenth century. Also, travel writers often followed in the wake of conquering armies capturing details of the topography, people, flora and fauna in their writings. The same holds true in the narratives of British personnel as varied in profession as the artist, missionary, military or the mere civilian. The element of superiority pervades in their literature as they write from a decisive position of power.

Many of these journals functioned as intelligence dossiers. In the wake of the nineteenth century, Alexander Hamilton had complained that

> the gleanings of information which [most Englishmen] may have collected respecting [India], are reposed in their minds, rather like exotic rarities in a museum, than as merchantable wares intended for use and circulation.[11]

The project of representing India was riven with internal debates in the imperial metropolis. Where, on the one hand, the "survey modality" was an accepted norm for the men of science, on the other there was a constant demand for expressions of first-hand sensory impressions. The latter camp was heavily critical of the Company men, who merely offered disinterested and objective descriptions of India of having "lost the European eyes on which its picturesque features stamp the most vivid impressions".[12] Travel literature in the nineteenth century therefore had to straddle a precarious path in trying to satisfy both sides. The narratives of the travellers discussed below strike a mean between the two camps.

Writing the picturesque

In a publishing formula that was only becoming popular in Europe, William Hodges, a painter by profession, combined graphic art with literary narrative in his *Travels in India, during the years 1780, 1781, 1782 and 1783* (1790).[13] William Hodges was not only one of the pioneers in painting Indian scenes, but also one of the first few to write a travelogue since the ascension to Diwani of the East India Company and after the range of cartographic activities of James Rennell. Though there were several other British travellers who wrote about their experiences in India, Hodges, being an artist by profession, definitely was the one to stabilise the vision of India according to the aesthetic norms of his time.[14] The format of combining text and image would also soon be adopted as the most viable medium for representing travel and conveying spatial characteristics. Moreover, the story about the fortunes Hodges made overseas made India seem a suitably lucrative field to be explored and exploited by generations of British travellers and artists later on. He is believed to have been trained by no less than the master landscape painter of Britain, Richard Wilson, who noticed his talent while he was enrolled in the Shipley's Drawing School, and took him as his apprentice. During the seven years that Hodges spent as Wilson's apprentice, he was not only trained in techniques of drawing and painting, but was also taught the principles of classical landscape tradition at the Royal Academy.

Having finished his training, Hodges embarked on projects of painting landscape scenes and architectural sites in both Britain and on the Continent. However, the most important of his projects, which launched his career and ripened him for the Indian journey, was his trip to the South Seas as a draftsman on Captain Cook in the latter's second expedition on board the *Resolution* to the South Pacific in 1772. Hodges had to adopt a new set of ideals and reinvent traditional modes of landscape depiction to suit the meteorological phenomena and the physiognomy of the tropics. Hodges had been working in close contact with naturalists and meteorologists located in the South Seas which finally earned him the accolade from no less than Cook himself:

> Mr. Hodges has made a very accurate view of the North and the South entrances, as well as of the other parts of the bay, and in these drawings he has represented the mood of the country with such a skill, that they will, without any doubt, give a much better idea than it is possible with words.[15]

Although from then on the task of describing all that was impossible to deal with in words passed on to the expedition artist, Hodges himself never doubted the ability of written words. As he wrote in the preface of his *Travels in India,* his immediate objective in writing a journal was "To supply in some slight degree, this hiatus in the topographical department in literature".[16] According to him, "It is only matter of surprize, that, of a country so nearly allied to us, so little should be known". For him, it was a matter of great regret that though a lot had been written and known about the "Laws and the Religion of the Hindoo tribes", and "the transactions of the Mogul government" little had yet been said about "the face of the country, its arts, and natural productions". Therefore Hodges' energies were directed to convey "the idea of the first impression which that very curious country makes upon an entire stranger", and which, the British gentlemen staying long in India, lose. As a travelling artist who came to do a series on India, he claims not to have let his "observations" be subsumed by "reasoning" and his "traveller" identity be usurped by that of the "philosopher".[17] Likewise, he informs readers that his journal:

> Consist[s] of a few plain representations of what I observed on the spot, expressed in the simple garb of truth, without the smallest embellishment from fiction or from fancy. They were chiefly intended for my own amusement, and to enable me to explain to my friends a number of drawings which I made during my residence in India.[18]

Hodges received the necessary permission on 18 October 1788 from the directors of the East India Company for passage to India as a professional artist with a mission to record, as the letter of introduction from John Macpherson to Warren Hastings mentioned, "the most curious appearances of nature and art in Asia".[19]

Hodges' narrative is closely linked with his artistic vision and is synchronised with his graphic art. It revolves around the various drafts he makes of places and sites. His paintings form demonstrative tableaux, as it were, of the places he visits and describes in the text. Rather, if we are to think of Hodges' narrative as Paul Carter sees Clark's historical narrative about the discovery of Australia, the "syntax creates the sense of diverse activities converging towards the single goal" with the "choice of events itself contribut[ing] to the illusion of growing purpose"; for Clark, the goal is forging a space for settlement, whereas for Hodges it is to reveal a space for picturesque viewing. Therefore, like Clark's, Hodges' descriptions do not simply reproduce events, he "narrates them, clarifies and orders them" in consonance with classic conventions formulated in distant Great Britain.[20] The spatial distance had to be transcended through aesthetic configuration. If the diction of the picturesque was what qualified domestic scenery, the colonised space "nearly allied to us" also had to receive similar appreciation. He resizes scenes as stage scenery, as it were, for the theatrical unfolding of colonial action and "advancement". (As has been discussed in Chapter 5, some of his paintings, in fact, were utilised as backdrops to theatrical performances.)

His route of travel from Calcutta along the Gangetic plains tracing significant architectural sites soon became, most certainly, the staple itinerary for numerous travellers and painters who followed. In fact, sketches and paintings of Indian religious sites, along with expository historical supplements, themselves became part of the necessary inventory of travel diaries and sketchbooks of most British travellers in India. He was definitely the first to extend the aesthetic ideas revolving around European ruins to Indian architecture, garbed in the language of the "sublime" and the "picturesque". Historians of art have often characterised Hodges' paintings as impressionist in nature, these impressions arising out of the prescient understanding of difference in climate and meteorology. On first arriving at the shore of Madras, he is already made aware of the existing polarity in climate:

> The clear, blue, cloudless sky, the polished white buildings, the bright sandy beach, and the dark green sea, present a combination totally new to the eye of an Englishman, just arrived from London, who, accustomed to the sight of rolling masses of clouds floating in a damp atmosphere, cannot but contemplate the difference with delight: and the eye being thus gratified, the mind soon assumes a gay and tranquil habit, analogous to the pleasing objects with which it is surrounded.[21]

The first encounter with India is expressed through a combination of sensory registers, visual and aural:

> Some time before the ship arrives at her anchoring ground, she is hailed by the boats of the country filled with people of business, who come in crowds on board. This is the moment in which an European feels the great distinction between Asia and his own country. The rustling of fine linen, and the general hum of unusual conversation, presents to his mind for a moment the idea of an assembly of females.[22]

The stimuli make the writer conscious of an unknown nebulous space which gradually solidifies as a cerebral image with the amorphous mob steadily gaining black faces thus implying racial identity, and the vague drone amplifying and manifesting into identifiable gestures:

When he ascends upon the deck, he is struck with the long muslin dresses, and black faces adorned with very large gold ear-rings and white turbans. The first salutation he receives from these strangers is by bending their bodies very low, touching the deck with the back of the hand, and the forehead three times.[23]

Steadily, he is introduced to various cultural practices of the region which surprise him. The delicately framed Hindoo, "the original inhabitants of the peninsula", with hands like women, men variedly clad, women carried on the shoulders of men, the native boats and boatmen "excite the strongest emotion of surprise".[24] All that he saw was so unexpected and so different from what he had ever heard of the country that he needed to explore it for himself and therefore venture further into the country. The war which broke out between Haider Ali and the British upset Hodges' plans in the south and therefore he had to sail eastward to the newly developed presidency of Calcutta to proceed with his ambition. On reaching Calcutta, he is overwhelmed by the architectural wonder that the British have constructed. What the place actually was and what it had been transformed to by the British is succinctly contrasted in celebrating the beauty of the sight:

> The appearance of the country on the entrance of the Ganges, or Houghly River ... is rather unpromising; a few bushes at the water's edge, forming a dark line, just marking the distinction between sky and water are the only objects to be seen. As the ship approaches Calcutta the river narrows; that which is called Garden Reach, presents a view of handsome buildings, on a flat surrounded by gardens. The vessel has no sooner gained one other reach on the river than the whole city of Calcutta bursts upon the eye ... superior to any in India.[25]

He was as overwhelmed by the plethora of white sparkling Palladian architecture as with the hospitality of people and "freedom of admission" making him not in the least uncomfortable with the foreign place. He already imagines the scene as a beautiful prospect in a picture neatly dividing the foreground from the background and pointing the angular perspective in the style of Canaletto, thus rearranging the optical space into a geometrical one:

> On the foreground of the picture is the water-gate of the fort, which reflects great honour on the talents of the engineer – the ingenious Colonel Polier. The glacis and esplanade are seen in perspective, bounded by a range of beautiful and regular buildings; and a considerable reach of the river, with vessels of various classes and sizes from the largest Indiamen to the smallest boat of the country, closes the scene.[26]

With subsidy from the Company, Hodges painted several landscapes and architectural sites in Bengal and through North India that caught his fancy, many of which were commissioned by Warren Hastings himself, his fervent patron. He eventually submitted three volumes of 90 pencil and wash drawings to Hastings. By now, the British trade settlement was increasingly involving itself in the affairs of some of the regional states, pushing their trade further inland, seeking the right to collect taxes from territories around their settlements and making alliances with princes who would favour them. Travelling at a time when the Company Raj was gradually expanding its territorial hold and steadily taking over several kingdoms, Hodges was witness to certain strategic military and civil operations. It is interesting to see how the spatial reordering under colonisation gets represented in Hodges' narrative.

In 1781, Hastings asked Hodges to join him in his entourage for a diplomatic visit to Cheyt Sing (Chait Singh), the zamindar of Benaras, to collect his dues to the Company. However, with the outbreak of war as a result of Chait Singh's arrest, the entire group took shelter in the fort of Chunar which the Company had usurped. These incidents gave Hodges the opportunity to write about the war as a first-hand witness, and also observe Asian warfare and native forts from close quarters. The paintings of the forts of Chunar and Vijaigarh which had been central scenes of the campaign, were the outcome of this episode. The choice of Hodges' vantage point for a high-altitude view of the action of war is worth mentioning here:

> Seated on the top of a high mountain, covered from its base to its summit with wood. This the last of a long range of mountains, which, at this place, rudely decline to the plain. Here I enjoyed an opportunity which falls to the lot of but few professional men in my line: I mean that of observing the military operations of the siege. The camp was formed nearly four miles from the fort: there was, however, a rock about the height of the top of the mountains, and within gunshot, commanding one face of the fort, which was square. From this station, the walls were battered; and, after a practicable breach was made, the garrison thought fit to surrender.[27]

After the successful suppression of Chait Singh's troops, Hastings returned back to Calcutta, but Hodges accepted the invitation from Augustus Cleveland, then collector of Rajmahal and Bhagalpur, to travel into the interior "through a part of the country called the Jungle Terry, to the westward of Bauglepoor".[28] On visiting the tribal region, Hodges describes the geographical terrain and gives an exposition of the inhabitants of the region who he heard were the aboriginal natives of the place. In his view, of course, it is otherwise:

> I could not help suspecting that these may have been formerly no other than the outcasts from the Hindoo tribes, who after having been driven out, formed themselves into society, and taking post in the more moun- tainous parts, to prevent being surprized, have occasionally issued to commit depredations on the defenceless peoples on the plains.[29]

Hodges describes a little later a savage ritual to which they were invited: of an animal sacrifice which was demonstrative of the violent nature of the tribes residing in the area. The policing of these tribes had become a necessity for the "Hindoo, the Moorish, and afterwards the English governments".[30] Cleveland's mission here, of course, was two-pronged: of disciplining the errant tribes and to bring them under the Company's revenue regime:

> It was the humanity of that gentleman, added to the desire of improv- ing the revenue of this part of his district for the Company's benefit, that induced him to venture into the hills, alone and unarmed.[31]

With the utter failure of chastising strategies, Cleveland adopts artifice in win- ning over the tribes:

> After the fullest assurance of his most peaceable intentions and good- will towards them, he invited them to visit him at his residence at Bauglepoor. The confidence which he manifested in their honour, by trusting to it for his personal safety, effectually gained their esteem, and some time after a deputation of their Chiefs waited on him.[32]

Moreover, he sent gifts for their wives, caressed their children and presented them with beads; more importantly, to the chiefs themselves he gifted medals "as a mark of friendship, and as a reward for their improving civilization".[33] Once he found them "civilized" enough, "when he found them prepared for the accomplishment of his plan", he ordered clothes for some, "like those of the Sepoys in the Company's service ... furnished them with fire-locks and they became regularly drill'd".[34] As the ultimate effect of the regimentation: " Vain of their newly acquired knowledge, these new soldiers soon imparted the enthusiasm to the rest of the nation, who earnestly petitioned for the same distinction".[35]

Here Hodges spent several months, making sketches of the desolate, rugged Jungleterry, and its adjacent plains, which, according to him, was once a flourishing farmland but had been deserted after the famine of 1770, which cast a mood of gloom over the terrain. The silence and the desolation "spreads a melancholy over the mind of the traveller, and for miles together, nothing is heard but the screams of the cormorant, nor is the trace of any footsteps found but those of the wild elephant".[36]

Hodges' topographical descriptions, which it was his original intention to portray through his writing, are not far removed from scientific discourse and the geological understanding of the times. The composition of the soil or vegetation is expressed in precise terms. In this, he was surely dictated by emerging European Enlightenment consciousness. For example, he describes the way to "Mootejerna", a cataract on the hills, in a precise language of science and mensuration:

> the falls of Mootejerna in the hills, about four coss, or eight English miles inland from the river. From the height of the hills, these cascades are clearly seen, in the time of the rains, the river being then near thirty feet higher than in the dry season, and the falls considerably increased. ... when rain has fallen in the hills, the noise of the cataract is distinctly heard at the distance of two English miles. It consists of two falls ... the perpendicular height measures one hundred and five feet. ... In the interior of the cave, which may be thirty feet from the front of the rock, the base appears to be a mixture of rock and charcoal.[37]

He even collects two large pieces of this rock to show in Calcutta for further study and satisfaction of his curiosity. It should also be pointed out here, that the extensive region bounded by the Ganges, the Sone, Mahanadi and the Bay of Bengal, and drained by the Damoodar and other tributaries of the Ganges, had received very close investigation on its northern side. Here the mineral deposits of the Rajmahal Hills had long attracted attention in the colonial era.[38] Hodges' narrative, and his subjection of the area to a picturesque gaze, (for he also paints several scenes of Rajmahal Hill), is only a precursor to the geological and topographical survey and mapping which took place in the later years of colonial rule. This is in Edney's language "the geographical traveller engaged in reasoned observation". Reason was meant to dictate the selection and observation of phenomena found on site, in the same manner as practised by natural historians.[39] The place which held religious significance for the local people, is first transformed and rationalised in a language of scientific discourse and later redrafted in accordance with the aesthetic principles of the picturesque into a painting, finally exhibited at the Royal Academy in 1787.

In 1783, he made a fresh journey from Calcutta, where he had returned after the Bhagalpur trip, via Benaras and Lucknow to Etawah. Here, he was joined by Major Browne and his troops who were on their way to Agra, which provided a promising field for Hodges' artistic enterprises. In tune with his mission of painting architectural sites, he sketched the Taj Mahal, the forts and ruins of Agra and Akbar's tomb at Sikandra. All of these harmonised with his artistic vision of the picturesque ruins already popular in the European landscape tradition. They fell in place with the general view of decay of the land. Many of the subjects selected either for his drafts or finished paintings achieved iconic status with later artists trying their hand at their representation, or travellers making it a point to talk about them in their narratives. Some such sites were the Ghats of Varanasi, the fort of Chunar, the ruins of Rajmahal, the Taj Mahal and forts of Agra, or practices like the Sati, and more general features and natural specimens like the banyan tree. Together, these are able to recreate the varied physiognomy of an "antique" land.

Characteristically, throughout his account, Hodges maintains a language of strict accuracy, always sharing minute calculations about distances and locations quite in the manner of a topographical map. He likewise ends his *Travels* with prescriptive passages on the prospect and utility of colonial representations of India in art, extolling the eschewal of imagination and adherence of realism:

> the imagination must be under the strict guidance of cool judgment, or we shall have fanciful representations instead of the truth, which, above all, must be the object of such researches. Everything has a particular character, and certainly it is the finding out the real and natural character which is required; for should a painter be possessed of the talents of a Raphael, and were he to represent a Chinese with the beauty of a Grecian character and form, however excellent his work might be, it would still have no pretensions to reputation as characteristical of that nation.[40]

Hodges' eyewitness account dramatises his act of painting and, in employing the long distance view, actually reduces the distance for the readers, making the foreign physical space a natural and proximate theatrical stage in which every movement of the artist could be followed.

Moral geography

The combination of land description and religious expansion presents a fascinating example of the coexisting aims of mapping and missionising, even if land surveyors and missionaries rarely envisioned themselves as involved in the same enterprise. Both provided a crucial link between their sponsoring agencies located in the West and the actual sites of expansion in the colonies in the East.[41]

In the special valedictory address of the Society for Promoting of Christian Knowledge, delivered by the bishop of Bristol on 13 June 1823, he conferred upon Reginald Heber the sacred duty of "communicating the blessings of Christianity to the *nations* of Hindostan" (my italics).[42] To the above, in acceptance of the position of the episcopal see of Calcutta, Heber remembers that it was "this society which administered the wants, and directed the energies of the first Protestant missionaries to Hindustan; that, under its auspices, at a later period, Shwartz and Gericke, and Kolhoff, went forth to sow the seeds of light and happiness in that benighted *country*, and that, still more recently, within these sacred walls ... Bishop Middleton made adieu to that *country* which he loved"(my italics). It was in this country that Heber would "conduce to, and accelerate the triumph of the Gospel among the Heathen" .[43] While the bishop of Bristol refers to an amorphous conglomerate of multiple regionalities with his use of the phrase "*nations* of Hindostan" especially, when he talks later about the practical difficulty of "procuring translations into the dialects of Hindostan"[44], Heber reveals a more unambiguous understanding of Hindustan as a unified terrain, a "*country*" of the heathen, an absolute space endowed with absolute (negative) connotations. Biblical meanings are superimposed onto a geography charging it with religious and moral significances. This is by no means, as the above examples of travel records should show, unexceptional. Nor was it the first time that physical and moral landscapes were conflated. As Harvey expounds, the Christian notion of geographical space is teleological and acts as the "revolving stage of a temporal drama" till it is symbolically incorporated into Christendom through "a progressive annexation of the inner space of the human soul".[45] In Bishop Reginald Heber's travel journal, *Narrative of a Journey through the Upper Provinces of India, from Calcutta to Bombay, 1824–1825, with Notes upon Ceylon, an Account of a Journey to Madras and the Southern Provinces, 1826, and Letters Written in India* (1828), one finds a duality: the interesting admixture of representations of space which existed "outside and before experience" and that which was experienced by him while he existed in it.[46] India, in his view, was a Stygian space with sublime possibilities.

The first few poems in Heber's collection spring out of a spatial understanding which is morally charged. The first poem in the collection, which he is said to have recited in the theatre of Oxford and had been honoured for, titled "Palestine", as well as the two next in the sequence, called "Europe" and "The Passage of the Red Sea", are consanguineous to the same consciousness of the moral universe in which he places Hindustan. "Palestine", an elegy mourning the fall of "Judea" attaches notions of death, decay and darkness to the spatial body captured by Islam:

Reft of thy sons, amid thy foes forlorn,

Mourn, widowed queen, forgotten Sion mourn.

Is this thy place, sad city, this thy throne,

Where the wild desert rears its craggy stone?

While suns unblest their angry lustre fling,

And way-worn pilgrims seek the scanty spring?

Where now thy pomp, which kings with envy view'd?

Where now thy might, which all those kings subdued?

No martial myriads muster in thy gate;

No suppliant nations in thy temple wait;

No prophet bards, thy glittering courts among,

Wake the full lyre, and swell the tide of song:

But lawless Force, and meagre Want is there,

And the quick-darting eye of restless Fear;

While cold Oblivion, 'mid thy ruins laid,

Folds his dank wing beneath the ivy shade.[47]

The images invoked in the quoted stanza are much akin to the antiquarian perspective through which many colonial painters and British romantic poets defined India. The idea generated in his poems is that of an upright yet isolated and vulnerable continent (Europe) circumscribed by morally debased and violent bands:

The manliest firmness in the fairest form –

Save, Europe, save the remnant! – Yet remains

One glorious path to free the world from chains.

Clear ideas of Europe and its other are evoked where the Euro-Christian self is endowed with the sacred duty of wresting space from its other, to enlighten the "benighted" world. However, where Christianity in general deploys metaphors of spatial liberation, "free the world from chains", Christianity under imperialism, with its connotation of conquest, becomes complex. The bishop's journal, especially, adhering to a romanticist framework, juggles the two competing notions, but in the end conciliates imperialism as divine providence.

Heber's acceptance of the bishopric of Calcutta came after numerous deliberations and apprehensions. When he was offered the position for the first time in 1822 after the death of Bishop Middleton, in spite of his eagerness to accept it, he had to decline it, based on various personal considerations. He was also discouraged by the opinion of an eminent physician who had recently arrived in England after a stint in Bengal. This gentleman thought that the weather in India could affect his daughter's health adversely. It was his wife, who shared his missionary zeal, (and it was she, who after his death put together his journal for publication), ultimately persuaded him into taking up the position. He finally accepted the post believing "that he should be doing God more acceptable service by going to India than by staying at Hodnet."[48]

Hailing from an old Yorkshire family and, after having distinguished himself as a scholar at Oxford, he had, by this time, already preached at two places, Hodnet, a parish in Shropshire, and at Lincoln's Inn. His position as a priest did not tarnish his literary genius. J.G. Lockhart, his contemporary and fellow litterateur, described him as an extraordinary churchman with the "eye of a painter, and the pen of a poet ... mind stored with the literature of Europe, both ancient and modern".[49] His literature was not merely personal and private in nature. He was very much a part of the British literary circuit, frequently publishing reviews of poetry or travelogues in the *Quarterly Review,* including the travelogues of Robert Ker Porter, E.D. Clarke and also that of Abu Taleb Khan, the first Indian writer to have written about his travels in Europe. He was well read in romantic poetry and had been especially influenced by Southey, to whom he was also acquainted. His personality has been aptly summed up by his twentieth-century editor, M.A. Laird, as "one imbued with the central traditions of Anglican Christianity together with the spirit of the early nineteenth-century Romantic and missionary movements."[50] His reflection of India, too, as captured in his journal, is imbued in this tradition. According to K.K. Dyson, with "Bishop Heber's report on India we move into a period when Romanticism becomes a prominent feature of the Indian journal."[51] With his *Narrative,* the travel genre definitely leaves the Wellesleyan era of survey narratives, and enters a more ornate realm of the "literary picturesque".

Not only did Heber review travel literature, he was also acquainted with the activity of travel. In 1805, following the cult of the Grand Tour, he set out as a young man on a tour of north Europe which extended to distant Russia and included Norway, Sweden, Finland, Crimea, Ukraine, Hungary, Austria, Prussia and Germany. In Dyson's view, Heber's Eurasian travels foreshadow his later ones undertaken in India. His encounter with Byzantine and Indo-Saracenic architecture in the East European and Russian regions made him put its Indian counterpart in a comparative format. He seems to have traced regions into a cultural schema and thereby could conclude in his memoir, despite William Jones's endeavours, that the Hindus "had made no great progress in art, and took all their notions of magnificence from ... Mahommedan conquerors".[52] Heber belonged to a time when travel and exploration literature was at its peak of popularity. He was aware of India before he actually visited it. It is very evident that Heber was as aware of travels in India as with James Rennell's map, or with Captain Cook's diaries. One of his biographers writes about his enthusiasm in carrying out missionary activities in India:

Looking over the map of India, with his wife, as Heber did occasionally, while tracing there the provinces which he afterwards visited ...

he has often avowed that had he no one's interest but his own to study, he would immediately devote himself to the missionary work.[53]

On reaching India, he felt the need for touring the entire Indian diocese in his capacity as the bishop in order to bring it under a manageable compass. Heber's tour encompassed almost the whole circumference of the Indian sub-continent, demarcating through his trail the Company's realm. Under the patronage of Governor-General Lord Amherst, he followed roughly the oft-taken route that Cohn talks about. He took his journey in two parts. From Calcutta he proceeded to Dacca, then on through what was called the Upper Provinces, to the Himalayan foothills, descending to the plains to visit Delhi and Agra, moving southward via Rajasthan and Gujarat to Bombay. From here he went to Ceylon by sea, from where, after a brief stay, he returned to make a tour of the southern region of the peninsula, where he met his demise. In this sense, the episcopal see conceptualised Calcutta as the physical centre with journeys outside as forays into the spatial edges of civilisation.

On 15 June 1824, he began his journey from Calcutta by river escorted by his chaplain, Mr Stowe, and some native servants. He described the onset of the journey in a quaint boat in a letter to his wife, Amelia Heber:

> We set out, attended by two smaller boats of very rude construction, with thatched cabins, and huge masts and yards of bamboo, something like the canoes of the Friendly Islands, as Cook has presented them.[54]

Clearly, Heber was no exception to the British enthusiasm for exploration narratives, especially that of Cook. Heber had not only gorged on the iconic text, but was even able to recall passages at the correct instances, identifying similarities, to remind him of "the drawings of Otaheite and the Friendly Islands".[55] Similarly, Rennell's map, too, was as firmly inscribed in his memory. After touring a few villages in Bengal, he headed towards Dacca, "the first station of [his] visitation"[56] of the route of which he says:

> Our way was through the heart of Lower Bengal, by the Matabunga, the Chundna, and those other branches of the Ganges, which make so tortuous a labyrinth in Rennell's map. The Sunderbunds would have been a nearer course, but this was pleasanter, and showed us more of the country which along the whole line of the river was fertile, well cultivated, and verdant to a great degree, and sometimes really beautiful.[57]

He seems to be asking for the greatest possible visibility of natural scenery of the region while travelling. Moreover, he needs to accustom and train his British eye to the foreign scenery for a duration before he can appreciate and develop a suitable aesthetics to map the tropics. Often, he exhibits a keen awareness of the existing iconography about specific places he visits, produced by professional and amateur British painters undertaken before his visit, such as that of William Hodges, George Chinnery or Sir Charles D'Oyly. Heber's seems to be an initiative to superscribe onto Rennell's factual map a piquant pictographic trail, which not only refers to natural scenery he encounters, but simultaneously gives a large amount of ethnographic and naturalistic information:

> The banks are generally covered with indigo, and beyond are wide fields of rice or pasture, with villages each under a thicket of glorious trees, banyans, palms and plantains, and bamboos, and though we here and there passed woods of a wilder character, their extent did not seem to be more than in one of our English counties. The villages are all mud and bamboos.[58]

While he often tries to stimulate imagination in the reader by providing a touch of the artist's palette to the narrative to make it accessible, he also points out flaws in Rennell's map, thereby not only contributing to the existing archive of knowledge resources about India but also to allay allegations of the narrative being entirely imaginative. In the entry dated 18 June, he writes:

> Our course from Ranaghat was up a wider and deeper stream, and chiefly to the N.W. – a circumstance irreconcileable with Rennell's map, unless the discrepancy can be accounted for by an extraordinary alteration of the river's channel.[59]

Again, a little while later:

> About half-past five we brought-to for the night, at a place which our crew called Sibnibashi, but so differently situated (being further to the south, and on a different side of the river) from the Sibnibas of Rennell, that I first thought they must be mistaken.[60]

Even Rennell's map and researches are not necessarily secure guides to places less frequented by travellers. Significantly, Rennell's map becomes the fundamental premise onto which new cartographic data could be assimilated with in-the-field inscriptive practices. The earlier map simultaneously provided grounds for claims about certainty for other kinds of non-scientific "authorcraft".[61] Moreover, map literacy and scientific knowledge here is placed at the service of religion. In this sense, Heber's textual scrutiny becomes a deeper scientific investigation of the geographical, antiquarian and historical relations of the foreign space.

The strength of his narrative are the passages which reflect a sentimental vein and romanticist view of nature, which had pervaded the age in which he was a child. Such passages recur through the narrative depicting the changes of scenery throughout the progress of his journey, differentiating one locality from another and thus representing a dynamic space:

241

The river continues a noble one, and the country bordering on it is now of a fertility and tranquil beauty, such as I never saw before. Beauty it certainly has, though it has neither mountain, nor waterfall, nor rock, which all enter into our notions of beautiful scenery in England. But the broad river, with a very rapid current, swarming with small picturesque canoes, and no less picturesque fishermen, winding through fields of green corn, natural meadows covered with cattle, successive plantations of cotton, sugar, and pawn, studded with villages and masts in every creek and angle and backed continually (though not in a continuous and heavy line like the shores of the Hoogly) with magnificent peepul, banian, bamboo, betel, and coco trees, afford a succession of pictures the most radiant that I have seen, and infinitely beyond anything which I ever expected to see in Bengal.[62]

A resounding nostalgia of double exile continues through his narrative, firstly from his home in England and secondly, in his separation from his wife and daughter who could not take the journey with him and had to remain behind in Calcutta (as represented in the poem "An Evening Walk in Bengal"). The haunting sentimental strain resurges in his narrative over and over again through the infinite comparisons and references he makes with his own native land. Since most of the narrative is also a correspondence with his wife, the descriptions are posed in a way as to make them vivid and amenable to his wife, who, as was the case with him so far, was only familiar with European registers. In talking about the alien geography, therefore, he selects in the manner of most British travel writers, European images which bear the greatest structural, organic and naturalistic similitude. He records on his journey after Cawnpur (Kanpur):

The country, as we advanced, became exceedingly beautiful and romantic. It reminded me most of Norway. ... It would have been like some part of Wales .... We saw some interesting plants and animals, black and purple pheasants, a jungle hen, some beautiful little white monkeys gambolling on the trees; and what pleased me most, we heard the notes of an English thrush. I also saw some very large nettles and some magnificent creepers, which hung their wild cordage as thick as a ship's cable, and covered with broad, bright leaves from tree to tree, over our heads. After about an hour and a half's ascent, we saw some dog-roses, a good many cherry trees of the common wild English sort in full blossom, some pear trees with fruit, and a wild thicket of raspberry and bilberry-bushes on either side of the road.[63]

Benaras, the epicentre of Hindu faith and practices, is an old town with narrow streets, which, "like those of Chester, are considerably lower than the ground-floor of the houses, which have mostly arched rows in front, with little shops behind them".[64] Curiously, Heber, in his review in 1809 of Abu Taleb Khan's *Travels* (the first travelogue written by a non-European on his travels in Europe), had commented on Taleb's overt references to his homeland when citing from Taleb's book: "He noticed, on his way from Holyhead, Conway [Castle], with its ancient walls resembling Allahabad; and Chester, with the *verandahs* which line the principle streets".[65] The parallelism between the two passages, the contiguity of the regions compared (Benaras and Allahabad, both sites of religious significance) and their comparison to Chester, is remarkably significant. Is Heber adopting the Indian perspective to view and compare native and foreign places? The view of Benaras might have recalled to his mind the passage from Taleb's *Narrative* that he himself had commented on earlier. Where he usually would cite the source of his associations and recollections of canonical authors, he significantly fails to name the Indian Abu Taleb. Moreover, the physical layout of roads, towns and buildings were indicators of the morals and habits of the inhabitants. In this sense, spaces become sacred or profane according to the reflections of socio-religious desires and anxieties.

His nostalgia nearly transports him to a realm of imagination where he seems to see and hear British images and notes. In Gour, ascending a steep place, passing a little brook, he almost fancies he sees gravel, a phenomenon which he had not seen in India yet,[66] and in Bengal, he seems to hear songs from the boatmen and children in the villages which remind him of Scottish melodies.[67] To prevent repetitions and stasis in his written descriptions, he often takes recourse to poetry as a literary strategy. As Nigel Leask points out, Heber occasionally adopts literary associations which provide the scene with an incremental interest. He evokes the British literary canon and frequently quotes romantic poetry to overcome the narrative stasis. The flooded villages by the Delaserry river in Dacca make him think of Gray's poetic representation of the Egyptian Delta:

On their frail boats to their neighbouring cities glide,

Which rise and glitter o'er the ambient tide.[68]

A practice of scaring away birds from the corn fields, Heber expresses by citing Southey's oriental romance *The Curse of the Kehama* (1810).[69] But the most significant of these is perhaps his breaking into rapture at the first glimpse of the Himalayas with the famous lines from Coleridge's "Kubla Khan":

But oh! that wild romantic chasm which slanted

Down the green hill athwart a cedarn cover!

A savage place! As holy and enchanted

As e'er beneath a waning moon was haunted

By woman wailing for her demon lover![70]

The awe inspiring inscrutable Himalayan landscape exudes connotations of the primitive atavism that the poem associates with the great Oriental despot, Kubla Khan. The significant reference to the poem imbues the sublime picturesque of the snowy ranges with a moral paradox: of a darkness beneath the white peaks: "a savage place". It is a place fraught with dualities in which resides both the sacred and the profane in being the periphery of civilisation. Heber's mysticism on viewing the towering mountain ranges springs from his overpowering moral topoi.

For all his efforts to believe and convey to the British public the similarity between Britain and Hindustan, there is always the lurking anxiety about the depravity clouding the space. At places where he "could have fancied [himself] inside in England", the "dark, naked limbs, and the weapons of [his] companions" jolt the evangelist into realisation "that [he] was in a far distant land", as "forlorn" as Keats finds himself at the end of his ecstatic reverie in the poem "Ode to the Nightingale".[71] The landscape itself was inscribed with signs of heathen superstitions which recalled back the establishment evangelist to his mission. A ramble amidst nature which could evoke the highest spiritual uplift, suddenly turns into a space conquered by monstrous abominations:

> A noble grove ... succeeded to the pawn rows at our village this evening, enbosoming the cottages together with their little gardens ... in greater perfection ... their little green meadows and homesteads. We rambled among these till darkness warned us to return. We saw a large eagle seated on a peepul tree near to us. On the peepul an earthen pot was hanging, which Abdullah said was brought thither by some person whose father was dead, that the ghost might drink.[72]

The ritualised space and the lived landscape of its native inhabitants are interpreted by its European observer as symptomatic of its being possessed by evil and ignorance. The same is most obscenely obvious in his view of the native part of Calcutta, "the City of Palaces" which obtrudes reason and defies logic in its haphazard, sloppy construction, devoid of any civic sense or discipline. The Black Town is a space which epitomises heathenism at its worst. The "clamour of voices", "thumping and jingling of drums and cymbals emerging from nearby temples of monstrous deities", the "villainous smell of garlic and rancid coconut oil", "sour butter", "stagnant and dirty ditches", the "breeding ground for malaria carrying mosquitoes" are constant reminders of the lurking depravity in the backyard of the magnificent and grand White Town. The visceral registers of sounds, sights and smells, penned down in language, could recreate the horrific and chaotic topography of the Black Town. The innumerable venomous creatures ranging from serpents to vermin which infest the geography, are manifestations of the same moral degradation:

> Within these few days all the vermin part of Noah's household seem to have taken a fancy to my little ark. To the scorpions, the cockroaches,

the ants, and the snake, were added this morning two of the largest spiders I ever saw, and such as I regretted afterwards I did not preserve in spirits. In a bottle they would have made monsters fit for the shelf of any conjurer of Christendom.[73]

However, Heber holds himself from complying completely with Rousseauvian views. The enervating heat, the putrid miasma and the torrential monsoon were not decisive factors shaping indolent natives. All that was bad about the place and its inhabitants seemed to be external to their otherwise innately gentle character:

All that is bad about them appears to arise either from the defective motives which their religion supplies, or the wicked actions which it records of their gods, or encourages in their own practice.

Any good Hindus who exist and that he is aware of are, "in no instance ... connected with, or arising out of, their religion, since it is in no instance to good deeds or virtuous habits of life that the future rewards in which they believe are promised".[74] In a letter to his friend R.J. Wilmot Horton, he is blatant in his disapproval of Hinduism which he finds "the worst" among "all the idolatries" he has "ever read or heard of". Among the reasons he provides for his antipathy is the system of caste: "a system which tends, more than any thing else the Devil has yet invented, to destroy the feelings of general bene-volence, and to make nine tenths of mankind the hopeless slaves of the remainder", and generally "the total absence of any popular system of morals, or any single lesson which the people at large ever hear, to live virtuously and do good to each other".[75] Dyson notes that Heber's intolerance is more obvious and rabid in his letters to friends and relatives in his clerical circuit than in his own travelogue. For example, in one of his letters to his friend Watkin Williams Wynn, who helped him acquire the position of bishop of Calcutta, he writes:

Of the people, so far as their natural character is concerned, I have ... a very favourable opinion. They have, unhappily, many of the vices arising from slavery, from an unsettled state of society, and immoral and erroneous systems of religion. But they are men of high and gallant courage, courteous, intelligent, and most eager after knowledge and improvements, with a remarkable aptitude for the abstract sciences, geometry, astronomy etc. and for the imitative arts, painting and sculpture. They are sober, industrious, dutiful to their parents, and affectionate to their children, of tempers almost uniformly gentle and patient, and more easily affected by kindness and attention .... Their faults seem to arise from the hateful superstitions to which they are subject, and the infavourable state of society in which they are placed.

245

But if should please God to make any considerable portion of them Christians, they would, I can well believe, put the best of European Christian to shame.[76]

The British Government could give the region its much-required stability and Christian missionary endeavours under its aegis could guarantee a moral transformation. Together, the two could reform the fallen state of the land. The onus is as much on the colonial government to herald and put in place an advanced system of learning at the core level; for to uproot Hinduism from the society would also mean the induction of an enlightened scholastic structure. He is severely critical of the existing government-aided schools which impart Hindu training to its students as the "Vidalaya" he inspects in Benaras:

> The astronomical lecturer produced a terrestrial globe, divided according to their system, and elevated to the meridian of Benaras. Mount Meru he identified with the north pole, and under the southern pole, he supposed the tortoise, "Chukwa", to stand, on which the earth rests. The southern hemisphere he imagines to be uninhabitable; but on its concave surface, in the interior of the globe, he placed Padalon. He then showed me how the sun went round the earth once every day, and how, by a different, but equally continuous motion, he also visited the signs of the zodiac. The whole system is precisely that of Ptolemy; and the contrast was very striking between the rubbish which these young men were learning in a government establishment, and the rudiments of real knowledge, which those whom I had previously visited, in another school in the very same city.[77]

The existing sense of space and cosmology of the inmates is discarded as Hindu chicanery. The same needed to be overhauled through the inception of Enlightenment foundations of scientific learning at the pedagogic level to tap impressionable young minds before they fall prey to Hinduism's irrational and immoral machinations. European spatial models and "scientistic" ideology were not only at the heart of British economic and political conquest but was a crucial element in its drive to rationalise and modernise the native mind. Heber repeatedly talks of Christian missions degenerating to Sisyphean hopelessness, especially in Ceylon and southern India, in its failure to intervene into a convert's private life and daily practices, who often continued to perform the same rituals as earlier. Therefore indoctrination of European rationality was crucial to Christian proselytism's success in India. Interestingly, Christian evangelism took recourse to science as an apparatus to promote and establish faith in itself and in the Christian godhead. Later in the nineteenth century, numerous schools and colleges were founded in India with this very objective. It is the voice of the establishment evangelist which speaks of the moral responsibility of the British in its colony which, through its maxim of science, rationality and progress, would be an agent of transformation of the space itself. The two ideologies completely synchronised with each other and spoke in perfect unison the mutually accepted language of the "white man's burden": Christianity would bloom, holding hands with the imperial order and imperialism would thrive through Christianity's unstoppable missionary zeal. Evidently, by Bishop Heber's time, British evangelism and colonialism in India had both come a long way since its troubled history in the eighteenth century, when the East India Company, then, mainly a mercantile concern, resisted the mission's entry citing it as unpopular with the natives and thus potentially detrimental to trade.[78]

What is interesting is the blending of the *Narrative* unobtrusively into the general discourse about India as a space fraught with danger, desultoriness and moral decadence. Simultaneously, it also reaffirms the bounds of the Company's territory and the regions under the supervision of the Church of Calcutta. Despite Heber's attempt to maintain the face of a liberal and a tolerant cleric, his journal formed a crucial text to justify expansion in Christian missionary activities in the nineteenth century. A sonnet by G.A. Vetch represents Bishop Heber during his travels through India:

Bright with the dews of pure Castalian springs,

See Heber gladdens now our sultry plains;

...

Hail, then, and Heaven speed thee on thy way,

Illustrious pilgrim of our distant shore:

Rous'd by thy call, enraptur'd by thy lay,

May nations learn their saviour to adore.

For thee the fairest garland shall be twin'd,

The Christian's palm and poet's wreath combin'd.[79]

The spatial metaphors are typecast with certain qualifying adjectives signifying remoteness of the place and the spatio-temporal distance and difference. The brightness of the "Castalian spring", or European Enlightenment would light up "our sultry plains" of "our distant shore": the physical touch or the benign presence of the godly figure of Heber would illuminate the dark recesses now owned by the imperial state. The white man's penetration into the land would reveal to light, the otherwise obscure space.

Although Heber tried his best to maintain a liberal and tolerant face in his *Narrative,* his work consolidates the discursive construction of India as a space ridden with primitive and savage practices justifying British (missionary) intervention into the region. Numerous works on India which surfaced after the publication of Heber's *Narrative* distilled out a totalising image of India as absolute and undiluted barbarism. The sylvan landscape that Heber celebrates mutates into a synoptic space essentially denoting evil. For example, James Chambers' *Bishop Heber and Indian Missions* (1846) before dealing with Bishop Heber's biography and works, provides an introductory chapter which focusses the view of its readers onto signs of Satanic practices inscribed on its picturesque landscape: the "gnawed corpses" gushing down a mighty river with "the black vulture settled on them", the scarecrow-like figure of the "fanatic fakir" in a "wild mountain glen", the "frantic yell of demoniacal worship" tearing the deep blue sky, and the "fitful glare" in which a widow is "food to flames".[80] First fixing the territorial expanse of the region into a numeric grid and then within defining natural boundaries, the chapter goes on to specify essences through the rhetorics derived from Christian myths and allegories:

> It is in such an hour, and in a scene of so fair a beauty, that the contrast between the moral and physical aspect of Hindoostan forces itself most strongly on the thoughtful mind. All external nature is rich in so surpassing a grandeur and loveliness, that the fond fancy might well deem it some long lost relic of Eden's bowers where sin and sorrow had found no place, and on which the primeval curse had not descended.
>
> Alas! Over this land, so abounding in the choicest beauties and blessings of nature, there broods a moral gloom of almost impenetrable obscurity. ... Even on the fairest works of creation, sin and error have impressed their foul marks, and when the excitement of imagination has passed away, hill and dale, wood and water, alike teem with signs of man's fall from his first estate.[81]

Tapping into regional nostalgia and scriptural longings, Heber's *Narrative* reveals deep connections between mapping and missionising that together participated in the unfolding of God's benevolent design for the region. The thinly veiled Christian rhetoric was employed to "convert" space into readable text by granting it visibility and opening it up for proselytising.

Frontiers and fixing boundaries

Moving beyond the model of theological construction of sacred and profane space, we can see applications with similar designs in claiming, controlling and disciplining frontier spaces. Military malefaction joining hands with cultural counteractive processes, war zones have a special ability to reveal to sight long unknown and neglected spaces. Just as the Highlands in Scotland became an important site for knowledge gathering and control after the final successful curbing of the Highland uprising, the Himalayan highlands became a field for intelligence garnering after the Anglo-Gurkha War. In order to draw unambiguous connections, tracing similar linkages related to the geographical construction of the British Empire in India and its frontiers is rather interesting. This also marks the inauguration of the movement from zonal frontiers to linear boundaries essential for the conceptualisation of the modern nation state.[82] Recent analysis states that frontiers originated in centrifugal forces and have proved to be an integrating factor by being open to both directions, the centre and the margins: "The Frontier is outward-oriented, and the main attention of the centre is concentrated on treasured space gained under calculated danger."[83] In contrast, the border could be seen in an inward-looking way as a consolidation of territory, a limitation to the expansion of colonisation and a clear definition of those beyond as well as within the polity.[84] The production of the borderland which is an integrating factor of all that is inside and an excluding factor of all that is the "other", required a great deal of scrutiny for recognising the social, cultural, and scientific registers which marked similarities and the discontinuities. Geographic knowledge was put to the aid of imperial power in carving this well-defined space. Travellers and travelogues worked concertedly to craft the confines of imperial space which also entailed a clear demarcation of a border. The continuous Himalayan belt in the north posed as a mighty wall reflecting the integrity and the strength of the empire inside. There were very many ways of reaffirming this and were achieved by numerous British professionals who chose the mountain ranges for imperial careering. Strikingly, many of them were Scots or had a Scottish Enlightenment education, which played a significant role in imagining the Himalayas as similar to the Scottish Highland fringes.

The fall-out of the Gurkha War of 1814–16 was that it disclosed to the Company officials the importance of knowledge about a hostile landscape, of the Taraee and the Himalayas and its potentially rebellious local residents.[85] Hereby, one may easily identify the replication of the Anglo-Scottish formula with similar effects, together with the noteworthy coalition of imperial injunctions and vociferous Scottish rejoinder. After the Gurkha War of 1814–16, a spate of Himalayan travel narratives appeared, as a passage from the *Edinburgh Review* proclaims:

> The Gorkha War subjected to us a large extent of these mountains; and the smaller Seikh chieftains on the south of the Sutlej having placed themselves under the British protection, the range of our influence has been widely enlarged; the farthest western boundary of our dominions now corresponding with the farthest eastern advance of Alexander the Great – a striking proof of the superiority maintained by the nations of Europe at an interval of two thousand years.[86]

The interest in this region had been steadily increasing, driven on the one hand by the rationale of trade, accession, defence and on the other by scientific knowledge. The length and breadth of the region was brought under rigorous academic scrutiny, whereas painters framed the mountainscape in terms of the picturesque. Numerous entries in the *Journal of the Asiatic Society of Bengal* from the time indicate specialised interest in the various facets of the Himalayan region. In 1788, *An Account of a Journey to Tibet* came out. In 1793, William Jones, who was then the president of the Society, brought out *On the Borderers, Mountaineers, and Islanders of Asia*. *Asiatic Researches* alone was replete with numerous works and excerpts based on the Himalayas at the end of the century, like those by John Eliot, William Dunkin and Thomas Hardwicke and many more.[87] In the early nineteenth century, a significant number of dossiers by Lambton in his conceptualisation of the Trigonometrical Survey and its progress, make repeated references to the Himalayas. After having defeated the Gurkhas in 1815, the Company forced the local chiefs to relinquish all land west of the River Sharada. The newly acquired territory was thereupon divided into three administrative divisions: British Kumaon, British Garhwal and a small princely province called Tehri Garhwal. The move spurred and encouraged many more invested professional and academic visitors ranging from naturalists and surveyors to artists and missionaries. In 1819, Francis Hamilton Buchanan's (who was in charge of the Institute for Promoting Natural History of India) highly systematic survey account of collecting and enumerating many plant species found across the Nepal region was published.[88]

The Himalayas provided a vast stage where Europeans explored, measured, sought and acquired scientific renown. As suggested earlier in this book, British interest in the river system was immense and held the key to journeys and routes inland. While the Trigonometrical Survey operated with aim of measuring the heights of various mountain summits, traveller after traveller undertook arduous expeditions to trace the course and sources of the rivers and lakes in the laps of the ranges. One of the first of these was James Baillie Fraser's (1783–1856) *Journal of a Tour through Part of the Snowy Range of the Himala Mountains and to the Sources of the Rivers Jumna and Ganges* (1820).[89] There were a host of others some of which were: *On the Sources of the Ganges, in the Himadri or Emodus* (1810) by H.T. Colebrook; *Narrative of a Survey for the purpose of discovering the sources of the Ganges* (1810) by F.V. Raper; *A Journey to Lake Manasarovara* (1818) by William Moorcroft; *Journal of a Survey to the Heads of the Rivers, Ganges and Jumna* (1822) by J.A. Hodgson; and *An Account of a tour made to lay down the course and levels of the River Setlej or Satudra* (1819) by J.D. Herbert.

Fraser mentions in his *Journal*, that a "belt of low, wooded, and marshy, but rich land, known by the name of the Turrāee or Turreeānā"[90] first fell into the notice of the British as a result of the Gurkha War, and:

> The conduct of this war, with its consequences, offered to us sources of information regarding Nepal and the countries contained in the mountainous belt that confines Hindostan, of which heretofore there was but little known.[91]

Fraser also accompanied his brother, William Fraser, who was responsible for overseeing the new land settlement in the Garhwal Hills after the incursion. At the end of his preface, Fraser almost foresees a tradition of colonial enterprises in the region, in which he puts himself at the forefront for having revealed the space as a rich field for investigation:

> the author will be gratified and proud if the effort at all succeed in satisfying or in awakening curiosity and inducing those who are better qualified than himself, to explore the field on which he has barely gazed from a distance.[92]

Being Scottish by birth and having travelled in the highlands intermittently, Fraser's modes of deriving essences and of drawing comparisons spring from his Scottish sensibilities, while making him suitable for the mountainous journey. He could spontaneously make deductions on land rights and authority based on his understanding of Scottish history. In one place, he says:

> On the whole, there seems at least a strong resemblance to be traceable between the state of this country and that condition of things which existed in the highlands of Scotland during the height of the feodal [sic] system, where each possessor of a landed estate exercised the functions of a sovereign, and made wars and incursions on his neighbours, as a restless spirit of ambition or avarice impelled him.[93]

There is a constant search for familiar signs amidst the unfamiliar:

> Here we found a birch-tree for the first time, precisely similar to that of Scotland in all respects. The bark, leaf, twig, and buds were quite the same; the leaf was somewhat larger, but seemed to possess no fragrance; yet we had been struck at a scent exactly like that of the birch after a shower ... we did not pluck even a bough, although ancient recollection almost tempted us to do so. We found sweet briar in great plenty, and giving a perfume perfectly the same as that from the home plant.[94]

The visual legacy of the "picturesque" which transformed the Scottish Highland landscape reverberates throughout Fraser's own narrative. Along with the *Journal*, 20 aquatints of Fraser's watercolours done from "on the spot" sketches, were published in a folio edition entitled *Views in the Himala Mountains* the same year. Being excessively conscious of portraying an alien landscape, he took special training from renowned professional artists like William Havell and George Chinnery in Calcutta during the interim four years until he felt fully confident to publish them.[95] Not only did he consult Company artists but there are ample indications that he took help from indigenous artists. He was engaged in producing a series of portraits of several servants and Gurkha locals, a skill he used in depicting native figures populating the landscape in his landscape paintings of the Himalayas.[96]

The sensory ecstasy of the revelation of familiar characteristics, that Fraser delineates, was the reason behind the establishment of numerous "hill stations" on the foothills and higher altitudes of the Himalayas, constructed to make the Britons feel at home in a picturesque locale in an otherwise alien land. The region, however, was still undeveloped. It was still clasped within petty rivalries and disputes among small-time feudal lords, a state which Scotland had long overcome and therefore managed to see its natural spatial apotheosis through a "modern" spatial reorganisation, namely that of a nation state. Therefore, the lack of an overarching political unity among these tribal societies, to which all of them would bow down, was the main difference with Scottish structure:

> Indeed, the chief political dissimilarity between this country and those in which the feudal system obtained seems to have been in this – that there did not exist even a nominal sovereign in this mountainous district to whom these independent barons acknowledged a feudal subjection.[97]

An overlord was what was required, to whom all would bow down to maintain peace in the region. By subordinating themselves to the British, the region could see its natural culmination. James Baillie Fraser's own work was determined by his brother William Fraser's (1784–1835) professional role as a surveyor posted near Delhi, in creating a visual record or a topographical vocabulary of local land settlements. Likewise, a cartographic dimension informs the landscape compositions by Fraser, where the below eye-level foreground is framed by the shaded dark green or brown hills and faded mountain peaks in the distance, demarcating a foreclosure to the expansive view. The region was otherwise comparable to European landscapes worthy of capturing, in pictures of no less quality than a master painter's. The cultural diversity and the antiquity of the people in their ethnic costumes would make the painting suitably picturesque while abiding by the dictates of maintaining a visual record:

> Around us the fantastic forms of the old trees, their rich masses of foliage contrasting with the gray bare crags, and the blasted pines and withered oaks, formed a foreground for a picture worthy of the pencil of a Salvator. Nor would our attendants, the Ghoorkhas, the hill-men, and the Patans, formed into groups reclined around, or loitering on the rocks and cliffs, have disgraced the composition.[98]

The version of the "sublime" and the "picturesque" that developed based on travels in the Celtic fringes played its part in articulating and appropriating the Indian highlands in similar terms. For a Scotsman, comparison with his native place was a means "of an associative ligature which mapped the exotic scene onto a nostalgic landscape of childhood and of home".[99] This became a popular and established idiom especially with artists, so much so that when Fanny Parks tours the Himalayan region "in search of the picturesque", she initially finds not having travelled to the Scottish Highlands a definite handicap in appreciating the great mountain belt of India:

> when we arrive at the hills, I hear we are to be carried back, in imagination, to the Highlands of Scotland. I have never been there; *n'importe*, I can fancy as well as others.[100]

An avid reader of Scott, her power of imagination soon overcomes her deficiency. For example, borderer Dr John Leyden, discovers in Coorg "the grotesque and savage scenery, the sudden peeps in romantic ridges of mountains bursting through the bamboo bushes, all contributed strongly to recall to memory some very romantic scenes in the Scottish Highlands";[101] while for Bishop Heber, the Himalayan landscape constantly brings back memories of Wales.

Though parts of the natural surroundings might have seemed familiar and even "sublime", it was sooner or later realised as essentially "other". Fraser adopts the trope of Terry (as talked about earlier in the chapter), in outlining a negative character lying beneath the illusory exalted visual landscape. The idea of the Arcadian space he had inadvertently generated had to be morally maligned to restrain it from attaining a utopian status. Through his travels "throughout those regions now fallen, or likely to fall into the British power",[102] he unravels a dominant moral landscape which unifies and binds spaces as varied as the plains and the hills:

> Such was the conduct of the hill people on this occasion, and it will probably be found of a piece with the whole tenor of that uncertain, vacillating, mean, and narrow policy, which marks and stains the Asiatic character. From such men no steady or good course of conduct can be looked for; on them no reliance can be placed. Even the tie of interest seems unsteady when viewed through so uncertain a medium.[103]

On the whole, these people were not different from the people to be encountered in the rest of the known regions of Hindustan. Also, an abundance of superstitions, heathen mythologies, idol worship, caste system, Brahmin priests, temples and numerous other practices in the region (all familiar Eurocentric distinguishing tropes to view the primordial society of India with), are evidence enough to relate and unify spaces across a topographically varied area.

Needless to point out, the area he traversed was one of utmost importance in Hindu/Tantrik *Teertha* (religious pilgrimage) tradition, an indigenous paradigm of travel. Once at Badrinath he hardly masks his contempt for the pilgrims and the shrine which he finds in a too dilapidated, ruinous condition to make any such claims to greatness:

> The structure and appearance of this edifice are by no means answerable to the expectations that might be formed of a place of such reputed sanctity, and for the support of which large sums are annually received.[104]

It should be pointed out that Badrinath is one of the four dhams marking the four directions of an early imagined geography. Mountains, and particularly the Himalayas themselves, hold a symbolic meaning in Hindu cosmogony, mountain peaks and hill tops always being deemed sacred. Teerthas are often conceived as geographical centres; in Badrinath Dhaam, other important pilgrimage destinations were strewn across the peninsula in the forms of rivulets of river Alakananda named after five great tirthas in India: Prabhasa, Pushkara, Gaya, Naimisa and Kurukshetra.[105] Fraser, in this sense, fails to understand or include the significance of the polycentric mythologically and narratively connected landscape linked with the tracks of pilgrimage in his European-styled topographical vocabulary. For Fraser, "in a country so rude where records are so loosely if at all preserved, and where the genius of the people runs so much upon fable and superstition, there is little faith to be placed in any details of history".[106] The layers of narratives which construe the land find no place in the logical and scientific construction of the new frontier geography which has travelled overseas from Great Britain.

Despite Fraser's shamefacedness at not having received proper training in the field sciences, his work nevertheless aspired to the textual status of a geographical narrative – an encyclopedic digest about the Himalayan region. He sent specimens of minerals, insects and ethnographical curiosities collected during his Himalayan travels to the Geological Society in London, together with notes about them. These were subsequently published first in *Transactions of the Geological Society* in 1819 and in many other issues later on.[107] He was soon elected a fellow of the society on the merit of his Himalayan travel, the first among many other journeys he undertook, later especially in the central Asian belt (another strategic area for the British about which little or nothing was yet known).[108] For Fraser, therefore, mountain climbing was as much symbolical of social ascension as was the conquest of mountain summits symbolical of political conquest of the entire land.

Pratt traces the prevalence of scientific scrutiny in travel narratives as being a part of the emergent "planetary consciousness" during the time period, 1750–1800 and subsequently, much work had already been done by that date to reveal the spirit of scientific enquiry, pursuit of mathematical precision and establishment of collaborative academies during this age and after. The trend continues all through the nineteenth-century Victorian British narratives about its empire.

These not only attempted to systematise, catalogue, and classify all existing species occupying the surface of the earth "into sequence of a descriptive language"[109] but, in being engaged in discovering the so-long hidden system of nature, the natural historians led "a new field of visibility being constituted in all its density".[110] Such expeditions completely overwrote indigenous cultural representations of space which embraced local stylised aesthetics and narrative expressions of religious, historical and moral topophilia and gave rise to disenchanted hodological space. In fact, from the 1790s onwards, a new optic emerged focussed entirely on foregrounds based on a new practice of botany.[111] Roots, tussocks, mosses, lichens, ferns, rocks, soil strata, trailing plants, pot herbs, bird nests, and bee-hives came to be inspected and studied in all their profusion and diversity, as parts of the local ecology. In Britain at this time, a commitment to agriculture and forestry merged with specialised interests in horticulture and botany. An interest in horticulture spread over to varied public and private spaces. Well-managed orchards, kitchen gardens, fields and plantations were linked not only to a national aesthetics, but also to a larger spectrum of twin ideas of profit and patriotism, for a well-managed plot semiotically referred to a well-managed nation. Globally, plant collection was aimed at understanding the hidden laws of nature, clinching questions on long-conjectured connections between a region's physical geography, such as its climate, rainfall, elevation, soil type and characteristic vegetation.[112] The Himalayan elevation and its rich, diverse characteristic vegetation was posited as a locus for deep enquiries into naturalistic laws which had wide-ranging implications, from the economic to the theological, from concerns about trade-worthy cash crops to debates about multiple centres of origin/creation. The ordering aesthetic with which the region was chalked and pictured also incorporated the task of segregating: making the differences observable and diversities knowable. Writing about hunting for typical genera, at elevations of many thousands of feet, not found anywhere else on the subcontinent, corresponded with the spatial analogy of the Himalayas's separateness being conceptualised as margins to the subcontinent's essential cohesion.

Joseph Dalton Hooker (1817–1911), surgeon, naturalist and traveller, embodied in himself the quintessential nineteenth-century Victorian scientific spirit and played a key role in conceptualising geobotany and elevated the otherwise descriptive subject to the realm of scientific scrutiny involving mathematical precision, aided by instruments for enhanced observation and collection: the Wardian case, vasculum, paper plant press, microscope and magnifying glass. He imbibed the Linnaean taxonomic impulse, as well as Humboldtian vision, producing numerous works on natural sciences in his lifetime. He was born into an English family of scientists. His grandfather was a fervent enthusiast in cultivating rare plants, whereas his maternal grandfather, Dawson Turner, had published works on Irish and English ferns and mosses. His father, William Jackson Hooker, was a professor of botany at Glasgow University and later became the director of the Botanical Gardens of Kew. Hooker himself was educated at Glasgow High School and Glasgow University. He completed his medical studies successfully but his passion remained the natural sciences and especially botany as he called himself "a born muscologist". He joined the medical department of the Royal Navy. Hooker's first major opportunity came with his appointment as assistant surgeon aboard H.M.S. *Erebus* on the Antarctic expedition commanded by James Ross from 1839 to 1843. The voyage gave him scope to study vegetation in places of which he had read in Cook's journals: New Zealand, Australia, Tasmania and the Falkland Islands. Huge numbers of plant specimens were collected from these places, which went to consolidate a scientific archive of these places.

Returning from the expedition, Hooker travelled throughout the Continent to collect and exchange specimens for Kew, for by now his father was the director of the Kew Gardens. After lecturing on botany for a few intervening years, he was appointed botanist to the Geological Survey of Great Britain in 1846, his work being in the field of palaeobotany, searching for plant fossils in the coal-beds of Wales. It seems to have been an archaeological enterprise to unravel, typify and categorise land and vegetation of the temperate zone of Europe. His conceptualisation of the tropics would in turn emerge from such an understanding. In the meantime, Hooker had become acquainted with Charles Darwin, and had become profoundly interested in Darwin's early ideas of "transmutation of species" and "natural selection". Later on, he also read the manuscript of *Origin of Species*. In the same vein as Darwin's theory of "survival of species", Hooker was to proclaim:

> Plants, in a state of nature, are always warring with one another, contending for the monopoly of the soil, – the stronger ejecting the weaker, – the more vigorous overgrowing and killing the more delicate. Every modification of climate, every disturbance of the soil, every interference with the existing vegetation of an area, favours some species at the expense of others.[113]

His theory is as much based on the hierarchy of species and genus as is Darwin's own. Curiously, his proposition held true in the nineteenth-century world order in terms of the land grab initiated by European colonialism for systematic exploitation of lands across the globe. The existing vegetation of an area, in turn, would be a stable indicator of a whole host of factors: climate, soil and meteorological conditions, by tabulation of which, conceptual and academic spatial categories could be configured and formulated.

In 1847, his father, noting his keenness and fervour in botanical research, nominated him to travel to India to collect botanical specimens for Kew. He states in the preface to his journal:

> On hearing about the kind interest taken by Baron Humboldt in my proposed travels, and at the request of my father (Sir William Hooker),

the Earl of Carlisle (then Chief Commissioner of Woods and Forests) undertook to represent to her Majesty's Government the expediency of securing my collections for the Royal Gardens at Kew.[114]

On 11 November 1847, Hooker left on his three-year long Himalayan travels, of which his *Himalayan Journals: or Notes of a Naturalist in Bengal, the Sikkim and Nepal Himalayas, the Khasia Mountains etc* (1854) was the outcome. He dedicated this journal to his friend Darwin. In the preface to *Himalayan Journal,* he elaborates upon his choice of India as a suitable field for his pursuits in natural sciences:

> Having accompanied Sir James Ross on his voyage of discovery to the Antarctic regions, where botany was my chief pursuit, on my return I earnestly desired to add to my acquaintance with the natural history of the temperate zones, more knowledge of that of the tropics than I had hitherto had the opportunity of acquiring.[115]

While his choice for field research laid "between India and the Andes", he chose the former on the premise of logistic support promised by Dr Falconer, then superintendent of the Botanic Garden at Calcutta. Falconer also:

> drew my [Hooker's] attention to the fact that we were ignorant of the geography of the central and eastern parts of these mountains, while all to the north was involved in a mystery equally attractive to a traveller and the naturalist.[116]

In fact, he was so preoccupied and overwhelmed by his Indian travel, that he even turned down a proposed trip to Borneo to report on the capabilities of Lebuan with reference to cultivation of cotton, tobacco, sugar, indigo, spices, gutta-percha etc., and extended the Himalayan tour for a third year. Sikkim was recommended by both Lord Auckland and Dr Falconer as "the portion of Himalayas best worth exploring ... as being ground untrodden by traveller or naturalist" and also because its ruler was a British dependent.[117] Moreover, no scientific exploration had till then been undertaken "of the snowy Himalaya eastward of the northwest extremity of the British possessions". He would be the first European to visit these parts since Samuel Turner's diplomatic embassy to Tibet in 1789. His own ideas regarding these regions were shaped by the works of two travellers, Cook and Turner, which incited in him fantasies about Lama worship, Chumulari and Kerguelen's Land, as described in those texts.

On reaching Calcutta, he:

> acquainted [himself] with the vegetation of the plains and hills of Western Bengal, south of the Ganges, by a journey across the

mountains of Birbhoom and Behar to the Soane valley, and thence over the Vindhya range to the Ganges, at Mirzapore ... to Bhagalpore.[118]

From there he proceeded north to Sikkim in the Himalayas. The proceedings of his meteorological observations during this tour are published in the *London Journal of Botany*. Much influenced by Brian Houghton Hodgson, who had accumulated a number of native informants of varied races and tongues as translators, collectors, artists, shooters and stuffers to assist him in establishing a knowledge resource on the Himalayan east, Hooker too outfitted his expedition in the same fashion as a polyglot group. Hooker wrote:

> My party mustered fifty six persons. These consisted of myself and one personal servant, a Portuguese half caste, who undertook all offices and spared me the usual train of Hindu and Mahomedan servants. My tent and equipment, instruments, bed, box of clothes, books and papers required a man for each. Seven more carried my papers for drying plants and other scientific stores. The Nepalese guard had two coolies of their own. My interpretor, the coolie sirdar (a headsman) and my chief collector (a Lepcha) had a man each. Mr. Hodgson's bird and animal shooter, collector and stuffer with their ammunition and indespensables [*sic*] had four more. There were, besides, three Lepcha lads to climb trees and change the plant papers, and the party was completed by fourteen Bhutan coolies laden with food.[119]

Hooker admired Hodgson's scholarly rigour in building up a consolidated archive with his ethnological and zoological enterprise in "unveiling the mysteries of Booddhist religion, chronicled the affinities, languages and customs, and faiths of the Himalayan tribes ... and natural history of birds and animals of this region".[120] Similarly, Hooker incorporated in his naturalist inspection, multi-facets of geology, geodesy, meteorology, botany, zoology and anthropology to architect a discursive space called the Indian tropics. However, Hooker packaged his scientific revelations in a language akin to the "literary picturesque" not different to that of Bishop Heber, as was popular at the time. His scientific Latin diction added to the poesis of his descriptions:

> It is difficult to conceive a grander mass of vegetation:– the straight shafts of the timber-trees shooting aloft, some naked and clean, with grey, pale, or brown bark; others literally clothed for yards with a continuous garment of epiphytes, one mass of blossoms, especially the white Orchids Caelogynes, which bloom in a profuse manner, whitening their trunks like snow. More bulky trunks were masses of interlacing climbers, *Araliaceae*, *Leguminosae*, Vines, *Menispermeae*, Hydrangea and Peppers, enclosing a hollow, once filled by the now strangled supporting tree, which has long ago decayed away. From the

sides and summit of these, supply branches hung forth, either leafy or naked; the latter resembling cables flung from one tree to another, swinging in the breeze, their rocking motion increased by the weight of great bunches of ferns or Orchids, which were perched aloft in the loops. Perpetual moisture nourishes this dripping forest: and pendulous mosses and lichens are met with in profusion.[121]

His scenic descriptions celebrate the beauty of natural diversity and paint a picture of the rich vegetation in its profusion. The region as constituted by the naturalist's eye would possibly, according to Hooker, be translatable through an artist's eye, as he expressed in a letter to Darwin about the Himalayan scenery:

> I am above the forest region, amongst grand rocks and such a torrent as you see in Salvator Rosa's paintings vegetations all a scrub of rhododods. with pines below me as thick and bad to get through as our Fuegian Fagi on the hill tops, and except the towering peaks of P.S. [perpetual snow] that, here shoot up on all hands; there is little difference in mt. scenery – here however the blaze of Rhod. flowers and various coloured jungle proclaimed a differently constituted region in a naturalist's eye and twenty species here, to one there, always are asking me the vexed question, where do we come from?[122]

However, it is the naturalist's eye which scores over the artist's as it is through the former that the space is made accessible and amenable to the European scopic order. His initiative was to chalk out differential realms within the Himalayan tropics, a recurring idea in all his works, which gets reflected as well in his *Flora Indica: being a systematic account of the plants of British India,* co-authored with Thomas Thomson:

> An attempt to divide the area embraced in the Flora Indica into physico-geographical or geographico-botanical districts. This is intended to serve the double purpose of giving a slight sketch of the physical characters and vegetation of these provinces, and of adopting such a carefully-selected system of nomenclature, as shall be available for assigning intelligible localities to the species in the body of the Flora and such as may be easily committed to memory, or found with little trouble on any map.[123]

The mission was to articulate conceptual and identifiable terrains by locating and segregating species and genus in the plant kingdom. His proposed map of "Botanical Provinces" in *Flora Indica* (1855) emerges from the idea that the entire Himalayan belt is posited as a continuous arboreal corridor with its internal divisions based on subtle variations.

His heteroglot companions were as much under his scrutiny as the plant kingdom. As racial types, they became the link to frame associative features as racial characteristics of the multiple hill tribes he encountered. The Lepchas, he found "rude but not savage, ignorant and yet intelligent" and could make loyal and diligent attendants.[124] The Gurkhas were also faithful as servants as they were used to working under an overruling power and had no pretensions as landowners. The Khasias, however, could not attain his appreciation and a show of their insubordination resulted in an incident of a bitter clash and the ruthless killing of the Khasia natives. Hooker's practice and examination of regional botany was fundamentally connected to the academic cult of local ecology. There was a practical dimension to this stream of academics. An understanding and management of an ecological part of a whole was an esteemed model generally believed to lead to good governance of a country as a whole.

There is another dominant strain which runs through his work and must not be overlooked in the context of this book. In his reminiscences of his childhood and adolescent botanical dabblings, he recalls incidents where the Himalayas and the Scottish Highlands figure as contiguous memory:

> when still in my early teens, I took up the study of these beautiful objects (mosses from his father's collection), and formed a good collection of the Scottish species, in the Highlands and elsewhere; and my first effort as an author was the description of three new mosses from the Himalaya.[125]

And a little later, in reminiscing about his early days of botanising with his father, he mentions his interest in arranging the specimens he gathered:

> A little older, and when still a child, my father used to take me [for] excursions in the Highlands, where I fished a good deal, but also botanised; and well I remember on one occasion, that, after returning home, I built up by a heap of stones a representation of one of the mountains I had ascended, and stuck upon it specimens of the mosses I had collected on it, at heights relative to those at which I had gathered them. This was the dawn of my love for geographical botany.[126]

Likewise, later in his life, mountain scenery below the snow line is compared in his *Himalayan Journals* to the Scottish Highlands. For example, "in the Tambur Valley, is an old lake bed, outspread under lofty hills, through which meandered a rippling stream, fringed with alder". Elsewhere, "the mountain rising out of the sea of valley mists" are like the mountains by "the Scotch salt water lochs". "A little lake, a rarity in these valleys" recalls "the turn at the entrance of Glencoe". It was a "home-like delight to espy abundance of a common Scotch fern, Cryptogramma crispa", growing in clefts of a rocky moraine under the

Choonjerna Pass at an elevation of 13000 feet. High on the Wallanchoon Pass, again, the same lichens coloured the rocks, as in Scotland and the "dwarf rhododendrons" and "masses of little Andromeda imitated Scottish heathery hill side".[127] Such findings stir his imagination, leading him to contemplate:

> Along the narrow path I found the two commonest of all British weeds, a grass (Poa annua), and the shepherd's purse! They had evidently been imported by man and yaks, and as they do not occur in India, I could not but regard these little wanderers from the north with the deepest interest. Such incidents as these give rise to trains of reflection in the mind of the naturalist traveller; and the farther he may be from home and friends, the more wild and desolate the country he is exploring, the greater the difficulties and dangers under which he encounters these subjects of his earliest studies in science, so much keener is the delight with which he recognises them, and the more lasting is the impression which they leave. At this moment these common weeds more vividly recall to me that wild scene than does all my journal, and remind me how I went on my way, taxing my memory for all it ever knew of the geographical distribution of the shepherd's purse, and musing on the probability of the plant having found its way thither over all Central Asia, and the ages that may have been occupied in its march.[128]

This was related to debates on distribution and to the problematising of sporadic species, or the existence of identical species in disconnected localities, which posed questions to the narrative of biblical creation in a singular centre and their descent from a single pair of progenitors. Naturalists like Darwin, Hooker and Forbes came up with multiple propositions regarding distribution. Hooker was vehemently against the theory of multiple centres of creation which answered questions about existence of similar but not identical species in disconnected physical surroundings with a convenient explanation. He tended to side with Forbes's hypothesis, aided by geological surveys and stratigraphy, that suggested geological unity of apparently disconnected lands at an anterior epoch. However, he believed that plant distribution studies depended on better classifications.

The scientific code underlining most of the narratives was designed to produce and establish what they represented. Encouraged by scientific institutions, the language of science immediately ratified the document's veracity. Apart from the purely scientific documentation, as has been seen in the course of the present chapter, the narratives are punctuated by several cultural discourses, which work in complicity with the dominant Enlightenment gaze of scientism. It is ultimately the voice of the narrator-author which unifies the imperial gaze in all its multifarious dimensions such as that of the aesthetic, visceral and the scientific to construct the space of the subject's location. The "I" of the narrator merges seamlessly and becomes the eye of imperialism which uses European frames of reference to translate and re-present a space which, by this process, is opened up for prospects of invasion and occupation. In providing a cohesive sense of geography and history of the land, the colonial travel narratives orchestrated an image of the British Empire's imagined geo-body.

Notes

1 Marlowe, Christopher. *The Tragical History of the Life and Death of Doctor Faustus.* (1604). Act II, sc. i. Project Gutenberg, 2009.
2 Barrow, Ian J. "Moving Frontiers: Changing Colonial Notions of the Indian Frontiers". <http://asnic.utexas.edu/ asnic/sagar/fall.1994/ian.barrow.art.html> 10 July 2005.
3 Cohn, Bernard S. *Colonialism and its Forms of Knowledge.* p. 6
4 Oaten, E.F. *European Travellers in India: During the Fifteenth, Sixteenth and Seventeenth Centuries; the Evidence afforded by them with respect to Indian Social Institutions, and the Nature and Influence of Indian Governments by Edward Farley Oaten.* London: Kegan Paul, Trench, Trubner and Co. Ltd., 1909. pp. 104–17.
5 Once in Agra, Mildenhall went through many a trial and tribulation against Jesuits who maligned England as "a nation of thieves" before the Great Mogul so much that his own interpreter deserted him in favour of the Jesuits who bribed all officials. Ultimately, he was forced to learn Persian in order to communicate with Akbar and finally bagged the desired signed and sealed document. However, Oaten is doubtful about the story of the treaty as no effective trade relations were established during this time.
6 Teltscher, Kate. *India Inscribed: European and British Writing on India 1600–1800.* Delhi: Oxford University Press. 1997. pp. 16–20.
7 Purchas, Samuel. *Hakluytus Postumus or Purchas His Pilgrimes,* iii, 49. Quoted in Kate Teltscher. *India Inscribed.* p. 17.
8 Teltscher, Kate. "India/ Calcutta: city of palaces and dreadful night". *The Cambridge Companion to Travel Writing.* Cambridge: Cambridge University Press, 2002. p. 191.
9 The Mughal grant to collect land revenues and administer civil justice in Bengal.
10 Cohn, Bernard S. *Colonialism and its forms of knowledge.* pp. 7–11.
11 ER, 6 (1805), 462. Quoted in Leask, Nigel. *Curiosities and the Aesthetics of Travel Writing, 1770–1840.* Oxford: Oxford University Press, 2002. p. 163.
12 John Gibson Lockhart in *Quarterly Review,* 37, January 1828. Quoted in Leask, Nigel. p. 158.
13 In the 1780s, there emerged a number of travel texts which carried illustrations by artists and were generally called "Voyage pittoresque". See Greppi, Claudio. "On the Spot: Travelling Artists and the Iconographic Inventory of the World, 1769–1859". In Felix Driver and Luciana Martins eds. *Tropical Visions in an Age of Empire.* Chicago: Chicago University Press, 2005. pp. 24–39.
14 Jemima Kindersley's *Letters from the Island of Teneriffe, Brazil, the Cape of Good Hope, and the East Indies* (1777) is an important memoir written during this period.
15 Cook's journal entry on 11 May 1773. Quoted in Greppi, Claudio. "On the Spot: Travelling Artists and the Iconographic Inventory of the World, 1769–1859". In Felix Driver and Luciana Martins eds. *Tropical Visions in an Age of Empire.* Chicago: Chicago University Press, 2005. p. 26.

16 Hodges, William R.A. "Preface". *Travels in India, during the years 1780, 1781, 1782 and 1783*. London: J. Edwards, Pall Mall, 2nd Edition,1794. P. iv.
17 Ibid.
18 Ibid. p. v.
19 Letter dated 31 December 1788. Quoted in Stuebe, Isabel. "William Hodges and Warren Hastings: A study in eighteenth century patronage". *The Burlington Magazine*, Vol. 115, No. 847 (October, 1973). pp. 657–66.
20 Carter, Paul. *Road to Botany Bay*. London: Faber and Faber, 1987. pp. xiv–xv.
21 Hodges, William R.A. *Travels in India*. p. 2.
22 Ibid. p. 2.
23 Ibid. p. 3.
24 Ibid. p. 5.
25 Ibid. p. 14.
26 Ibid.
27 Ibid. p. 85.
28 Ibid. p. 86.
29 Ibid. p. 88.
30 Ibid.
31 Ibid. p. 89.
32 Ibid.
33 Ibid.
34 Ibid. p. 90.
35 Ibid.
36 Ibid. p. 95.
37 Ibid. p. 24.
38 Other overtly geographical researches dealing with the area are R.E. Sherwill's "Notes upon a tour through the Rajmahal hills" in his *Geological Memoirs* and Hooker's journal and his paper in the *Journal of the Asiatic Society of Bengal*, xvii, part II. See Markham, Clements R. A *Memoir on the Indian Surveys*. London: Allen and Co., 1871. p. 262.
39 Edney, Matthew H. "Reconsidering Enlightenment Geography and Map Making: Reconnaissance, Mapping, Archive". Livingstone, David N. and Charles W.J. Withers eds. *Geography and Enlightenment*. Chicago: Chicago University Press, 1999. p. 177.
40 Hodges, William R.A. *Travels in India*. p. 155.
41 DeRogatis, Amy. *Moral Geography: Maps, Missionaries, and the American Frontier*. New York: Columbia University Press, 2003. p. 19. This book makes an interesting case for a collaboration between the missionaries and land surveyors/ map makers on the American frontiers.
42 Heber, Reginald. *Sermons Preached in India by the Late Right Reverend, Reginald Heber D.D. Lord Bishop of Calcutta*. London: John Murray, 1829. p. xxii.
43 Ibid. p. xxxi–xxxii.
44 Ibid. p. xxviii.
45 Harvey, David. *Justice, Nature and the Geography of Difference*. Oxford: Basil Blackwell, 1996. p. 214–15.
46 Ibid. p. 214.
47 Heber, Reginald. *Poems by the Late Rt. Rev. Reginald Heber D.D. Lord Bishop of Calcutta*. Hingam, C. and E.B. Gill, 1830. p. 11. ll. 1–16.
48 Chambers, James. *Bishop Heber and Indian Missions*. London: John W. Parker. 1846. p. 101.
49 *Quarterly Review*, 27 Jan. 1828, 102. Quoted in Leask, Nigel. p. 184.
50 Laird, M.A. *Bishop Heber in Northern India, Selections from Heber's Journal*. 1971. p. 36. Quoted in K.K. Dyson. *A Various Universe: A Study of the Journals*

and Memoirs of British Men and Women in the Indian Subcontinent, 1765–1856. Delhi: Oxford University Press, 1978. p. 225.

51 Dyson, K.K, *A Various Universe.* p. 225.

52 Reginald, Heber. *Narrative of a Journey through the Upper Provinces of India, from Calcutta to Bombay, 1824–1825, with Notes upon Ceylon, an Account of a Journey to Madras and the Southern Provinces, 1826.* 2 Vols. Vol. 2. London: John Murray, (2nd. Edition). 1828. p. 392.

53 Taylor, Thomas. *Memoirs of the Life and Writings of the Right Reverend Reginald Heber, D.D. Late Lord Bishop of Calcutta.* London: John Hatchard and Son, Piccadilly, 1835. p. 112.

54 Heber, Reginald and Amelia Shipley Heber. *The Life of Reginald Heber.* 2 Vols. Vol. 2. New York: Protestant Episcopal Press, 1830. p. 192.

55 Reginald, Heber. *Narrative of a Journey through the Upper Provinces of India, from Calcutta to Bombay, 1824–1825, with Notes upon Ceylon, an Account of a Journey to Madras and the Southern Provinces, 1826.* 2 Vols. Vol. 1. London: John Murray, (2nd. Edition), 1828 Ibid. p. 169.

56 Ibid. p. 107.

57 Heber, Reginald and Amelia Shipley Heber. *The Life of Reginald Heber.* 2 Vols. Vol. 2. New York: Protestant Episcopal Press, 1830. p. 206.

58 Ibid. p. 193.

59 Reginald, Heber. *Narrative of a Journey through the Upper Provinces of India, from Calcutta to Bombay, 1824–1825, with Notes upon Ceylon, an Account of a Journey to Madras and the Southern Provinces, 1826.* 2 Vols. Vol. 1. London: John Murray, (2nd. Edition), 1828 p. 118.

60 Ibid p. 119.

61 Various kinds of authors require various kinds of styles, called authorcrafts, for their writings; for example, professional scientists adopted the style of ordered facts. However, popular and non-scientist authors also borrowed this style for claiming veracity.

62 Taylor, Thomas. *Memoirs of the Life and Writings of the Right Reverend Reginald Heber.* p. 83.

63 Reginald, Heber. *Narrative of a Journey through the Upper Provinces of India, from Calcutta to Bombay, 1824–1825, with Notes upon Ceylon, an Account of a Journey to Madras and the Southern Provinces, 1826.* 2 Vols. Vol. 1. London: John Murray, (2nd. Edition), 1828 p. 389.

64 Ibid. p. 245.

65 *Quarterly Review.* 4 (August 1810), p. 87. Quoted in Leask, Nigel. p. 188.

66 Taylor, Thomas. *Memoirs of the Life and Writings of the Right Reverend Reginald Heber.* p. 118 .

67 Heber, Reginald. Letter to the Rev. Cholmondoley and Mrs. Cholmondoley. In *Narrative of a Journey through the Upper Provinces of India, from Calcutta to Bombay, 1824–1825, with Notes upon Ceylon, an Account of a Journey to Madras and the Southern Provinces, 1826.* 2 Vols. Vol. 2. London: John Murray, 1844 p. 214.

68 Heber, Reginald. *Narrative of a Journey through the Upper Provinces of India, from Calcutta to Bombay, 1824–1825, with Notes upon Ceylon, an Account of a Journey to Madras and the Southern Provinces, 1826.* 2 Vols. Vol. 1. London: John Murray, (second edition), 1828. p. 207.

69 Ibid. p. 269.

70 Heber, Reginald. *Narrative of a Journey through the Upper Provinces of India, from Calcutta to Bombay, 1824–1825, with Notes upon Ceylon, an Account of a Journey to Madras and the Southern Provinces, 1826.* 2 Vols. Vol. 2. London: John Murray, (second edition), 1828. p. 194.

71 Heber, Reginald. *Narrative of a Journey through the Upper Provinces of India, from Calcutta to Bombay, 1824–1825.* Vol. 1. p. 276.

72 Ibid. p. 164.

73 Ibid. p. 319.

74 Heber, Reginald. Letter to R.J. Wilmot Horton. In *Narrative of a Journey through the Upper Provinces of India, from Calcutta to Bombay, 1824–1825, with Notes upon Ceylon, an Account of a Journey to Madras and the Southern Provinces, 1826.* 2 Vols. Vol. 2. London: John Murray, 1844. p. 229.

75 Ibid. Also see Dyson, K.K. *A Various Universe.* p. 230.

76 Heber, Reginald. Letter to Charles Williams Wynn. p. 220.

77 Heber, Reginald. *Narrative of a Journey through the Upper Provinces of India, from Calcutta to Bombay, 1824–1825, with Notes upon Ceylon, an Account of a Journey to Madras and the Southern Provinces, 1826.* 2 Vols. Vol. 1. London: John Murray, (second edition), 1828. p. 295.

78 Teltscher, Kate. *India Inscribed.* p. 76.

79 Taylor, Thomas. *Memoirs of the Life and Writings of the Right Reverend Reginald Heber.* p. 282.

80 Chambers, James. *Bishop Heber and Indian Missions.* pp. 1–10.

81 Ibid. p. 10.

82 See Michael, Bernardo A. *Statemaking and Territory in South Asia: Lessons from the Anglo-Gorkha War (1814–1816).* London: Anthem, 2014.

83 Moschek, Wolfgang. "The Limes: Between open Frontier and Borderline". In Fryde, Natalie ed. *Walls, Ramparts and Lines of Demarcation: Selected studies from Antiquity to Modern Times.* Berlin: Lit Verlag, 2009. p. 15.

84 Ray, Keith and Ian Bapty. *Offa's Dyke: Landscape and Hegemony in Eighth Century Britain.* United Kingdom: Oxbow, 2016. n.p.

85 The accession of Awadh brought the Company-occupied territory in close proximity with the kingdom of Nepal. The intervening narrow strip of fertile lowlands called the Tarai fell under border dispute as a zone of combat between the two self-seeking warring factions with expansive aspirations. The two-year-long Gurkha War saw several mountainous campaigns until, in 1816, Nepal was forced to cede its occupied provinces of Kumaon, Garhwal and Tarai and accept a British Resident in Kathmandu, the capital of Nepal. Despite the defeat, the spectacular show of Gurkha martial prowess impressed the British and Gurkha regiments have played a crucial role in British military campaigns throughout their imperial tenure at various places on the globe.

86 *Edinburgh Review.* Vol. 57. July 1833. p. 360.

87 Eliot, John. "On the Inhabitants of the Garrow Hills". AR. Vol. 3. 1793. pp. 21–45; Dunkin, William. "Extract from a diary of a journey over the Great Desert, from Aleppo to Bussora, in April 1782". *Asiatick Researches.* Vol. 4. 1795. pp. 399–402; Hardwicke, Thomas. "Narrative of a Journey to Srinagur". AR. Vol. 6. pp. 309–81.

88 Hamilton, Francis Buchanan. *An Account of the Kingdom of Nepal: and of the territories annexed to this dominion by the House of Gorkha.* Edinburgh: A. Constable, 1819.

89 For a more detailed analysis of Fraser's travelogue and paintings and its Scottish connections, see Mukherjee, Nilanjana. "Scottish Resonances in James Baillie Fraser's Travel Accounts". In Sutapa Dutta and Nilanjana Mukherjee eds. *Mapping India: Transitions and Transformations.* New Delhi: Routledge, 2020..

90 Fraser, James Baillie. *Journal of a Tour through Part of the Snowy Range of the Himala Mountains and to the Sources of the Rivers Jumna and Ganges.* London: Rodwell and Martin, 1820. p. 3.

91 Fraser, James Baillie. *Journal of a Tour through Part of the Snowy Range of the Himala Mountains and to the Sources of the Rivers Jumna and Ganges*. London: Rodwell and Martin, 1820Ibid. p. 4.

92 Ibid. p. Ix.

93 Ibid. p. 4.

94 Ibid. p. 158.

95 He also made some excellent drawings on Calcutta, 24 of them being published in eight parts between 1824–6.

96 James' Diary, Fraser Papers Vol. 8. As cited in Archer, Mildred and Toby Falk. *India Revealed: The Art and Adventures of James and William Fraser 1801–35* (The Passionate Quest: The Fraser Brothers in India). London, 1989. pp. 45, 59. Also see Sharma, Yuthika. (Unpublished dissertation) *"Art in between empires: Visual culture and artistic knowledge in late Mughal Delhi 1748–1857'"*. (Order No. 3563677). Available from ProQuest Dissertations & Theses Global. (1400005965). Retrieved from https://search.proquest.com/docview/1400005965?accountid=49663. pp. 195–6.

97 Fraser, James Baillie. *Journal of a Tour through Part of the Snowy Range of the Himala Mountains and to the Sources of the Rivers Jumna and Ganges*. London: Rodwell and Martin, 1820 . p. 4.

98 Ibid. p. 158.

99 Leask, Nigel. *Curiosities and the Aesthetics of Travel Writing* p. 176.

100 Parks, Fanny. *Wanderings of a Pilgrim in Search of the Picturesque*. Quoted in Leask. Nigel. *Curiosities and the Aesthetics of Travel Writing*. p. 176.

101 Quoted in Leask, Nigel. *Curiosities and the Aesthetics of Travel Writing*. p. 176.

102 Fraser, James Baillie. *Journal of a Tour through Part of the Snowy Range of the Himala Mountains and to the Sources of the Rivers Jumna and Ganges*. London: Rodwell and Martin, 1820. p. 167.

103 Ibid. p. 129.

104 Ibid. p. 373.

105 As narrated in the *Skanda Purana*, these teerthas came here staggering beneath the load of sin they had removed from pilgrims and were finally relieved here after which they returned to their purifying duties, to their assigned places, but also remained here. This is the place where at the end of Mahabharata, the last of the Pandavas, Yudhisthir, finally left the earth, and also the place where Urvashi, the celestial danseuse was born of Brahma's thighs.

106 Fraser, James Baillie. *Journal of a Tour through Part of the Snowy Range of the Himala Mountains and to the Sources of the Rivers Jumna and Ganges*. London: Rodwell and Martin, 1820. p. 382.

107 Wright, Denis. "James Baillie Fraser: Traveller, Writer and Artist (1783–1856)". p. 126.

108 James Baillie Fraser wrote *An Historical and Descriptive Account of Persia* published in 1834 as Volume XV in the Edinburgh Cabinet Library. Apart from this he was the author of numerous romances set in central Asia and often based on his grandfather's tract *History of Nadir Shah*. Some of these are *The Khuzzilbash, A Tale of Khorasan* (1828), and its sequel, *The Persian Adventurer* (1830), *The Highland Smugglers* (1832), *The Tales of the Caravanserai* and *The Khan's Tale* (1833).

109 Foucault, Michel. *The Order of Things*. Quoted in Pratt, Mary Louise. p. 28.

110 Pratt, Mary Louise. Imperial Eyes: Travel Writing and Transculturation. London: Routledge, 1992. p. 29.

111 Daniels, Stephen, Susanne Seymour and Charles Watkins. "Enlightenment, Improvement, and the Geographies of Horticulture in Later Georgian England". In David N. Livingstone and Charles Withers ed. *Geography and Enlightenment*. Chicago: University of Chicago Press, 1999. p. 348; Endersby, Jim. "From having

no Herbarium. Local Knowledge versus Metropolitan Expertise: Joseph Hooker's Australasian Correspondence with William Colenso and Ronald Gunn." *Pacific Science* 55.4 (2001): 343–58.

112 Endersby, Jim. *Imperial nature: Joseph Hooker and the practices of Victorian science*. Chicago: University of Chicago Press, 2008. p. 17.

113 Hooker, Joseph Dalton and Thomas Thomson. *Flora Indica: Being A Systematic Account of the Plants of British India*. London: W. Pamplin, 1844. p. 41.

114 Hooker, Joseph Dalton. *Himalayan Journals, or, Notes of a Naturalist, in Bengal, the Sikkim and Nepal Himalayas, the Khasia Mountains*. 2 Vols. Vol. 1. London: John Murray, 1854. p. i.

115 Ibid.

116 Ibid.

117 Ibid. p. ii.

118 Ibid. p. xi.

119 Ibid. Vol. I, p. 179

120 Ibid. p. ii.

121 Ibid. p. 110–11.

122 24 June, 1849. Quoted in Burkhardt, Frederick and Sidney Smith eds. *The Correspondence of Charles Darwin*. Vol. 4 1847–50, Cambridge: Cambridge University Press. p. 242.

123 Hooker, Joseph Dalton and Thomas Thomson. *Flora Indica: Being A Systematic Account of the Plants of British India*. London: W. Pamplin, 1844. p. 2.

124 Hooker, Joseph Dalton. *Himalayan Journals, or, Notes of a Naturalist, in Bengal, the Sikkim and Nepal Himalayas, the Khasia Mountains*. 2 Vols. Vol. 1. London: John Murray, 1854. p. 175.

125 Huxley, Leonard, and Joseph Dalton Hooker. *Life and letters of Sir Joseph Dalton Hooker: Based on materials collected and arranged by Lady Hooker*. Vol. 1. London: John Murray, 1918. p. 5

126 Ibid.

127 Ibid. p. 281.

128 Hooker, Joseph Dalton. *Himalayan Journals*. Vol. 1. p. 221.

POSTSCRIPT

hai kahan tamanna ka doosra qadam ya rab
  hamne dashte imkan ko ek naqsh-e-pa paya
  Where will the next step of desire be, O Lord,
  I/We found the whole desert of possibility to be just one footprint.

Mirza Ghalib[1]

I put a clamp on yearning, shun latitudes, renounce form.
  And turn my eye to the far kingdom
  of bloodless Kalinga battling with a storm.
  Dampen your fires, turn from lighthouse, spire, steeple.
  Forget maps and voyaging, study instead
  the parched earth horoscope of a brown people.

The Map-maker: Keki N. Daruwalla[2]

Imagination is a key productive force behind processes of spatialisation. Therefore, investigating imagination, too, is a productive approach in understanding processes of place-making. Why that should be important is because it helps us understand and critique spatial matrices constituted by the forces of colonialism, imperialism and nationalism. Michelet, the historian of the French Revolution, famously declared: "l'histoire est d'abord toute geography" (history is first of all completely geography).[3] Histories take place and for that reason, every past is a place: an imagined place. Michelet talked of how France, even having existed in a particular space, did so as an organic whole in which the universal spirit of the country replaced the local spirit.[4] This "universal spirit" is imaginary, I argue, a unique product of the cartographic reasoning of modernity which promotes geographic persuasion and control through spatial representations. This book opens up an investigation of how the modern state, in this particular case, India, stems from a practice of cartographic reasoning, fashioned in Britain. Modernity, here, is not merely a temporal or a periodising signifier, devoid of a political basis, but related to colonial histories of power and control exerted over people and spaces. This book, therefore, is about the cartographic method which underlies models of representations, time–space-based narrative strategies or visual designs: one that operates on axiomatics, on normative science, on geometrical idealities and exhaustive deductivity.[5] On the other hand, it is not as though Enlightenment's *esprit geometrique* went uncritiqued. Many Romantic era texts and visionaries including Edmund Burke condemned the reorganisation of administrative boundaries (such as that in revolutionary France) into a rational structure as the malady of "a geometrical and arithmetical constitution" for which "nothing more than an accurate land surveyor, with his chain, sight, and theodolite is requisite".[6] Rousseau too implicated cartographic reason as a culture of instrumentation which thrived on distrust in natural human faculties:

> a theodolite dispenses with our estimating the extent of angles; the eye, which is capable of measuring distances with great exactness, gives up the task to the chain ... the more ingenious and accurate our instruments, the most unsusceptible and inexpert become our organs.[7]

It has been the purpose of this book to reflect on the assumption of spatial identity of India under the British with precisely this kind of instrumentalised vision. In this context, as India became the staging ground for numerous exploratory and survey activities, the entire region which had fallen under the administrative control of the British also received the defining feature of boundaries. The forging of British India as a political entity, however, had far-reaching consequences. As the colony was brought under cartographic surveillance, it also received an autonomous existence as a geographical unit on the globe. In later years, the geo-space thus shaped from geometric fantasies, became a core edifice for nationalist movements. Studies by Benedict Anderson, Manu Goswami and Sumathi Ramaswami show the later sacralisation of the boundaries and the consolidation of the geo-body through print images of the map. If one is to trace the continuity of the colonial imagination, that certainly is the afterlife of the effects of the cultural practices discussed in the chapters of this book. Clearly then, sentiments were tied to, but were also products of, the myriad practices which evoked the sense of the space. It is in this context, that I have looked at the construction of the idea of India through the cultural apparatus wielded by the empire. In trying to test the philosophico-epistemological basis of the construction of India, we find the dominance of the mental realm, enveloping the social and physical dimensions. We also see the theoretical practices which gave birth to this space in the mental realm, first in the imperial consciousness and later in the collective awareness of the inhabitants of the region. This mental realm may seem extra ideological or simply organic, but as has been seen from the study, it is invariably linked to ideologies. This is, thus, also the study of the dominant ideas of the dominating order of the empire, which becomes the key reference point of circulation of knowledge. The process is circular: where epistemology provided the basis for knowledge, knowledge defined the basis for epistemology. Attached to this were the artistic and scientific credentials promoted by the age. It is through these theoretical practices that mentally constructed spaces steadily acquired identity with physical and social space. The articulation of colonial space which worked with representations became a crucial feat in sustaining the bases of power, control and governance, first of the empire and later for the nation. I have in my work tried to look at processes which sought to transform space from without, by imposing an external design to it, a design which was imbued with the codes and conventions of the socio-cultural climate of the place of its origin. Yet, any production of space is not complete and present if not changed and transformed from within. By this I mean those practices which are not merely theoretical. The two kinds of constructions are not two separable ideas. The next step in the process of production is to study how space is transformed within through the direct actions upon it, which irrevocably alter existing space and establish a new place.

Though landscape paintings, travel narratives and maps were important, cultural artefacts which consolidated the idea and appearance of the colony, when confronted with the question of colonial space one cannot ignore the aspect of architecture, which ultimately transformed native space irrevocably. British architecture, spatial planning and urbanisation, especially in the Indian presidencies, ushered the final eradication of the lived space of the inhabitants to mark the complete take over, occupation and super-imposition of the representational space of the empire. The symbolic space thus constructed, also marked the hierarchisation of space and imported the relations of the urban–rural divide which characterised the contemporary European societies. The initial basis of that transformation is, of course, the natural or physical space, which acts as the material on which the empire inscribes its power. Upon these are superimposed networks tangible in form such as roads, paths, telegraph posts and railways along with state buildings and other such structures, acting as properties to reorder and to redefine space. This new colonial space is an expression of power.

Having said that, it should also be pointed out that no existing space is completely annihilated even with its transformation. Traces of each space survive even after repeated reworking, forming time layers or sedimented palimpsests.[8] It is exactly as Derrida suggests:

> The present appears neither as the rupture nor the effect of a past, but as retention of a present past, i.e. a retention of a retention, and so forth. Since the retentional power of living consciousness is finite, this consciousness preserves significations, values and past acts as habitualities (*habitus*) and sedimentations.[9]

There is thus also, the space of latent violence and revolt, of struggle over space which bodies inhabit and which ideologies aspire to root out. Therefore, even though colonial space sought to be absolute, it could never be, and left out pockets for challenging the construction. As Lefebvre points out, rhythms in all their multiplicity interpenetrate one another through a body's inventiveness deployed in space. It can therefore be validly pointed out that during the time period with which I have dealt, there have been interventions of another sort, which sought to challenge and disrupt the figurative and discursive constructions of space as undertaken by the imperial order. The translation could never be complete and absolute and left out pockets of dissension and resistance. A study of the policy decisions which varied from area to area in the entire South Asian region only demonstrates that. It only followed that in their implementation, the ideologies which were tested and put into practice in Great Britain got refigured and configured in unprecedented ways, often resulting in implementation of completely new policies.

I began this book with a reference to cartographic knowledge being deeply embroiled with questions of language and writing, especially with the surge of English education and pedantic learning. Until they have been written about, places, societies, people and cultures stay largely amorphous. The same applies for a geographic location. This is reminiscent of what the Kenyan writer, Ngugi Wa Thiong'o, famously observed:

> Language as communication and as culture are then products of each other. Communication creates culture: culture is a means of communication. Language carries culture, and culture carries, particularly through orature and lit-erature, the entire body of values by which we come to perceive ourselves and our place in the world.[10]

As has been indicated in this book, this intersection between language, literature and knowledge systems, affirms the primacy of geography and the ideology of domination and power over territory. The fact that all of the discussed media went on to be printed, make for its intersection with book history and processes of control and dissemination of knowledge. As G.N. Devy suggests, the linkage between print, knowledge and power with questions of language is all the more visible and abnormal in the Indian subcontinent because of the region's history of colonial domination. His work has categorically reminded us of the history of bibliocentric literate culture as it rose in the West and its perilous consequences on modern India. He details the ways in which worldviews of the disempowered have been historically destroyed in India through a deliberate sidelining and silencing of their own languages and therefore from local and living knowledge systems altogether. According to him, colonial education and English literacy turned mainstream Indian intelligentsia into a society of amnesiacs, removed from their own native culture and driving tribal and oral traditional knowledge systems out of circulation. In this, Western, primarily English medium print and book culture, is to be squarely blamed for monopolising knowledge and hegemonising reality. Similarly, alternatives to contemporary knowledge are hard to revive if local languages and their literatures die unceremonious deaths. The loss of *bhasha* is in effect a loss of history and realities. As has been raised by Devy, geographical knowledges in many tribal and nomadic dialects/languages have innumerable terms to describe geographical features. The People's Linguistic Survey of India enlists nearly 200 terms describing different varieties of snow among 16 Himachali dialects, whereas a matching number of terms describe the desert and barrenness among nomadic tribes in desert Rajasthan. If investigated further, *bhashas* could have the potential of representing an alternate reality, or an alternate cartographic imagination very different from the global "one world" framework that our present adheres to.[11] The urban centric social location of production of knowledge and its pedagogic proliferation has utterly alienated marginal socio-linguistic groups and their realities, superimposing another imaginary on their habitats.

Print and the medium of generation of knowledge is today at the threshold of transformation as we enter the digital age. As mainstream knowledge continues to overlook the margins, there is worry that an already stabilised format of knowledge in the earlier age, with its continued neglect of local alternative ideas and tribal knowledge systems, is being transferred to this new media without review and revision. Moreover, in the age of remote sensing GIS–GPS (Geographic Information System–Global Positioning Satellite) digital mapping, we merely cross over from a writerly web to a readerly web:

> A readerly web [is one] in which knowledge is presented to viewers who are asked to consume but not alter its conditions, meaning making strategies, and modes of production. The readerly web is ultimately just that: a web of pages to be read, a web governed by a singularity of meaning achieved through a common set of naturalized practices of interaction and consumption.[12]

Visual media by and large disallows arguments and therefore shuns critical thinking, which spoken or written word very often still permits. Therefore, as the entire concept of the map is changing from being static and stylised paper representations found in books or hung on a wall, to being carried around in smart phones and portable devices, we have reasons for concern when Michael Jones, Google's chief technologist announces his plans "to geographically organise the world's information and make it universal and useful" through Google Maps/Google Earth. "Knowledge" and "information" or even "statistics" of yesteryears are today's data which, in satellite-generated digital maps, have either utilitarian goals, such as aids for navigation, or portend ominous designs, like micro-level surveillance and population monitoring. With apparent democratisation of digital technology, as maps such as these become part and parcel of daily lives, we need to question who controls organisation and quality of daily lives: is it the financiers or developers, or the people?

More recently, critical cartography studies, as advocated by Gregory, Harvey and others, have deepened their interaction with visual studies and digital humanities at the human–digital technology interface investigating the nexus of mapping technologies, politics and power.[13] A number of them, such as Laura Kurgan, Trevor Paglen, Anne Knowles, Annette Kim, Bill Rankin and T. Presner, have also shown hope in the new media and its developments.[14] The search for a high-altitude aerial view in mimetic mapping in the age of Enlightenment perhaps came to its decisive apotheosis once the earth was photographed from space back in the 1970s. The remote-sensed satellite images of today reduce earth to a digital globe without start or finish and more importantly without a centre. For some, as the digital map is zoomed in, panned out, it allows for endless possibilities to explore, contribute, alter, retrieve and relocate data with applications like archaeological coring through which temporal and historical layers of a place can be captured.

The question remains: can alternate spatial imaginations, systematically ousted through an onslaught of cartographic modernity, be mappable?

Notes

1 Transliteration of Ghalib's Sher by Tauseef Warsi. https://www.quora.com/What-is-the-translation-of-the-sher-I-have-attached-in-the-question-detail
2 Daruwalla, Keki N. *The Map-maker*. Delhi: Ravi Dayal Publisher, 2002.
3 Osborne, Peter. *The Politics of Time: Modernity and Avant-garde*. London: Verso, 1995. p. 13.
4 Kaufmann, Thomas DaCosta. *Toward a Geography of Art*. Chicago: University of Chicago Press, 2004. p. 52.
5 See Derrida, Jacques. *Edmund Husserl's "The Origin of Geometry": An Introduction*. Tr. John P. Leavey Jr. Lincoln: University of Nebraska Press, 1989.
6 Hewitt, Rachel. "Mapping and Romanticism". The Wordsworth Circle, 42:2. 2011, pp. 157–65. p. 160–1.
7 Ibid. p. 161.
8 Presner, T., D. Shepard and Y. Kawano. *Hypercities: Thick Mapping in the Digital Humanities*. UCLA Open Access Publication, 2014. pp. 53–4.
9 Derrida, Jacques. *Edmund Husserl's "The Origin of Geometry": An Introduction*. Tr. John P. Leavey Jr. Lincoln: University of Nebraska Press, 1989. p. 57.
10 Wa Thiong'o, Ngugi. *Decolonising the Mind: The politics of language in African literature*. Nairobi: East African Publishers, 1992. pp. 15–16.
11 Kaushik, Martand. "The Centre Cannot Hold: How GN Devy challenges our concept of knowledge". *The Caravan*. 1 July, 2018. People's Linguistic Survey of India (PLSI). Vols. 11, 12, 25. Hyderabad: Orient Blackswan, 2014–15.
12 Presner, T., D. Shepherd and Y. Kawano. *Hypercities Thick Mapping in the Digital Humanities*. UCLA, 2014. https://escholarship.org/uc/item/3mh5t455, accessed on 5.5. 2019.
13 See Gregory, Derek. "Gabriel's Map: Cartography and Corpography in Modern War". In *Geographies of Knowledge and Power, Knowledge and Space*. Vol. 7. Springer, 2015, pp. 89–121. DOI: 10.1007/978-94-017-9960-7_4; Gregory, Derek. "Tahrir: Politics, Publics and Performance of Space". *Middle East Critique*. 22: 3, September, 2013. pp. 235–46. DOI: 10.1080/19436149.2013.814944. Harvey, David. *The Ways of the World*. US: Oxford University Press, 2016; Harvey, David. Youtube

Video. "Technology and Post Capitalism". https://www.youtube.com/watch?v=g18JoOZsoEM. Harvey, David. Youtube Video "Anti Capitalist Chronicles: Social Media and Internet as a Powerful Organising Tool" Part 2. https://www.youtube.com/watch?v=rUA0xbNMCeQ. Accessed on 06.05.2019.

14 Kurgan, Laura. *Close Up at a Distance: Mapping, Technology and Politics*. New York: Zone Books, 2013; Paglen, Trevor. *The Last Pictures*. California: University of California Press, 2012; Knowles, Anne. ed. *Placing History: How Maps, Spatial Data, and GIS are Changing Historical Scholarship*. California: ESRI. 2008; Kim, Annette. *Sidewalk City: Remapping Public Space in Ho Chi Minh City*. Chicago: University of Chicago Press, 2015; Rankin, Bill. *After the Map: Cartography, Navigation, and the Transformation of Territory in the Twentieth Century*. Chicago: University of Chicago Press. 2016; Presner, T., D. Shepherd and Y. Kawano. *Hypercities Thick Mapping in the Digital Humanities*. UCLA, 2014. https://escholarship.org/uc/item/3mh5t455, sourced on 5.5. 2019.

BIBLIOGRAPHY

Books and journals

Ahuja, Ravi. *Pathways of Empire: Circulation, Public Works and Social Space in Colonial Orissa, C. 1780–1914*. Hyderabad: Orient BlackSwan, 2009.

Alam, Muzaffar and Sanjay Subrahmanyam. *Indo-Persian Travels in the Age of Discoveries 140–1800*. New Delhi: Cambridge University Press, 2008.

Allen, Brian. "East India Company's Settlement Pictures: George Lambert and Samuel Scott". PaulineRohatgi and Pheroza Godrej eds. *Under the Indian Sun: British Landscape Artists*. Delhi: Marg Publications, 1995. pp. 7–11.

Alpers, Svetlana. *The Art of Describing: Dutch Art in the Seventeenth Century*. Chicago: John Murray, 1983.

Anderson, Benedict. *Imagined Communities: Reflections on the Origin and Spread of Nationalism*. London: Verso, 1991.

Archer, Mildred. *British Drawings in the India Office Library*. Vol. 1: Amateur Artists. London: Her Majesty's Stationary Office, 1969.

Archer, Mildred and Toby Falk. *India Revealed: The Art and Adventures of James and William Fraser 1801–35 (The Passionate Quest: The Fraser Brothers in India)*. London: Cassell, 1989.

Archer, Mildred. *British Drawings in the India Office Library*. Vol. 1: Amateur Artists. London: Her Majesty's Stationary Office, 1969.

Archer, Mildred. *Early Views of India: The Picturesque Journeys of Thomas and William Daniell 1786–94*, London: Thames and Hudson, 1980.

Archer, Mildred and Toby Falk. *India Revealed: The Art and Adventures of James and William Fraser 1801–35 (The Passionate Quest: The Fraser Brothers in India)*. London: Cassell, 1989.

Armitage, David. *Ideological Origins of the British Empire*. Cambridge: Cambridge University Press, 2000.

Arnold, David. *The Tropics and the Traveling Gaze: India, Landscape, and Science 1800–1856*. Delhi: Permanent Black, 2005.

Avery, I. Emmett and Arthur H. Scouten. *The London Stage 1600–1700: A Critical Introduction*. Carbondale: Southern Illinois University Press, 1968.

Axelby, Richard. "Calcutta Botanic Garden and the colonial re-ordering of the Indian environment." *Archives of Natural History* 35. 1 (2008): 150–163.

Bacon, Francis. *The Essays and Counsels, Civil and Moral, of Francis Ld. Verulam Viscount St. Albans*. 1601. New York: E.P. Dutton, 1900.

Bacon, Francis. *The Works of Francis Bacon*. James Speddinget al. eds. 7 Vols. London: Longmans, 1862–72. Vol. 7.

Baigent, Elizabeth. "Lambton, William". *Oxford Dictionary of National Biography*. Oxford: Oxford University Press, Sept 2004; online edn. Jan 2008. http://www. oxforddnb.com/view/article/15948, 24 Oct 2009

Barker, Hugh. *Hedge Britannia: A Curious History of a British Obsession*. London: Bloomsbury, 2012.

Barnes, Trevor J. and James S. Duncan eds. *Writing Worlds: Discourse, Text and Metaphor in the Representation of Landscape*. London: Routledge, 1992.

Barrell, John. *The Dark Side of the Landscape: The Rural Poor in English Painting, 1730–1840*. Cambridge: Cambridge University Press, 1980.

Barrell, John. *The Idea of Landscape and the Sense of Place 1730–1840*. Cambridge: Cambridge University Press, 1972.

Barrow, Ian J. "Moving Frontiers: Changing Colonial Notions of the Indian Frontiers". http://asnic.utexas.edu/asnic/sagar/fall.1994/ian.barrow.art.html, accessed 10 July 2005.

Barrow, Ian J. *Making History, Drawing Territory. British Mapping in India c. 1765–1905*. Oxford: Oxford University Press, 2003.

Barrow, Ian J. "India for the Working Classes: The maps of the Society for the Diffusion of Useful Knowledge". *Modern Asian Studies*, 38: 3, 2004. pp. 677–702.

Barthes, Roland. *The Responsibility of Forms: Critical Essays on Music, Art and Representation*. Tr. Richard Howard, Oxford: Basil Blackwell, 1986.

Baylis, Gail. "England, Seventeenth and Eighteenth Centuries". *Literature of Travel and Exploration: An Encyclopedia*. ed. Jennifer Speake ed. 3 Vols. Vol. 1. London: Fitzroy Dearborn, 2003.

Bayly, Chris. *Empire and Information: Intelligence Gathering and Social Communication in India, 1780–1870*. Cambridge: Cambridge University Press, 1996.

Bearley, K.G. "Mapping them 'out': Euro-Canadian cartography and the appropriation of the Nuxalk and Ts'ilhqot'in First nations" territories, 1793–1916". *The Canadian Geographer*, 39, 1995. pp. 140–156.

Benjamin, Roger Harold. "The Decorative Landscape, Fauvism, and the Arabesque of Observation". *The Art Bulletin*, 75:2, 1993. pp. 295–316.

Berger, John. *Ways of Seeing*. London: Penguin, 2008.

Bergmann, Sigard. *Religion, Space and the Environment*. New Brunswick: Transaction Publishers, 2014.

Bermingham, Ann. *Landscape and Ideology: The English Rustic Tradition, 1740–1860*. Berkeley: University of California Press, 1989.

Blanton, Casey. *Travel Writing: The Self and the World*. London: Routledge, 2002.

Blomley, Nicholas. "Making private property: enclosure, common right and the work of hedges." *Rural History*, 18. 1, 2007. pp. 1–21.

Bollnow, Otto Friedrich. *Human Space*. Tr. Christine Shuttleworth. London: Hyphen, 2011.

Bowen. H.V. *Elites, Enterprise and the Making of the British Overseas Empire, 1688–1775*. New York: St. Martin's Press, 1996.

Bratton, J.S.et al. *Acts of Supremacy: The British Empire and the Stage 1790–1930*. Manchester: Manchester University Press, 1991.

Bravo, Michael T. "Precision and Curiosity in Scientific Travel: James Rennell and the Orientalist Geography of the New Imperial Age (1760–1830)". In Jas Elsner and Joan Pau Rubies eds. *Voyages and Visions: Towards a Cultural History of Travel*. London: Reaktion Books, 1999.

Breckenridge, Carol A. "The Aesthetics and Politics of Colonial Collecting: India at World Fairs". *Comparative Studies in Society and History*, Vol. 31, no. 2, 1989, pp. 195–216.

Brewer, Daniel. "Lights in Space". *Eighteenth Century Studies*. Vol. 37 No. 2, 2004. pp. 171–186.

Brewer, John. *The Pleasures of the Imagination: English Culture in the Eighteenth Century*. London: Harper Collins, 1997.

Broglio, Ron. *Technologies of the Picturesque: British Art, Poetry and Instruments 1750–1830*. Lewisburg: Bucknell University Press, 2008.

Burkhardt, Frederick and Sidney Smith eds. *The Correspondence of Charles Darwin*. Vol. 4. (1847–1850). Cambridge: Cambridge University Press, 2017.

Butlin, Robin A. *Geographies of Empire: European Empires and Colonies c. 1880–1960*. Cambridge: Cambridge University Press, 2009.

Buzard, James. "The Grand Tour and after (1660–1840)". In Peter Hulme and Tim Youngs eds. *The Cambridge Companion to Travel Writing*. Cambridge: Cambridge University Press, 2002.

Carter, Paul. *The Road to Botany Bay: An Essay in Spatial History*. New York: Knopf, 1988.

Chakrabarti, Pratik. *Western Science in Modern India: Metropolitan Methods, Colonial Practices*. Delhi: Permanent Black, 2004.

Chambers, James. *Bishop Heber and Indian Missions*. London: John W. Parker, 1846.

Chatterjee, Indrani. "Connected Histories and the Dream of Decolonial History". *South Asia: Journal of South Asian Studies*. 41:1, 2018. pp. 69–86.

Chatterjee, Indrani. *Forgotten Friends: Monks, Marriages, and Memories of Northeast India*. Delhi: Oxford University Press, 2013.

Chatterjee, Kumkum. *Merchants, Politics, and Society in Early Modern India: Bihar, 1733–1820*. Leiden: Brill, 1996.

Chattopadhyaya, B.D. *The Concept of Bharatavarsha and Other Essays*. New Delhi: Permanent Black, 2017.

Close, Charles F. *The Early Years of the Ordnance Survey*. Great Britain: David and Charles, 1926.

Cohn, Bernard. *An Anthropologist Among Historians. And Other Essays*. New Delhi: Oxford University Press, 1987.

Cohn, Bernard. *Colonialism and its Forms of Knowledge*. Oxford: Oxford University Press, 1997.

Colley, Linda. *Britons: Forging the Nation, 1707–1830*. New Haven: Yale University Press, 1992.

Colville, Berres Hoddle. "Robert Hoddle: Pioneer Surveyor, 1794–1881." *The Globe*. 57, 2005. pp. 17–26.

Comment, Bernard. *The Panorama*. London: Reaktion, 1999.

Conner, Patrick. *George Chinnery 1774–1852: Artist of India and the China Coast*. Suffolk: Antique Collectors Club, 1993.

Conner, Patrick. "The Poet's Eye: The Intimate Landscape of George Chinnery" in Pauline Rohatgi and Pheroza Godrej eds. *Under the Indian Sun: British Landscape Artists*. Bombay: Marg Publication, 1995.

Cosgrove, Denis. "Maps, Mapping, Modernity: Art and Cartography in the Twentieth Century". *Imago Mundi*. 57:1, 2005. pp. 35–54.

Cosgrove, Denis. "Landscape and landschaft." *German Historical Institute Bulletin* 35, 2004. pp. 57–71.

Cosgrove, Denis. "Prospect, Perspective and the Evolution of the Landscape Idea". *Transactions of the Institute of British Geographers*, Vol. 10, no. 1, 1985, pp. 45–62.

Cronin, Nessa. "Lived and Learned Landscapes. Literary Geographies and the Irish Topographical Tradition". In Marie Mianowski ed. *Irish Contemporary Landscapes in Literature and the Arts*. London: Palgrave Macmillan, 2012.

Cronin, Nessa. "Writing the 'New Geography': Cartographic Discourse and Colonial Governmentality in William Petty's The Political Anatomy of Ireland." *Historical Geography*. Vol. 42, 2014. pp. 58–71.

Cronin, Nessa. *The Eye of History: Spatiality and Colonial Cartography* (unpublished Ph.D. thesis). Galway: NUIZ, 2007.

Curley, Thomas M. *Samuel Johnson and the Age of Travel*. Athens: University of Georgia Press, 1976.

Dalrymple, William. *White Moghuls: Love and Betrayal in Eighteenth Century India*. London: Penguin, 2002.

Daniels, Stephen. "Re-visioning Britain: Mapping and Landscape Painting, 1750–1820". In Katherine Baetjer ed. *Glorious Nature: British Landscape Painting 1750–1820*. London: Zwemmer, 1994.

Daniels, Stephen. "The Culture of Cartography". In Geoffrey Cubbit ed. *Imagining Nations*. Manchester: Manchester University Press, 1998.

Daniels, Stephen, "Thresholds and Prospects". In Nicolas Alfrey, Stephen Daniels and Martin Postle eds. *Art of the Garden*. London: Tate, 2004.

De Bolla, Peter. *The Education of the Eye: Painting, Landscape, and Architecture in Eighteenth-Century Britain*. Stanford: Stanford University Press, 2003.

de Certeau, Michel. *Practice of Everyday Life*. Vol. 1. Berkeley: University of California Press, 1984.

de Saussure, Ferdinand. *Course in General Linguistics*. Trans. R. Harris. Chicago: Open Court Publishing, 1998.

Delano-Smith, Catherine and Roger J.P. Kain. *English Maps: A History*. London: The British Library, 1999.

Deleuze, Gilles. *Essays Critical and Clinical*. Trans. Daniel W. Smith and Michael A. Greco. London: Verso, 1998.

DeRogatis, Amy. *Moral Geography: Maps, Missionaries, and the American Frontier*. New York: Columbia University Press, 2003.

Derrida, Jacques. *Edmund Husserl's The Origin of Geometry: An Introduction*. Trans. John P. Leavey Jr. Lincoln: University of Nebraska Press, 1989.

Dirks, Nicholas B. "Guiltless Spoliations: Picturesque Beauty, Colonial Knowledge, and Colin Mackenzie's Survey of India". In Catherine B. Asher and Thomas R. Metcalf eds. *Perceptions of South Asia's Visual Past*. New Delhi: Oxford and IBH Publishing Co. Ltd., 1994. pp. 211–212.

Dirks, Nicholas B. *Castes of Mind: Colonialism and the Making of Modern India*. Princeton: Princeton University Press, 2001.

Distad, Merrill. "Scientific Travelling". In Jennifer Speake ed. *Literature of Travel and Exploration: An Encyclopedia*. pp. 1022–1023.

Douglas M. Peers. "Conquest Narratives: Romanticism, Orientalism and Intertextuality in the Indian writings of Sir Walter Scott and Robert Orme". In Michael J. Franklin ed. *Romantic Representations of British India*. London: Routledge, 2006.

Dove, Jane. "Geographical board game: promoting tourism and travel in Georgian England and Wales". *Journal of Tourism History*. Vol. 8, Issue 1. 2016. pp. 1–18.

Driver, Felix and Luciana Martins. *Tropical Visions in an Age of Empire*. Chicago: Chicago University Press, 2005.

Driver, Felix. *Geography Militant: Cultures of Exploration in the Age of Empire*. Oxford: Blackwell, 1999.

Duncan S.A. Bell, "Dissolving Distance: Technology, Space, and Empire in British Political Thought, 1770–1900", *Journal of Modern History*. Vol. 77, 2005. pp. 523–562.

Duncan, James and Derek Gregory eds. *Writes of Passage: Reading Travel Writing*. London: Routledge, 1999.

Dyson. K.K. *A Various Universe: A Study of the Journals and Memoirs of British Men and Women in the Indian Subcontinent, 1765–1856*. Delhi: Oxford University Press, 1978.

Eaton, Natasha Jane. *Imaging Empire: The Trafficking of Art and Aesthetics in British India c. 1772 to c. 1795. Vol. 1.* (Ph.D. Thesis). University of Warwick, 2000.

Eaton, Natasha. "Between mimesis and alterity: Art gift and diplomacy in colonial India". In Michael J. Franklin ed. *Romantic Representations of British India*. London: Routledge, 2006. pp. 98–99.

Eck, Diana L. "The Goddess Ganges in Hindu Sacred Geography". In John Hawley and Donna Wulff eds. *Devi: Goddess of India*. Berkeley: University of California Press, 1996.

Eck, Diana L. "The Imagined Landscape: Patterns in the Construction of Hindu Sacred Geography". *Contributions to Indian Sociology* 32:2, 1998. pp. 165–188.

Eck, Diana L. *India: A Sacred Geography*. New York: Harmony Books, 2012.

Edgerton, Samuel. *Renaissance Rediscovery of Linear Perspective*. New York: Basic Books, 1975.

Edney, Matthew. *Mapping and Empire: British Trigonometrical Surveys in India and the European Concept of Systematic Survey, 1799–1843* (Ph.D. thesis). University of Wisconsin-Madison, 1990.

Edney, Matthew. "Reconsidering Enlightenment Geography and Map Making: Reconnaissance, Mapping, Archive". In David N. Livingstone and Charles W.J. Withers eds. *Geography and Enlightenment*. Chicago: Chicago University Press, 1999.

Edney, Matthew. *Mapping an Empire: The Geographical Construction of British India, 1765–1843*. Chicago: Chicago University Press, 1997.

Egerton, Judy. "*National Gallery Catalogues: The British School.*" (1998): 218. pp. 80–86.

Elden, Stuart. *The Birth of Territory*. Chicago Scholarship Online, 2014.

Eliot, George. *Felix Holt, The Radical*. London: William Blackwood and Sons, 1866.

Ellis, Markman. "Spectacles within Doors: Panoramas of London in the 1790s". *Romanticism: The Journal of Romantic Culture and Criticism*. Vol. 14, no. 2, 2008, pp. 133–148.

Elsner, Jas and Joan-Pau Rubies. "Introduction". Jas Elsner and Joan-Pau Rubies eds. *Voyages and Visions: Towards a Cultural History of Travel*. London: Reaktion, 1999.

Endersby, Jim. "From having no Herbarium. Local Knowledge versus Metropolitan Expertise: Joseph Hooker's Australasian Correspondence with William Colenso and Ronald Gunn". *Pacific Science* 55. 4 (2001): 343–358.

Endersby, Jim. *Imperial Nature: Joseph Hooker and the Practices of Victorian Science*. Chicago: University of Chicago Press, 2008.

Eyre, Giles. "Foreword". In Shellim, Maurice. *Oil Paintings by Sir Charles D'Oyly, 7th Baronet 1781–1845.* London: Spink, 1989.

Feldmann, Doris. "Economic and/as Aesthetic Constructions of Britishness in Eighteenth-Century Domestic Travel Writing". *Journal for the Study of British Cultures.* Vol. 4/1–2, 1997. pp. 31–45.

Ford, Thomas F. *Wordsworth and the Poetics of Air.* Cambridge: Cambridge University Press, 2018.

Forrest, Lt Col. *A Picturesque Tour Along The Rivers Ganges and Jumna, In India.* New Delhi: Niyogi, 2016 [1824].

Foucault, Michel. "Of Other Spaces: Utopias and Heterotopias". In *Architecture/Mouvement/Continuite.* Trans. Jay Miskowiec. 1967.

Foucault, Michel. "Questions on Geography". In Colin Gordon ed. *Power/Knowledge: Selected Interviews and Other Writings 1972–1977.* Hempstead: Harvester Wheatsheaf, 1980.

Foucault, Michel. *Discipline and Punish: The Birth of the Prison.* (1975) New York: Random House, 1989.

Foucault, Michel. *The Order of Things: An Archaeology of the Human Sciences.* (1966) London: Routledge, 2005.

Franklin, Michael J. ed. *Romantic Representations of British India.* London: Routledge, 2006.

Friel, Brian. *Collected Plays.* Vols. 2 & 3. Ireland: The Gallery Press, 2016.

Fulford, Tim and Carol Bolton eds. *Travels, Explorations and Empires: Writing from the Era of Imperial Expansion 1770–1835.* 4 Vols. Vol. 1. London: Pickering and Chatto, 2001.

Gikandi, Simon. *Maps of Englishness: Writing Identity in the Culture of Colonialism.* New York: Columbia University Press, 1996.

Godlewska, Anne. "The Idea of the Map". In S. Hanson ed. *Ten Geographic Ideas that Changed the World.* New Brunswick: Rutgers University Press, 1997. pp. 15–39.

Gole, Susan. *India Within the Ganges.* New Delhi: Jay Prints, 1983.

Gombrich, E.H. *Art and Illusion: A Study in the Psychology of Pictorial Representation.* London: Phaidon, 1962.

Gordon, John E. "Rediscovering a sense of wonder: geoheritage, geotourism and cultural landscape experiences." *Geoheritage* 4. 1–2 (2012): 65–77.

Goswami, Manu. *Producing India: From Colonial Economy to National Space.* Chicago: Chicago University Press, 2004.

Green, Nicholas. "Looking at the Landscape: Class Formation and the Visual". In Eric Hirsch and Michael O'Hanlon eds. *The Anthropology of Landscape: Perspectives on Place and Space.* Oxford: Oxford University Press, 1995.

Gregory, Derek. *Geographical Imaginations.* Oxford: Blackwell, 1994.

Gregory, Derek. "Edward Said's Imaginative Geographies". In Mike Crang and Nigel Thrift eds. *Thinking Space.* London: Routledge, 2000.

Gregory, Derek. "Gabriel's Map: Cartography and Corpography in Modern War". In *Geographies of Knowledge and Power, Knowledge and Space.* Vol. 7. Springer, 2015, pp. 89–121. doi:10.1007/978-994-017-9960-7_4.

Gregory, Derek. "Tahrir: Politics, Publics and Performance of Space". *Middle East Critique.* 22: 3, 2013. pp. 235–246. doi:10.1080/19436149.2013.814944.

Greppi, Claudio. "'On the Spot': Travelling Artists and the Iconographic Inventory of the World, 1769–1859". In Felix Driver and Luciana Martins eds. *Tropical Visions in an Age of Empire.* Chicago: Chicago University Press, 2005. pp. 24–39.

Guha-Thakurta, Tapati. *Monuments, Objects, Histories: Institutions of Art in Colonial and Postcolonial India*. Delhi: Permanent Black, 2004.

Hardwicke, Thomas. "Narrative of a Journey to Srinagur". *Asiatick Researches*. Vol. 6. 1799. pp. 309–381.

Harley, J.B. "The re-mapping of England, 1750–1800". *Imago Mundi*. Vol. 19. 1965. pp. 56–123.

Harley, J.B. "The Society of Arts and the Surveys of English Counties 1759–1809". *Journal of the Royal Society of Arts*, Vol. 112, (1963–1964). pp. 43–46.

Harley, J.B. "Maps, knowledge and power". In D. Cosgrove and S. Daniels eds. *The Iconography of Landscape: Essays on the Symbolic Representation, Design and the Use of Past Environments*. Cambridge: Cambridge University Press, 1985.

Harvey, David. *The Condition of Postmodernity: An Enquiry into the Origins of Cultural Change*. Oxford: Basil Blackwell, 1989.

Harvey, David. *Justice, Nature and the Geography of Difference*. Cambridge: Blackwell, 1996.

Harvey, David. *The Condition of Postmodernity: An Enquiry into the Origins of Cultural Change*. Oxford: Basil Blackwell, 1989.

Harvey, David. *The Ways of the World*. New York: Oxford University Press, 2016.

Harvey, David C. "Ambiguities of the hedge: an exercise in creative pleaching – of moments, memories and meanings." *Landscape History* 38. 2, 2017. pp. 109–127.

Harvey, P.D.A. *The History of Topographical Maps: Symbols, Pictures and Surveys*. London: Thames and London., 1980.

Harvey, P.D.A. "English Estate Maps: Their Early History and Their Use as Historical Evidence". In David Buisseret ed. *Rural Images, Estate Maps in the Old and New Worlds*. Chicago: University of Chicago Press, 1996.

Heaney, G.F. "Rennel and the Surveyors of India". *The Geographical Journal*. Vol. 134, No. 3, 1968. pp. 318–325.

Helgerson, Richard. "The Land Speaks: Cartography, Chorography and Subversion in Renaissance England". *Representations*, 16, 1986. pp. 50–85.

Helgerson, Richard. *Forms of Nationhood: The Elizabethan Writing of England*. Chicago: Chicago University Press, 1994.

Hemingway, Andrew. *Landscape Imagery and Urban Culture in early 19th Century*. Cambridge: The University Press, 1992.

Hewitt, Rachel. "Wordsworth and the Ordnance Survey in Ireland: 'Dreaming O'er the Map of Things'". *The Wordsworth Circle*. Vol. 37. No. 2. 2006. pp. 80–85.

Hewitt, Rachel. "Mapping and Romanticism". *The Wordsworth Circle*, 42:2, 2011. pp. 157–165.

Hewitt, Rachel. *Map of a Nation: A Biography of the Ordnance Survey*. London: Granta, 2013.

Hirsch, Eric and Michael O'Hanlon eds. *The Anthropology of Landscape: Perspectives on Place and Space*. Oxford: Oxford University Press, 1995.

Hobbes, Thomas. *Leviathan*. 1660. London: Penguin, 2003.

Hodges, William. *Travels in India: During the years 1780, 1781, 1782 and 1783*. Delhi: Munshiram Manoharlal, 1999.

Hogan, Charles Beecher. *The London Stage 1776–1800: A Critical Introduction*. Carbondale: Southern Illinois University Press, 1968.

Hogan, Sarah. "Of Islands and Bridges: Figures of Uneven Development in Bacon's New Atlantis". *Journal of Early Modern Cultural Studies*. Vol. 12, Issue 3. 2012. pp. 28–59.

Huggan, Graham. *Territorial disputes: maps and mapping strategies in contemporary Canadian and Australian fiction.* (Dissertation.) University of British Columbia, 1989.

Hulme, Peter and Tim Youngs eds. *The Cambridge Companion to Travel Writing.* Cambridge: Cambridge University Press, 2002.

Hunt, John Dixon and Peter Willis eds. "Introduction". *The Genius of the Place: The English Landscape Garden, 1620–1820.* Cambridge, MA: MIT Press, 1988. pp. 1–45.

Hyde, Ralph. *Panoramania: The Art and Entertainment of the All-Embracing View.* London: Barbican Art Gallery, 1988–1989.

Inden, Ronald B. *Imagining India.* Oxford: Basil Blackwell, 1990.

"James' Diary, Fraser Papers". As cited in Archer, Mildred and Toby Falk. *India Revealed: The Art and Adventures of James and William Fraser 1801–35 (The Passionate Quest: The Fraser Brothers in India).* Vol. 8. London: Cassell, 1989. pp. 45, 59.

Jameson, Fredric. *The Geopolitical Aesthetic – Cinema and Space in the World System.* London: BFI, 1992.

Jarvis, Robin. "Romanticism". *Literature of Travel and Exploration: An Encyclopedia.* ed. Jennifer Speake ed. Vol 3. New York: Fitzroy Dearborn, 2003.

Jenkins, Ralph E. "'And I travelled after him': Johnson and Pennant in Scotland". *Texas Studies in Literature and Language.* Vol. 14. No. 3, 1972. pp. 445–462.

Jonson, Ben. *The Complete Masques.* Stephen Orgel ed. New Haven: Yale University Press, 1969.

Kalpagam, U. "Cartography in Colonial India". *Economic and Political Weekly.* Vol. 30, No. 30. 29 July 1995. pp. 87–98.

Kalpagam, U. *Rule by Numbers: Governmentality in Colonial India.* New York: Lexington Books, 2014.

Kaufmann, Thomas DaCosta. *Toward a Geography of Art.* Chicago: University of Chicago Press, 2004.

Kaushik, Martand. "The Centre Cannot Hold: How GN Devy challenges our concept of knowledge". *The Caravan.* 1 July 2018.

Kaviraj, Sudipta. "The Imaginary Institution of India". In Partha Chatterjee and Gyanendra Pandey eds. *Subaltern Studies,* Vol. VII, New Delhi: Oxford University Press, 1999. pp. 1–39.

Keay, J. *India Discovered: The Achievement of the British Raj.* Leicester: Windward, 1989.

Keay, John. *The Great Arc: The Dramatic Tale of How India Was Mapped and Everest Was Named.* London: HarperCollins, 2000.

Keighren, Innes M. and Charles W.J. Withers. 'The spectacular and the sacred: narrating landscape in works of travel'. *Cultural Geographies.* 19: 1, 2012. pp. 11–30.

Kemp, Martin. *The Science of Art: Optical Themes in Western Art from Brunelleschi to Seurat.* New Haven: Yale University Press, 1990.

Kiberd, Declan. *Inventing Ireland: The Literature of the Modern Nation.* London: Jonathan Cape, 1995.

Kim, Annette. *Sidewalk City: Remapping Public Space in Ho Chi Minh City.* Chicago: University of Chicago Press, 2015.

Kipling, Rudyard. *Kim.* (1901). Oxford: Oxford University Press, 2008.

Klein, Bernard, *Maps and the Writing of Space in Early Modern England and Ireland.* Hampshire: Palgrave, 2001.

Klein, Bernard. "Constructing the Space of the Nation: Geography, Maps and the discovery of Britain in the Early Modern Period". *Journal for the Study of British Cultures.* Vol. 4/1–2, 1997. pp. 11–29.

Klonk, Charlotte. *Science and the Perception of Nature: British Landscape Art in the Late Eighteenth and Early Nineteenth Centuries*. New Haven: Yale University Press, 1996.

Knowles, Anne ed. *Placing History: How Maps, Spatial Data, and GIS are Changing Historical Scholarship*. California: ESRI, 2008.

Kohl, Stephen. "Imagining the country as 'The Country' in the 1830s: William Cobett, William Howitt, William Turner". *Journal for the Study of British Cultures*. Vol. 4/1–2, 1997. pp. 113–127.

Korte, Barbara. *English Travel Writing: From Pilgrimages to Postcolonial Explorations*. Basingstoke: Macmillan, 2000.

Kramrisch, Stella. "Space in Indian Cosmogony and in Architecture". In Kapila Vatsyayan ed. *Concepts of Space: Ancient and Modern*. Indira Gandhi National Centre for the Arts. New Delhi: Abhinav Publications, 1991.

Kurgan, Laura. *Close Up at a Distance: Mapping, Technology and Politics*. New York: Zone Books, 2013.

Lahiri, Manosi. *Mapping India*. New Delhi: Niyogi Books, 2012.

Laird, M.A. *Bishop Heber in Northern India: Selections from Heber's Journal*. Cambridge: Cambridge University Press, 1971.

Lambert, David and Alan Lester eds. *Colonial Lives Across the British Empire: Imperial Careering in the Long Nineteenth Century*. Cambridge: Cambridge University Press, 2009.

Lamont, Claire and Michael Rossington eds. *Romanticism's Debatable Lands*. London: Palgrave Macmillan, 2007.

Leask, Nigel. "Introduction: Practices and Narratives". *Curiosity and the Aesthetics of Travel Writing, 1770–1840: From an Antique Land*. Oxford: Oxford University Press, 2002.

Leask, Nigel. *Curiosities and the Aesthetics of Travel Writing, 1770–1840*. Oxford: Oxford University Press, 2002.

Lefebvre, Henri. *The Production of Space*. Oxford: Blackwell, 1991.

Locke, John. "An essay concerning human understanding". In *Works of John Locke in Nine Volumes*. Vol. 1. London: Rivington, 1824.

Loomba, Ania: *Colonialism/Postcolonialism*. London: Routledge, 1998.

Losty, J.P. "A Career in Art: Sir Charles D'Oyly". Pauline Rohatgi and Pheroza Godrej eds. *Under the Indian Sun: British Landscape Artists*. Bombay: Marg, 1995.

Losty, J.P. "Sir Charles D'Oyly's Lithographic Press and His Indian Assistants". In Pauline Rohatgi and Pheroza Gandhi eds. *India: A Pageant of Prints*. Bombay: Marg, 1989.

Low, Martina. "The Constitution of Space: The Structuration of Spaces Through the Simultaneity of Effects and Perception". *European Journal of Social Theory*. 11:1, 2008. pp. 25–49.

MacDonald, Kenneth Iain. "Issues of Ethics". In Jennifer Speake ed. *Literature of Travel and Exploration: An Encyclopedia*. Vol. 1. London: Routledge, 2003. p. 404.

MacKenzie, John ed. *Imperialism and Popular Culture*. Manchester: Manchester University Press, 1986.

Mackenzie, John M. "Essay and Reflection: On Scotland and the Empire". *The International History Review*, 15: 4, 1993. pp. 714–739.

MacKenzie, John. *Propaganda and Empire: The Manipulation of Public Opinion, 1880–1960*. Manchester: Manchester University Press, 1984.

Majeed, Javed. "Nationalism, Travel and, Modernity". University of Delhi Lecture Series. Arts Faculty Building. Department of English, Delhi. 18 December, 2008.

Marin, Louis. *On Representation*. Stanford: Stanford University Press, 2001.

Markovits, Claude eds *Society and Circulation: Mobile People and Itinerant Cultures in South Asia, 1750–1950*. London: Anthem Press, 2006.

Markovits, Claude. "The Political Economy of Opium Smuggling in Early Nineteenth Century India: Leakage or Resistance?", *Modern Asian Studies*, 43, 2009. pp. 89–111.

McKenzie, John M. "Scottish Orientalists, Administrators and Missions: A Distinctive Scots approach to Asia?". In T.M. Devine, and Angela McCarthy eds. *The Scottish Experience in Asia c. 1700 to the Present: Settlers and Sojourners*. London: Palgrave Macmillan, 2017. pp. 51–73.

McLeod, Bruce. *The Geography of Empire in English Literature 1580–1745*. Cambridge: Cambridge University Press, 1999.

Metcalf, Thomas R. *An Imperial Vision: Indian Architecture and Britain's Raj*. London: Faber, 1989.

Michael, Bernardo A. *Statemaking and Territory in South Asia: Lessons from the Anglo-Gorkha War (1814–1816)*. London: Anthem, 2014.

Mitchell, Timothy. "The World as Exhibition". *Comparative Studies in Society and History*, Vol. 31. No. 2. 1989. pp. 217–236.

Mitchell, W.J.T. ed. *Landscape and Power*. Chicago: University of Chicago Press, 1994.

Mookerji, R.K. *The Fundamental Unity of India*. (1913) New Delhi: Chronicle Classics Series, 2003.

Moran, Joe. *Interdisciplinarity*. London: Routledge, 2007.

Moretti, Franco. *Graphs, Maps and Trees: Abstract Models for Literary History*. London: Verso, 2007.

Moschek, Wolfgang. "The Limes: Between open Frontier and Borderline". In Natalie Fryde ed. *Walls, Ramparts and Lines of Demarcation: Selected Studies from Antiquity to Modern Times*. Berlin: Lit Verlag, 2009.

Mukherjee, B.N. *The Concept of India*. Calcutta: Sanskrit Pustak Bhandar, 1998.

Mukherjee, Nilanjana. "Translations/Representations/Colonisations", *Language Forum*. Vol. 30, No. 2. July–Dec.2004.

Mukherjee, Nilanjana. "'A Desideratum More Sublime': Imperialism's Expansive Vision and Lambton's Trigonometrical Survey of India", *Postcolonial Studies*. 14, 2011, 429–447.

Mukherjee, Nilanjana. "Drawing Roads, Building Empire: Space and Circulation in Charles D'Oyly's Landscapes of India". *South Asia: Journal of South Asian Studies*. Vol. 37, II, 2014. pp. 339–355.

Mukherjee, Nilanjana. "Scottish Resonances in James Baillie Fraser's Travel Accounts". In Sutapa Dutta and Nilanjana Mukherjee eds. *Mapping India: Transitions and Transformations, 18th–19th Century*. New Delhi: Routledge, 2019.

Nagai, Kaori. *Empire of Analogies: Kipling, India and Ireland*. Cork, Ireland: Cork University Press. 2007.

Netzloff, Mark. *England's Internal Colonies: Class, Capital and the Literature of Early Modern Colonialism*. New York: Palgrave Macmillan, 2003.

Ogborn, Miles. "Writing travels: power, knowledge and ritual on the English East India Company's early voyages". *Transactions of the Institute of British Geographers*. Vol. 27 No. 2, 2002. pp. 155–171.

Ogborn, Miles and Charles W. J. Withers. "Introduction: Georgian Geographies" in Miles Ogborn and Charles W.J. Withers eds. *Georgian Geographies: Essays on Space,*

Place and Landscape in the Eighteenth Century. Manchester: Manchester University Press, 2004. pp. 2–23.

Ogborn, Miles and Charles W.J. Withers ed. *Georgian Geographies: Essays on Space, Place and Landscape in the Eighteenth Century*. Manchester: Manchester University Press, 2004.

Ogborn, Miles. "Geographia's pen: writing, geography and the arts of commerce, 1660–1760". *Journal of Historical Geography*, 30, 2004. pp. 294–315.

Olwig, Kenneth Robert. *Landscape, Nature and the Body Politic*. Wisconsin: University of Wisconsin Press, 2002.

O'Quinn, Daniel. *Staging Governance: Theatrical Imperialism in London, 1770–1800*. Baltimore: Johns Hopkins University, 2005.

Osborne, Peter D. *Travelling Light: Photography, Travel and Visual Culture*. Manchester: Manchester University Press, 2000.

Osborne, Peter. *The Politics of Time: Modernity and Avant-garde*. London: Verso, 1995.

Paglen, Trevor. *The Last Pictures*. California: University of California Press, 2012.

Panofsky, Erwin. *Perspective as Symbolic Form*. New York: Zone Books. 1991.

Paris, H.J. "English Water-colour Painters". In W.J. Turner ed. *Aspects of British Art*. London: Collins, 1947.

People's Linguistic Survey of India (PLSI). Vols. 11, 12, 25. Hyderabad: Orient Black-Swan, 2014–15.

Pickles, John. "Texts, Hermeneutics and Propaganda Maps". In Trevor J. Barnes and James S. Duncan ed. *Writing Worlds: Discourse, Text and Metaphor in the Representation of Landscape*. London: Routledge, 1992.

Pinney, Christopher. "Moral Topophilia: The Significations of Landscape in Indian Oleographs". In Eric Hirsch and Michael O'Hanlon eds. *The Anthropology of Landscape: Perspectives on Place and Space*. Oxford: Oxford University Press, 1995.

Pittock, Murray G. H. *Inventing and Resisting Britain: Cultural Identities in Britain and Ireland 1685–1789*. Basingstoke: Macmillan, 1997.

Porter, Dennis. *Haunted Journeys: Desire and Transgression in European Travel Writing*. Princeton: Princeton University Press, 1991.

Pratt, Mary Louise. *Imperial Eyes: Travel Writing and Transculturation*. London: Routledge, 1992.

Presner, T., D. Shepherd and Y. Kawano. *Hypercities Thick Mapping in the Digital Humanities*. UCLA, 2014. https://escholarship.org/uc/item/3mh5t455, on 5. 5. 2019.

Raj, Kapil. "Colonial encounters and the forging of new knowledge and national identities: Great Britain and India, 1760–1850." *Osiris* 15, 2000. pp. 119–134.

Raj, Kapil. *Relocating Modern Science: Circulation and Construction of Knowledge in South Asia and Europe, 1650–1900*. Basingstoke: Palgrave Macmillan, 2007.

Ramaswamy, Sumathi. "Maps and Mother Goddesses in Modern India". *Imago Mundi*. Vol. 53, 2001. pp. 97–114.

Rankin, Bill. *After the Map: Cartography, Navigation, and the Transformation of Territory in the Twentieth Century*. Chicago: University of Chicago Press, 2016.

Ray, Keith and Ian Bapty. *Offa's Dyke: Landscape and Hegemony in Eighth Century Britain*. London: Oxbow, 2016.

Rees, Ronald. "Historical Links between Cartography and Art". *Geographical Review*. Vol. 70, no. 1, 1980. pp. 61–78.

Robinson, A.H.W. *Marine Cartography in Britain, A History of the Sea Chart to 1855*. Leicester: Leicester University Press, 1962.

Rohatgi, Pauline and Pheroza Godrej eds. *India: A Pageant of Prints*. Bombay: Marg, 1989.

Rohatgi, Pauline and Pheroza Godrej eds. *Under the Indian Sun: British Landscape Artists*. Bombay: Marg, 1995.

Rorty, Richard. *Contingency, Irony and Solidarity*. Cambridge: Cambridge University Press, 1989.

Rose, Gillian. "Looking at landscape: the uneasy pleasures of power". In *Feminism and Geography: The Limits of Geographical Knowledge*. Cambridge: Polity Press, 1993. pp. 86–112.

Rosenfeld, Sybil. *Georgian Scene Painters and Scene Painting*. Cambridge: Cambridge University Press, 1981.

Rosenthal, Michael, Christiana Payne, and Scott Wilcox eds. *Prospects for the Nation: Recent Essays in British Landscape, 1750–1880*. New Haven: Yale University Press, 1997.

Roskill, Mark. *The Language of Landscape*. Pennsylvania: Pennsylvania State University Press, 1997.

Ryan, Simon. *The Cartographic Eye: How Explorers Saw Australia*. Melbourne: Cambridge University Press, 1996.

Said, Edward. *Culture and Imperialism*. London: Vintage, 1994.

Said, Edward. *Orientalism*. London: Routledge and Kegan Paul, 1978.

Scouten, Arthur H. *The London Stage 1729–1749: A Critical Introduction*. Carbondale: Southern Illinois University Press, 1968.

Sen, Sudipta. *Distant Sovereignty: National Imperialism and the Origin of British India*. London: Routledge, 2002.

Serlio, Sebastio. *Tutte l'Opera d'Architettura et Prospectivadi Sebastatediano Serlio Bolognese*. Vols 1–5, Venice, 1555. London: Dover, 1982.

Sharma, Yuthika. *Art in between empires: Visual culture and artistic knowledge in late Mughal Delhi 1748–1857* (unpublished dissertation). Columbia University, 2013.

Shellim, Maurice. *Oil Paintings by Sir Charles D'Oyly, 7th Baronet 1781–1845*. London: Spink, 1989.

Singh, Shiv Sahay. "Of Trigonometry and towers – and two centuries of history". *The Hindu*. Kolkata, 13 December2017.

Sinha, Amita. *Landscapes in India: Forms and Meanings*. New Delhi: Asian Educational Services, 2011.

Smiles, Sam. *Eye Witness: Artists and Visual Documentation in Britain 1770–1830*. Farnham: Ashgate, 2000.

Smith, David. *Antique Maps of the British Isles*. London: BT Batsford Ltd., 1928.

Soja, Edward. *Postmodern Geographies: The Reassertion of Space in Critical Theory*. London: Verso, 1989.

Soja, Edward. *Third Space: Journeys to Los Angeles and Other Real and Imagined Spaces*. Oxford: Blackwell, 1996.

Speake, Jennifer ed. *Literature of Travel and Exploration: An Encyclopedia*. Vol 2. London: Routledge, 2003.

Standring, Timothy J. "Watercolor Landscape Sketching during the Popular Picturesque era in Britain". In Katherine Baetjer ed. *Glorious Nature; British Landscape Painting 1750–1850*. London: Zwemmer, 1994.

StoneJr, George Winchester. *The London Stage 1747–1776: A Critical Introduction*. Carbondale: Southern Illinois University Press, 1968.

Stuebe, Isabel. "William Hodges and Warren Hastings: A study in eighteenth century patronage". *The Burlington Magazine*, Vol. 115, No. 847 (October1973). pp. 657–666.

Subrahmanyam, Sanjay. "Connected Histories: Notes Towards a Reconfiguration of Early Modern Eurasia". *Modern Asian Studies*. 31:3, 1997. pp. 735–762.

Teltscher, Kate. *India Inscribed: European and British Writing on India 1600–1800*. Delhi: Oxford University Press, 1997.

Teltscher, Kate. "India/Calcutta: city of palaces and dreadful night". In Peter Hulme and Tim Youngs eds. *The Cambridge Companion to Travel Writing*. Cambridge: Cambridge University Press, 2002. pp. 191–206.

Thomas, Adrian P. "The establishment of Calcutta Botanic Garden: plant transfer, science and the East India Company, 1786–1806." *Journal of the Royal Asiatic Society*. Vol.16 Issue 2, 2006: 165–177.

Tobin, Berth Fowkes. *Colonizing Nature: The Tropics in British Arts and Letters, 1760–1820*. Philadelphia: University of Pennsylvania Press, 2005.

Tolbert, Jane T. "Censorship and Travel Writing". In Jennifer Speake ed. *Literature of Travel and Exploration: An Encyclopedia*. Vol. 1. London: Fitzroy Dearborn, 2003. p. 207.

Tooley, R.V. *Maps and Mapmakers*. London: BT Batsford Ltd., 1952.

Traub, Valerie. "Mapping the Global Body". In Peter Erickson and Clarke Hulse eds. *Early Modern Visual Culture: Representation, Race, Empire in Renaissance England*. Philadelphia: University of Pennsylvania Press. 2000.

Tuan, Yi-Fu. *Topophilia: A Study of Environmental Perceptions, Attitudes, and Values*. New York: Columbia University Press, 1990.

Turner, Henry S. "Literature and Mapping in Early Modern England 1520–1628". In J.B. Harley and David Woodward eds. *The History of Cartography*. Vol. 6. Part 1. Chicago: University of Chicago Press, 2007.

Tyner, Judith. "Persuasive Cartography". *Journal of Geography*, 81, 1982. pp. 140–144.

Urry, John. *Consuming Places*. London: Routledge. 1995.

Van Noy, Rick. *Surveying the Interior: Literary Cartographers and the Sense of Place*. Nevada: University of Nevada Press, 2003.

Wa Thiong'o, Ngugi. *Decolonising the Mind: The Politics of Language in African Literature*. Nairobi: East African Publishers, 1992.

Wagoner, Phillip B. "Precolonial Intellectuals and the Production of Colonial Knowledge". *Comparative Studies in Society and History*. 45:4, 2003. pp. 783–814.

White, Daniel E. "Imperial spectacles, imperial publics: panoramas in and of Calcutta." *The Wordsworth Circle*. 41. 2, 2010: 71–81.

White, Hayden. *Metahistory: The Historical Imagination in Nineteenth Century Europe*. Baltimore: The Johns Hopkins University Press, 1973.

Whyte, Ian D. *Landscape and History since 1500*. London: Reaktion, 2004.

Williams, Raymond. *The Country and the City*. New York: Oxford University Press, 1975.

Williamson, Tom and Liz Bellamy. *Property and Landscape: A Social History of Land Ownership and the English Countryside*. London: George Philips, 1987.

Wilson, Jon E. "'A Thousand Countries To Go To': Peasants and Rulers in Late Eighteenth-Century Bengal". *Past & Present*, Vol. 189:1. 2005.

Wilson, Kathleen. *The Sense of the People: Politics, Culture and Imperialism in England, 1715–1785*. Cambridge: Cambridge University Press, 1998.

Winichakul, Thongchai. *Siam Mapped: A History of the Geo-Body of the Nation*. Honolulu: University of Hawaii Press, 1994.

Winlow, Heather. "Anthropometric cartography: constructing Scottish racial identity in the early twentieth century". *Journal of Historical Geography*, 27:4, 2001. pp. 507–528.

Wintle, Michael. "Renaissance maps and the construction of the idea of Europe". *Journal of Historical Geography*, 25:2, 1999. pp. 137–165.

Withers, Charles. "Memory and the history of geographical knowledge: the commemoration of Mungo Park, African explorer". *Journal of Historical Geography*. Vol. 30, 2004. pp. 316–339.

Withers, Charles W.J. "Authorizing landscape: 'authority', naming and the ordnance survey's mapping of the Scottish Highlands in the nineteenth century." *Journal of Historical Geography*. 26, 4, 2000. pp. 532–554.

Withers, Charles W.J. "Art, Science, Cartography and the Eye of the Beholder". *Journal of Interdisciplinary History*. Vol. 42, Issue 3. 2012. pp. 429–437.

Withers, Charles and Innes M. Keighren. "Travels into print: authoring, editing and narratives of travel and exploration, c.1815–c.1857". *Transactions of the Institute of British Geographers*. Vol. 36: 4. 2011. pp. 560–573.

Wolters, O.W. *History, Culture, and Region in Southeast Asian Perspectives*. Ithaca, NY: SEAP Publications, 1999.

Wood, Gillen D'Arcy. *The Shock of the Real: Romanticism and Visual Culture, 1760–1800*. New York: Palgrave, 2001.

Wood, John. *Personal Narrative of a Journey to the Source of the River Oxus, by the Route of the Indus, Kabool, and Badakshan*. London: John Murray, 1841.

Woodward, David ed. "The History of Renaissance Cartography". *Cartography in the European Renaissance*. Vol. 3. Part 1 The History of Cartography. Chicago: University of Chicago Press. 2007.

Wright, Denis. "James Baillie Fraser: Traveller, Writer and Artist (1783–1856)". *Journal of Persian Studies, Iran*, Vol. 32:1. 1994. pp. 125–134.

Wyld, Helen. "Re-Framing Britain's Past: Paul Sandby and the picturesque tour of Scotland". *The British Art Journal*. Vol. XII, No. 1. pp. 29–36.

Yang, Anand A. *Bazaar India: Markets, Society, and the Colonial State in Gangetic Bihar*. Berkeley: University of California Press, 1998.

Ziter, Edward. *The Invention of the Middle East in British Scene Painting and Mise en Scene: 1798–1853* (Ph.D. Thesis). University of California, 1997.

Zou, David Zumlallian and M. Satish Kumar. "Mapping a Colonial Borderland: Objectifying the Geo-Body of India's Northeast". *The Journal of Asian Studies*. Vol. 70. No. 1, 2011. pp. 141–170.

Online resources

https://www.walesonline.co.uk/whats-on/arts-culture-news/famous-paintings-inspired-tintern-abbey-7147126, accessed on 13. 07. 2018. "Famous paintings inspired by Tintern Abbey and Llanthony Priory to go on show". Wales Online, 22 May2014; https://www.bl.uk/picturing-places/articles/the-spectacle-of-the-panorama, accessed on 12. 07. 2018. Special online collections titled "Picturing Places" of the British Library.

Wordsworth's journey in Ireland: https://uploads.knightlab.com/storymapjs/05489351d2507e423715fc1443163ad3/wordsworths-irish-tour/index.html, sourced on 3. 8. 2018.

Harvey, David. YouTube Video *"Anti Capitalist Chronicles: Social Media and Internet as a Powerful Organising Tool"* Part2. https://www.youtube.com/watch?v=rUA0xbNMCeQ.

Harvey, David. YouTube Video. *"Technology and Post Capitalism"*. https://www.youtube.com/watch?v=g18JoOZsoEM.

Surveying Empires – Archaeologies of the GTS in West Bengalhttp://www.surveyingemp ires.org/ accessed 09. 06. 2018.

Victoria and Albert Museum of Childhood Collectionshttp://collections.vam.ac.uk/ item/O26289/tour-through-england-and-wales-board-game-wallis-john/, accessed 27. 06. 2018.

Primary sources: travelogues, paintings, maps, memoirs and personal journals

Anon. "Curious Account of the Island of Staffa (one of the Hebrides) communicated to Mr. Pennant, by Joseph Banks, Esq." *The Annual Register or a View of the History, Politics and Literature for the year 1774.* London: J. Dodsley, 1775.

Anon. "How to Understand Geography", *The Penny Magazine*, 16 June 1832. p. 112.

Beckford, William. *The Travel Diaries of William Beckford of Fonthill.* Edited with a memoir and notes by Guy Chapman. Cambridge: University Press for Constable and Company Limited & Houghton Mifflin Company, 1928.

Boswell, James. *The Journal of a Tour to the Hebrides with Samuel Johnson LL.D. 1775.* http://www.gutenberg.org/cache/epub/6018/pg6018-images.html. Date of access 6. 6. 2018.

British Library, Warren Hastings Papers, Add MSS 29232ff. 393–410.

Daniell, Thomas and William Daniell. *A Picturesque voyage to India by the Way of China.* London, 1810.

D'Oyly, Charles. *Tom Raw the Griffin.* V: XXV. London: R. Ackermann, 1823.

D'Oyly, Charles. *Behar Amateur Lithographic scrap book.* Patna: Behar Amateur Lithographic Press, 1828(?).

D'Oyly, Charles. *Sketches of the New Road in a Journey from Calcutta to Gyah.* Calcutta: Asiatic Lithographic Press, 1830.

D'Oyly, Charles. *Antiquities of Dacca/* Nos. 1–4. (With engravings by J. Landseer from drawings by Sir C. D'Oyly.) London: John Landseer, 1830(?).

D'Oyly, Charles. *Views of Calcutta and its environs.* London: Dickinson & Co., 1848.

Devereux, Walter Bourchier. *Lives and Letters of the Devereux, Earls of Essex: in the Reigns of Elizabeth, James I., and Charles I. 1540–1646.* London: John Murray, 1853.

Dibdin, Thomas Frognall. *Reminiscences of a Literary Life.* 2 Vols *Vol.* 1. London: John Major, 1836.

Dunkin, William, "Extract from a diary of a journey over the Great Desert, from Aleppo to Bussora, in April 1782". *Asiatick Researches.* Vol. 4. 1795. pp. 399–402.

Edinburgh Review. Vol. 57. July1833.

Eliot, John. "On the Inhabitants of the Garrow Hills". *Asiatick Researches.* Vol. 3. 1793. pp. 21–45.

Faden, William. *A Map of Scotland: drawn chiefly from the topographical surveys of Mr. John Ainslie and those of the Late General Roy, 1807.* London: J. Wyld, 1875. NLS MMSID: 9938361263804341

Falconer, H. and P.T. Cautley. "Note on the fossil Hippopotamus of the Siwalik Hills". *Asiatic Research*, 19, 1836. pp. 39–53.

Falconer, H. and P.T. Cautley. "Sivatherium Gigantium, a new fossil ruminant genus from the Valley of the Markanda in the Siwalik branch of the Sub Himalayan Mountains". *Asiatic Research*, 19, 1836. pp. 1–24.

Falconer, H. and P.T. Cautley. "On some fossil remains of Anoplotherium and Giraffe from the Siwalik Hills, in the north of India". *Proceedings of Geological Society.* Vol. 4, 1843–1844. pp. 235–349.

Fergusson, James. "On Recent Changes in the Delta of the Ganges". *Quarterly Journal of the Geological Society*, 19, 1 Feb.1863. pp. 321–354.

Fergusson, James. *History of Indian and Eastern Architecture*. (1876) Cambridge: Cambridge University Press, 2013.

Fiennes, Celia. *Journeys of Celia Fiennes*. London: Cresset Press, 1947.

Fraser, James Baillie. *Journal of a Tour through Part of the Snowy Range of the Himala Mountains and to the Sources of the Rivers Jumna and Ganges*. London: Rodwell and Martin, 1820.

Gilpin, William, *Observations on the Highlands of Scotland*. London, 1789.

Gilpin, William. *Observations relative chiefly to picturesque beauty, made in the year 1776: on several parts of Great Britain; particularly the Highlands of Scotland*. Vol. 2. London: R. Blamire. 1789.

Gilpin, William. Essay 1 "On Picturesque Beauty", *Three essays: on picturesque beauty; on picturesque travel; and on sketching landscape: to which is added a poem, on landscape painting*. London: R. Blamire, 1792.

Gilpin, William. *Observations relative chiefly to picturesque beauty, made in the year 1776: on several parts of Great Britain; particularly the Highlands of Scotland*. Vol. 2. London: R. Blamire. 1789.

Hamilton, Francis Buchanan. *An Account of the Kingdom of Nepal: and of the territories annexed to this dominion by the House of Gorkha*. Edinburgh: A. Constable, 1819.

Heber, Reginald. *Sermons Preached in India by the Late Right Reverend, Reginald Heber D.D. Lord Bishop of Calcutta*. London: John Murray, 1829.

Heber, Reginald. *Poems by the Late Rt Rev. Reginald Heber D.D. Lord Bishop of Calcutta*. Hingam, C. and E.B. Gill, 1830.

Heber, Reginald. *Narrative of a Journey through the Upper provinces of India*, 2 Vols. London: John Murray, 1844.

Hodges, William R.A. "Preface". *Travels in India, during the years 1780, 1781, 1782 and 1783*. London: J. Edwards, Pall Mall, 2nd Edition,1794.

Hooker, Joseph Dalton and Thomas Thomson. *Flora Indica: Being A Systematic Account of the Plants of British India*. London: W. Pamplin, 1844.

Huxley, Leonard, and Joseph Dalton Hooker. *Life and letters of Sir Joseph Dalton Hooker: Based on materials collected and arranged by Lady Hooker*. Vol. 1. London: John Murray, 1918.

Imray, James F. *The Bay of Bengal Pilot. A Nautical Directory for the Principal Rivers, Harbours, And Anchorages, Contained within the Bay of Bengal; also for Ceylon, Andaman and Nicobar Islands, and the North Coast of Sumatra*. London: James Imray and Son, Chart and Nautical Book Publishers, 1879.

Johnson, Samuel. *A Journey to the Western Islands of Scotland*. London: David Price, 1775.

Kippis, Andrew. *A Narrative of the voyages around the world, performed by capt. James Cook*. (1788) London: Bickers and Son, 1878.

Knowles, John ed. *The Life and Writings of J.H. Fuseli Esq., MA, RA*. Vol. II. London: Colburn and Bentley, 1831.

Lambton, William. 1811. *An Account of the Trigonometrical Operations in crossing the Peninsula of India, and connecting Fort St. George with Mangalore. Extract from Asiatic Researches*. Vol.10. pp. 290–384.

Lambton, William. 1811. *An Account of the Trigonometrical Operations in crossing the Peninsula of India, and connecting Fort St. George with Mangalore. Extract from Asiatic Researches*. Vol.10. pp. 290–384.

Lambton, William. *Memoir*. Ddn. 85, 14–12–10. Phillimore, R.H. ed. *Historical Records of the Survey of India*. Vol. 1 (18th Century), India: Office of Survey of India, 1950.

La Touche, E. ed. *The Journals of Major James Rennell, first surveyor general of India, written for the information of governors of Bengal during his surveys of the Ganges and the Brahmaputra rivers, 1764–1767*. Calcutta: The Asiatic Society, 1910.

"Map Plan of the trigonometrical operations carried on in the Peninsula of India from the year 1802 to 1814 inclusive, under the superintendence of Lt. Col.W.Lambton". London: J. Horsburgh, 1827. Cartographic Items Maps 52415.(25.) UIN: BLL01004862522

Markham, Clements R. *A Memoir of the Indian Surveys*. London: W.H. Allen & Co., 1878.

Markham, Clements R. *Major James Rennell and the Rise of Modern English Geography*. London: Cassell and Co., 1895.

Mudge, William. *An Account of the Operations Carried Out for Accomplishing a Trigonometrical Survey of England and Wales*. Vol. II. London: Philosophical Transactions, 1801.

Oaten, E.F. *European Travellers in India: During the Fifteenth, Sixteenth and Seventeenth Centuries; the Evidence afforded by them with respect to Indian Social Institutions, and the Nature and Influence of Indian Governments by Edward Farley Oaten*. London: Kegan Paul, Trench, Trubner and Co. Ltd., 1909.

O'Keeffe, J. *Recollections of the life of John O'Keeffe, written by himself*. Vol. 2. London.

Pennant, Thomas. *A Tour in Scotland*. (1769) London: John Monk, 1771.

Pennant, Thomas. *A Tour in Scotland and Voyage to the Hebrides*. (1772) London: John Monk, 1774.

Phillimore, R.H. ed. *Historical Records of the Survey of India*. Vol. 1. Dehradun, India: Survey of India, 1945.

Phillimore, R.H. ed. *Historical Records of the Survey of India*. Vol. 2 (1800–1815), Dehradun, India: Survey of India, 1950.

Pyne, W.H. *Etchings of Rustic figures for the Embellishment of Landscape*. London: James Rimmel and Son, 1815.

Quarterly Review, Vol. 27, January1828. p. 102.

Quarterly Review. Vol. 4, August1810. p. 87.

Radcliffe, Ann. *A Journey made in the Summer of 1794, through Holland and the West Frontier of Germany, with a return down the Rhine: to which are added observations during a tour to the Lakes of Lancashire, Westmoreland and Cumberland*. Dublin: P. Wogan, 1795.

Rennell, James. *A Description of the Roads in Bengal and Bahar*. London: East India Company, 1778.

Rennell, James: *Memoir of a Map of Hindoostan, or the Mogul Empire*. London: C. Nicol, 1793.

Rennell. James. *Memoir of a Map of Hindoostan*. (1793). New Delhi: Editions Indian, 1976.

Rennell, J. A. *Map of Hindoostan. Hindoostan. By J. Rennell, F.R.S., 1782 ... The map engd. by J. Phillips. The writing by W. Harrison. 100 Geographical miles. Memoir of a map of Hindoostan; or the Mogul Empire. With an examination of some positions in the former system of Indian Geography ... and a complete Index of names to the Map*. London: W. Faden, 1785.

Rennell, James, WilliamLambton and George Everest, "Nine copper printing plates for British maps and triangulation diagrams of parts of India". East India Company. Cartographic Items Maps 183.a.3.(1–9). UIN: BLL01017910057

Sandby, Paul. *Virtuosi's Museum Containing Select Views in England, Scotland and Ireland.* London: George Kearsley, 1778–81.

Sleeman, SirWilliam Henry. *Rambles and Recollections of an Indian Official.* Vol. 1. London: J. Hatchard, 1844.

Starke, Mariana. *The Widow of Malabar: A Tragedy as it is performed at the Theatre Royal, Covent Garden.* London: J. Barker, 1799.

Taylor, Thomas. *Memoirs of the Life and Writings of the Right Reverend Reginald Heber, D.D. Late Lord Bishop of Calcutta.* London: John Hatchard and Son, Piccadilly, 1835.

Wordsworth, William. "Upon a Stone on the Side of Black Comb". *Poems.* Vol. II. London: Longman, 1815.

Wordsworth, William. "Letter to Christopher Wordsworth, September, 1829". William Knight ed. *Letters of the Wordsworth Family From 1787–1855.* Vol. 2. New York: Haskell House, 1969 [1907].

INDEX

Absolon, John 145
Abul Fazl 4, 148
Act of Union 204, 214
Ain-i-Akbari 75, 148
Akbar 68, 75, 226, 227, 236, 262n
Akbar II 143, 182n
Alberti 36
Albertopolis 139
Allan, Alexander 156
amateur: amateur artists 153, 166, 167, 275n; amateur painters 178
Anburey, Thomas 156
Andamans 94
Anderson, Adam 70
Anderson, Benedict 8, 26, 33, 269
Andrada, Antonio di 91
Anglo-Gurkha War 248
anthropology 10, 13, 27, 213, 258
archaeological coring 273
Ashton, Lever 125, 131n
Asiatic Society of Bengal 151, 250, 263
Atkinson, James 156, 170
Aurangzeb 67, 170
Austen, Jane 213, 223
authorcraft 241, 264

Bacon, Francis 19, 33, 58, 60n, 63n, 196, 218
Badrinath 254
Baffin, William 46, 67
Baird, General 78, 141
Banks, Joseph 55, 77, 98, 113, 120, 131n
Barak Valley 92
Barron, Richard 156
Barthes, Roland 10, 27n
Bartholomew, John ix, 29n
Bartlet, William Henry 117

base line 48, 53
Bay of Bengal 67, 93, 98, 225
Bay of Bengal Pilot 93, 94, 98n
Beaufort, Francis 60
Beckford, William 211, 222
Behar 158, 165, 167, 171, 175, 178, 258
Behar School of Athens 158, 167
Behar Lithographic Scrapbook 165
Bellin 67
Benaras 227, 233, 235, 242, 246
Bentinck 83
Bernier, Francois 228
Bertelli 67
Bharatvarsha 4
Bibiena 122
bildungsroman 191
Black Combe 57
Black Hole 155, 183
Blanchard, Capt. 94
Blundeville, Thomas 197
body politic 3, 42, 69, 102, 104, 127n, 202
book history 271
Boswell, James 42, 202, 210, 220, 223
botany 13, 14, 213, 215, 255, 256, 257, 258, 260; botanical 14, 112, 113, 168, 215, 222n, 256, 259; botanical gardens 215, 222; Botanical Garden of Calcutta 151, 255; Botanic Gardens, Kew 215, 222n, 255; botanical Provinces 259; botanise, botanising 260; palaeobotany 256; regional botany 260
Bouchet, Father 68
Boudier, Father 68
boundary 36, 52, 58, 75, 89, 92, 95, 106, 109, 133, 155, 205, 249; Boundary Commission 91; in mapping 52, 90; as conception 248, 249; as fortification 92; cartographic 52

293